Advanced Sciences and Technologies for Security Applications

Indexed by SCOPUS

The series Advanced Sciences and Technologies for Security Applications comprises interdisciplinary research covering the theory, foundations and domain-specific topics pertaining to security. Publications within the series are peer-reviewed monographs and edited works in the areas of:

- biological and chemical threat recognition and detection (e.g., biosensors, aerosols, forensics)
- crisis and disaster management
- terrorism
- cyber security and secure information systems (e.g., encryption, optical and photonic systems)
- traditional and non-traditional security
- energy, food and resource security
- economic security and securitization (including associated infrastructures)
- transnational crime
- human security and health security
- social, political and psychological aspects of security
- recognition and identification (e.g., optical imaging, biometrics, authentication and verification)
- smart surveillance systems
- applications of theoretical frameworks and methodologies (e.g., grounded theory, complexity, network sciences, modelling and simulation)

Together, the high-quality contributions to this series provide a cross-disciplinary overview of forefront research endeavours aiming to make the world a safer place.

The editors encourage prospective authors to correspond with them in advance of submitting a manuscript. Submission of manuscripts should be made to the Editor-in-Chief or one of the Editors.

Umair B. Chaudhry · Hamid Jahankhani

Autonomous Revolution: Strategies, Threats and Challenges

Springer

Umair B. Chaudhry
Northumbria University
London, UK

Hamid Jahankhani
Information Security and Cyber Criminology
Northumbria University
London, UK

ISSN 1613-5113 ISSN 2363-9466 (electronic)
Advanced Sciences and Technologies for Security Applications
ISBN 978-3-031-96784-9 ISBN 978-3-031-96782-5 (eBook)
https://doi.org/10.1007/978-3-031-96782-5

This Springer imprint is published by the registered company Springer Nature Switzerland AG
The registered company address is: Gewerbestrasse 11, 6330 Cham, Switzerland

Acknowledgements We extend our deepest appreciation to the researchers who enriched this project with their valuable contributions. Our thanks go Connor Kevin Stevens, Nick Costello, Dhana Lakshmi Mariyala, Sanjay Kumar Kodur, Wayne Gill, Chisom Sylvia Onunta, Rob Patton, Clement Ayodeji Sanni, Enejda Torba, from Northumbria University London Campus for their insightful inputs and their significant expertise on cybersecurity. Their dedication and knowledge have greatly shaped this book's depth and direction.

We also owe a debt of gratitude to the countless scholars, researchers, and thought leaders whose prior work laid the foundation for this exploration. Their studies at the intersection of marketing, technology, psychology, and consumer rights provided essential context and inspiration, and we are honored to build upon their intellectual legacy.

To everyone who has contributed to this endeavor—thank you. Your support, encouragement, and collaboration have made this book both possible and meaningful.

Umair B. Chaudhry
Hamid Jahankhani

Competing Interests The authors have no competing interests to declare that are relevant to the content of this manuscript.

Contents

Chapter 1
A Security Analysis of the Vulnerabilities of Drones That Use the IEEE 802.11 (Wi-Fi) Standard by Using Simple Hacking Techniques, with Poor Governance Procedures

Connor Kevin Stevens and Hamid Jahankhani

Abstract Several studies have identified the way in which drones are used to communicate to the ground controlling station. However, conducting a security analysis on the vulnerabilities of drones, that use the IEEE 802.11 (Wi-Fi) standard (Islam in Securing your wireless network: an in-depth guide to Wi-Fi password best practices, 2024 [61]) with the use of simple hacking techniques is yet to be conducted. The research will include a literature review on the way drones are being used, operated and their communication techniques. The data analysis will identify the security vulnerabilities between the ground controlling station and the drone by conducting an ethical hack, which will use a man in the middle attack vector. The vulnerabilities will identify that if drones are not being used correctly, the vulnerabilities found could then be integrated with artificial intelligence, to create an offensive or defensive effect. The data analysis will create an awareness on how drones are currently not being governed correctly. It will then create a Drone Information Governance (DIG) framework, which will provide awareness to all drone operators. The drone information governance framework will identify the importance of data collection, which will provide guidance on how the data should be collected, protected, and stored correctly. Furthermore, this research will identify the vulnerabilities found throughout the analysis. Drones are growing within all environments, with no overarching governance to show a safe direction. It is pertinent that policies, frameworks and permissions are created to stop the identified risks growing. This research urges a DIG framework to be implemented at the earliest opportunity. Nevertheless, future work is needed.

Keywords Drones · Information governance · 802.11 · Hacking · Warfare

---

C. K. Stevens · H. Jahankhani (✉)
Northumbria University London, London, UK
e-mail: Hamid.jahankhani@northumbria.ac.uk

U. B. Chaudhry and H. Jahankhani, *Autonomous Revolution: Strategies, Threats and Challenges*, Advanced Sciences and Technologies for Security Applications,
https://doi.org/10.1007/978-3-031-96782-5_1

1.1 Introduction

The aim of the research is to produce a personalised awareness for cyber-attacks on drones, this will be implemented by low-cost simple hacking techniques. An assessment of the vulnerabilities will be conducted on a first person view unmanned air vehicle that utilises the IEEE 802.11 standard, which is also known as Wi-Fi. The vulnerability assessment will identify flaws, which will include how drones can be exploited by a man in the middle attack, by using the denial-of-service and brute force attack vectors. The research will then produce a Drone Information Governance framework. This will provide recommendation to alleviate these vulnerabilities being exploited and create a safety-first culture, to protect the data, drone, and drone operator. A governance framework at the national level will be created to help protect the drone operator's data.

The popularity and usage of drones has increased within a myriad of environments within recent years and the advancements of drones is growing every day. The first record of the use of drone's dates back to World War I, the United States developed the Kettering Bugg unmanned aerial vehicle, which was used as a rudimentary cruise missile [3]. Although the Kettering Bugg unmanned aerial vehicle never saw combat, it was the initial stage of drone technology [5]. Furthermore, drone technology evolved again during World War II, which saw fifteen thousand Radioplane OQ-2 unmanned aerial vehicles made for the United States Army. The Radioplane OQ-2 was developed for both target practice and reconnaissance [3]. The creation of drones in both World War I and World War II set the conditions for how militaries around the world use drones today.

The data analysis has identified multiple vulnerabilities for drones using the IEEE 802.11 standard to communicate from the ground controlling station to the drone. The researcher was able to infiltrate the communications link and manipulate the drone. The researcher was able to take control of the drone through a brute force attack, which could be used as both an offensive and defensive technique. Additionally, once the researcher had exploited the vulnerabilities a Drone Information Governance Framework was written to create awareness for how drones should be managed and controlled, within the United Kingdom's government. The results showed that if these communication vulnerabilities could be integrated with Artificial Intelligence, this could then be weaponised as an offensive or defensive capability. The drone Information Governance framework created awareness on the operation of drones and how they should be governed correctly. Additionally, the drone information governance framework recommended that both policies and permissions need to be created at the governmental level, enabling drone manufacturers to enforce training prior to drone operations within any environment need for future work.

1.2 Literature Review

1.2.1 Drones in Modern Warfare

The terrain of modern conflict has been transformed by the arrival of drones, this has adapted both tactics, strategy and new standard operating procedures (SOPs) for militaries around the world. Drones have been indispensable for surveillance, reconnaissance, intelligence collecting and offensive actions. Drones have been used within every environment. A primary example in the use of drones in the air, which includes first person view (FPV) drones or one way attack drones [7, 73]. Unmanned aerial vehicles (UAV) and affordable drones that have been brought of the shelf, have adapted the employment of drones within armed conflict, which has changed relationships and allowed technologically deprived armies to successfully have an effect against more powerful enemies. The terrorist organisation the Houthis that operate within the middle east, have demonstrated this by successfully launching UAVs at shipping vessels, which has disturbed trading routes across the Red Sea [32]. This shows that the use of drones has dramatically changed and governance at the national level ought to be enforced.

The Russia-Ukraine war is the most current major conflict within the modern day. This demonstrates the use of drones on a daily basis. Drones have been employed since 2014 which was the start of the conflict, this advanced considerably when Russia conducted the special military operation in 2022, which was Russia's full-scale invasion of Ukraine [92]. Ukraine has deployed UAVs for multiple uses, including reconnaissance, cyber operations, and attacks on the Russian military; drones have become crucial to offensive and defensive operations. The Ukrainian military have used drones to enable cyber operations or teams to breach surveillance systems in areas that have been occupied by Russia. Therefore, Ukraine is able to obtain intelligence regarding the movement of Russian troops [45]. FPVs have attached explosives to have detrimental effects on Russian infrastructure or personnel, this is shown within Fig. 1.1 the Hercules 13 FPV drone [87]. However, Russia is using the same effector to counter Ukraine and drones are playing a vital part in precision strikes, which shows that the communication between the ground controlling station (GCS), and the drone needs to remain secure, to mitigate the communication becoming compromised.

Drones that are now used for combat operations like the Hercules 13, are also being used within the intelligence, surveillance and reconnaissance (ISR) roles. This allows military forces to gain the upper hand, which will enhance situational awareness [50]. In modern conflicts this has been crucial by helping make tactical and strategic decisions, by painting a real time intelligence picture. Drones are aiding the forward line of troops (FLOT) in any conflict and helping chains of command have current and up to date situational awareness. This helps ease the fog of war and improves operational efficiency [31]. The dual-use capability of providing ISR and being used as a precise offensive weapon, has enabled militaries to reduce collateral damage by providing more focused attacks, which uses a smaller warhead to have a larger effect. This is enabling militaries to avoid a capability gap and permitting militaries

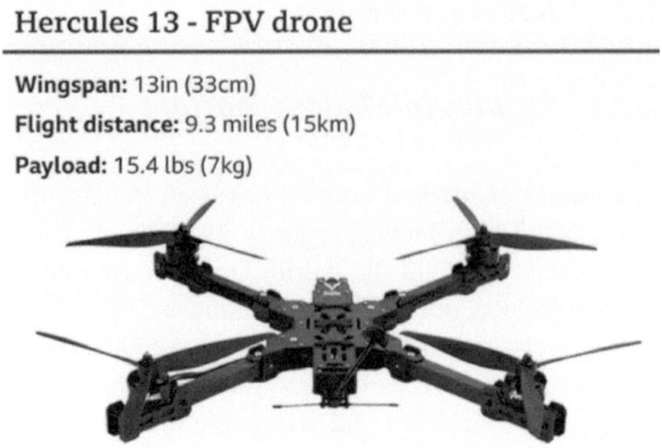

Fig. 1.1 Hercules 13 FPV drone [87]

to have the most up to date technologies in modern conflict. Small, cheap drones have let the Ukraine armed forces provide a great effect, which has nullified high-value Russian assets such as support vehicles, tanks and artillery at a fraction of the cost. The Ukraine armed forces are spending around 400 USD on the Wild Hornet FPV drone, which is manufactured in Ukraine and has been used on a daily basis within combat operations providing ISR capabilities, strike effects and payload dropping [101]. Additionally, the more commercially accessible drones like the Da Da-Jiang Innovations (DJI) Mavic 3 costs 2000 USD and upwards, which has fewer customising options for combat operations [16]. This has been demonstrated within a congested airspace; were both less expensive and smaller UAVs have a superior advantage in asymmetric battles. It is pertinent to be able to customise drones for different missions, this includes the changing of frequencies to evade electronic warfare (EW) [68], which has been demonstrated throughout the Russia-Ukraine war.

Additionally, this is mitigating risks to pilots that would usually have to fly in congested and dangerous airspace, which militaries around the world would have seen in previous conflicts. This risk has been shown in the Red Sea where two F18 Hornets were shot down [69]. The two aircraft were shot down due to friendly fire because the airspace was congested with missiles, drones and piloted aircraft. The United States (US) was trying to protect the trading routes within the Red Sea and the terrorist organisation the Houthis was launching one way attack drones at merchant shipping vessels. The future of drones is developing rapidly, which shows that the use of drones will mitigate the risk to life, due to a technology driven environment and it is critical that all elements of the drone remain secure.

1.2.2 The Use of Drones by Non-state Actors or Criminal Organisations

Building on the use of drones has rapidly changed how non-state actors fight, which has also equalled the playing field for how terrorist organisations, criminal organisations or non-state actors operate. The Islamic State (IS) demonstrated this in 2014 by using drones to guide suicide bombers onto targets, which included talking through a communication device and was guided onto the target by the drone operator [52, 93]. This was a steep learning curve for the rest of the world and emphasised the requirement for change. This showed that hardened defences were needed and if change was not made effects would be detrimental, which increased security concerns at the national level.

Non-state organisations that include terrorists, insurgents and criminal networks have all evolved with drone technology within all environments, which has enabled non-state actors to help achieve their goals. Drones brought off the shelf at low costs are being modified for myriads of purposes, specifically it has assisted criminal networks to achieve crime. For example, criminal groups have employed rotary wing drones, modified for dropping drugs at prisons, which has contributed to a 45% increase in violence in prisons; due to depts that have been obtained by the prisoners because of drug use [100]. Additionally, Mexican cartels are using the same effectors to smuggle drugs around the US [40]. Drones need to be controlled and all ground controlling stations (GCS) ought to be held accountable. GCS needs to adhere to correct governance frameworks with the relevant control measures, detection and tracking of drones needs to be set in place [51]. This will help mitigate criminal groups being able to gain access and create adverse effects on society.

Terrorist groups have been adapting their SOPs on how they operate drones, which has been seen internationally. IS have demonstrated the use of UAV drones in both Iraq and Syria by using UAVs with shape charges to launch at targets. Houthis rebels in Yemen and Israel have been using a one-way attack (OWA) uncrewed surface vessels (USVs) to launch at vessels transiting through the Red Sea [43]; a variant can be seen below that has been used in Ukraine [24]. Hezbollah's drone fleet of OWA UAVs are operating out of Lebanon and targeting Israel, which has caused detrimental effect to the Israeli forces because the OWA UAVs can sometimes remain undetected by slipping below air defence (AD) systems. This was seen in the middle of October 2024 where sixty-five Israeli soldiers were injured while eating [46]. Drones have become a factor threat to all state actors, which shows that drone technology has more control and power over modern warfare. Thus, all states need to ensure they are critically cyber and drone aware (Fig. 1.2).

In addition, a number of drones have been sighted over the UKs military bases, and this included the US bases that are located in the UK. The drones were spotted in Gloucestershire, Norfolk and Suffolk [82]. The drones were conducting ISR over military establishments and were most likely operated from another country, which shows the threat can be operated from geographically anywhere. Drones are causing

Fig. 1.2 USV drone [24]

adverse and complex problems for all states around the world and the spread of available, affordable technology is changing the environment of world conflict. Terrorist organisations, insurgents and criminal organisations are no longer dependent on conventional weapons such as tanks or fighter jets. Non-state actors are using tiny drones to conduct strike capabilities or ISR, these drones have let a wide spectrum of actors operate within complex operations at a reduced cost, which would have been impossible in the past.

1.2.3 Drone Technology's Possibilities and the Evolution

Technology within drones has grown considerably over the past few years. Initially drones performed reconnaissance operations within all sectors of the military or civil service [12]. However, drones are now multifunctional systems and highly sophisticated. Drones were initially used for ISR operations, which provided real time intelligence on enemy movements, and this provided situational awareness for military forces [50]. The developments within drone sensors, technology, data and weapon systems allowed them to become multifunctional. Drones were then enabled to perform armed reconnaissance, close air support (CAS) or communication support while conducting other tasks.

Advanced drone sensors are able to collect up to date intelligence by using electro-optical (EO) cameras, infrared (IR) cameras, synthetic aperture radar (SAR), and artificial intelligence to support piloted aircraft. These sensors can be seen within the Multi Role Remotely Piloted Aircraft System (MQ) 28 Ghost Bat and MQ-9 Reaper [15, 20], which are top of the range drones within modern conflict. Both drones are able to capture high density data which is used for both strategic planning and enables real time decision making. The MQ-9 has a fourteen-hour flight time and communicates through satellites so it can be controlled within a safe and controlled environment [10]. These aircraft can operate at multiple altitudes and can conduct operations within any weather (Fig. 1.3).

In addition, on top of the ISR capability, drones are being used for offensive or counter offensive operations. The MQ-9 Reaper carries a laser guided Air to Ground

Fig. 1.3 MQ9 AGM 114 [78, 104]

Missile (AGM) 114 Hellfire, which is used for low collateral damage strikes, anti-armour, and anti-personnel engagement [10, 94]. Also, a laser Guided Bomb Unit (GBU) 12 Paveway II is used for engaging larger targets requiring more effect. The weapon systems minimise collateral damage, which lowers risk to the civilian population [94]. Furthermore, within the US military drones are being used to employ weapons for counter terrorism [13], this has raised ethical concerns regarding the use of unmanned systems in targeting operations.

In the near future drones will be used for logistics and support tasks, which includes the MQ-25 Stingray, which is an aerial refuelling drone. The MQ-25 will be used within the US Navy and is situated on an aircraft carrier, which provides aircraft to aircraft refuelling [14]. Unmanned aerial refuelling drones are developing to allow crewed aircraft to have an extended range and maintain posture within the airspace for longer, which is due to become operational in 2026 [67]. This will enable militaries to have a major cutting operational edge, as the mission will cover a longer distance or endure longer. Additionally, resupply drones will be used to aid in ground resupply, this will provide supplies that are vital without creating a risk to life. At present time humanitarian missions have used this capability to get vital supplies to unreachable locations, within a dangerous environment or has been destroyed by a natural disaster [99]. As the capability grows this will be life changing and help save lives.

Artificial intelligence (AI) is advancing within drones, with machine learning this allows drones to grow autonomously and help by delivering arduous jobs without human involvement [22]. Drones will start becoming more software driven rather than hardware driven [38], which would give the drones room to overcome and adapt in difficult situations. Autonomous drones can help navigate around difficult terrain, which can recognise and follow targets, then prosecute if authorised. This would have a need for important ramifications within warfare because it lessens human error and human involvement for quicker decision making within combat situations [83]. However, this must have certain controls for ethical reasons, by ensuring that both policies and permissions are adhered to.

The capabilities of drone growth have also resulted in the acceptance by the civilian sector and emergency services. Drones have been used within the agriculture environment, which drones have monitored weed control or fertilizers. The use of drones has resulted in reduced costs within agriculture operations, increased the yield rate and has improved the quality of crops [37]. The emergency services are implementing drones within disaster response, search and rescue operations. In 2017 drones were saving one life a week on average [80], which shows the benefits that drones can add within the search and rescue sector. Logistics businesses such as Amazon, Dalsey, Hillblom and Lym (DHL), are testing drones for package or parcel delivery [36]. Amazon and DHL can see a fully functioning delivery service within the next five or ten years [36]. Drones are operating within many different areas and with the current influence drones are projected to increase, thus it is essential to ensure drones are secured.

Drone Operations: Communication and Security

The effective use of drones within both the civilian and military environments is essential. The communication link between the drone and its ground control station (GCS) is necessary for mission success, as the operator within the GCS maintains control and receives data from the drone's sensors [66]. However, the communication between the drone and GCS is the most crucial element of the drone operation, as it can be either disrupted or intercepted. Therefore, the confidentiality, integrity and availability (CIA) of the data must remain secure for anyone requiring the information [42]. In a hostile environment the data could get compromised and could cause serious damage to the operator, drone or target.

In addition, a range of communication strategies have been explored, predominantly within the decentralised and centralised control systems [79]. A centralised system is where an operator controls the drone from a command centre. This is commonly used within military operations where intelligence is time sensitive, and precision is required for decision making. Decentralised systems, allow the drones to operate autonomously or semi autonomously, which can include multiple drones. Decentralised drone systems don't need direct input from the human operator and are able to coordinate with each other or operate as a single entity. The approach is often used within the OWA or swarm operation and often deployed simultaneously to conduct wide area surveillance and overwhelm the target's defence systems [79]. Additionally, centralized and decentralised drone operations have been used within the Russian-Ukraine war that has caused detrimental effects on both states [49]. Thus, communication between the operator and the drone is pertinent, which could create a risk to life if not secured correctly.

Figure 1.4 shows the UAV architecture, which is broken down into the UAV structure, main computer, sensors or payload, communication link and the GCS [104]. The UAVs structure is commonly lightweight, small in size and has aerodynamic flexibility. The main computer is the most crucial part as for both the flight control and autonomous functioning, which provides the information to the operator within the control base. Sensors and payload can be adapted to the operator's need, to meet the requirement of the operational need and allow the operator to gain real time intelligence. The communication link can be fitted with either high quality wireless communication or cable tethered with a fibre optic cable to mitigate jamming [11]. Finally, the ground controlling station is mainly employed to operate the UAV and alter the mission parameters if required.

UAVs are equipped with high quality wireless communication and several different communication technologies are used to maintain links between the drone,

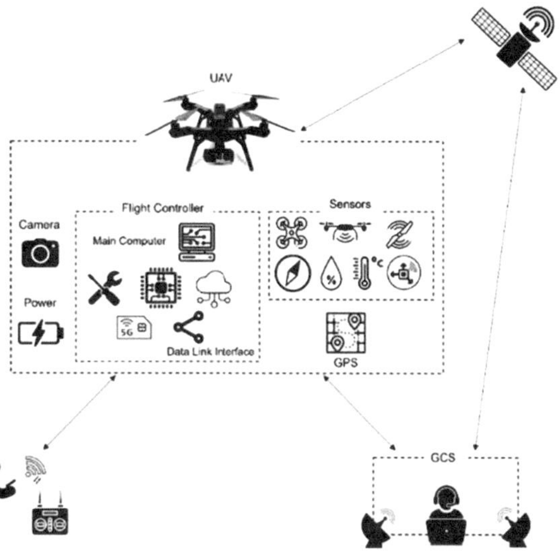

Fig. 1.4 Architecture of a UAV [104]

operator or ground controlling station. Communication technologies include Bluetooth, Zigbee, cellular networks Wi-Fi (2.4 GHz) and radio frequency identification (RFID) [104]. Cellular networks offer broader coverage but are less reliable in remote areas. Whereas Wi-Fi provides high bandwidth but less range [84]. Additionally, both optical communication and acoustic signals have been researched as alternatives for the operation of underwater drones or space drone's, which shows the vast number of drones that are being created for multiple different purposes. Therefore, these systems will have security vulnerabilities that can be infiltrated by cyber criminals. Furthermore, the advances within UAVs communication technology makes the system vulnerable to cyberattacks. The most common form of cyberattacks is a DoS attack, in which the communication link between the GSC and the drone becomes overwhelmed with excessive traffic, causing the connection to break down [64]. This within any environment can cause the drone to become unworkable or cause the system to crash, depending on the magnitude of the cyber-attack. In military applications, the loss of drones can have adverse consequences if the drone is carrying sensitive data or expensive equipment. This has been seen within the Russia-Ukrainian war where an £11,500,000 stealth drone has been shot down [35] and the debris would have been collected for further research.

The risk of DoS or cyber-attacks can be mitigated with encryption, which will secure the link between the drone and the GCS. This will enable the operator who is controlling the system to have a secure connection and have a larger impact on the mission [61]. In addition, encrypted communications protocols make it challenging for the adversaries to disrupt or intercept the transmission signal and enable the operator to have full control over the system. The more developed drones have an alternate means of communication, which allows them to switch to a different amplitude, frequency or wavelength [89]. Amplitude modulation (AM) sets patterns of disturbance as the wave travels from its undisturbed position at the waves furthest range. Wavelength of a wave is the distance between the point of one wave and the same point on the following wave. Frequency modulation (FM) of a wave is the number of waves produced by the controlling systems each second and is also the number of waves that pass through a certain point within each second. This allows the operator or drone to switch to a different form of communication if the link has been compromised. Advances within artificial intelligence and machine learning are also being utilised for when the connection becomes compromised [83], which allows the drone to operate autonomously when the communication link has been lost with the operator.

The security of communication transmission is also not only a concern for military forces but all public sectors or civilian industries. Drones are becoming more integrated to all commercial applications, which can include package delivery [36], infrastructure monitoring, and the pressure washing of buildings [70]. Therefore, the more drones in operation will increase the chance of cyberattacks on any system. This has been seen within a corporate network and cyber criminals used modified DJI drones and attached a Wi-Fi Pineapple device to the drone, which enabled them to spoof the company's Wi-Fi network. Then a MitM attack was conducted through a rogue access point, where the cyber criminals intercept the transmission and then relay messages between two parties, which from the company's understanding they are communicating with one and other. The outcome was that the cyber criminals were able to intercept the data which included login credentials and Wi-Fi information [88]. Drone manufacturers must ensure they are fitted with robust security protocols and safeguard the systems from being modified, to prevent disruption to public safety and critical infrastructure.

1.2.4 Governance, Policies, and Ethical Concerns

The rapid development and spread of drones within all sectors have driven a necessity for rules, regulations and governance structures. Drones being used within civilian and military sectors has raised concerns regarding their ethical or legal limits. There is a growing concern about the possible abuse that drones are capable of conducting and some of these concerns can be seen below within Fig. 1.5 [1]. In addition, just some of the ethical concerns include, controlling signals, spoofed signals, malicious communications, recording, manipulating inputs and signal jamming, which are all broken down within the below figure. The image below breaks down some of the reasons why governments should be enforcing strict policies and correct governance procedures to ensure that drones are not being used for the immoral reasons.

Guidelines are aiming at regulating the usage of drones across many different countries and are being developed by international agencies such as the International Civil Aviation Organisation (ICAO). Furthermore, guidance is looking to address security challenges, data privacy and the safety of airspace [59]. Rules and regulations will ensure that drones mitigate endangering civilian air traffic, which is one of the

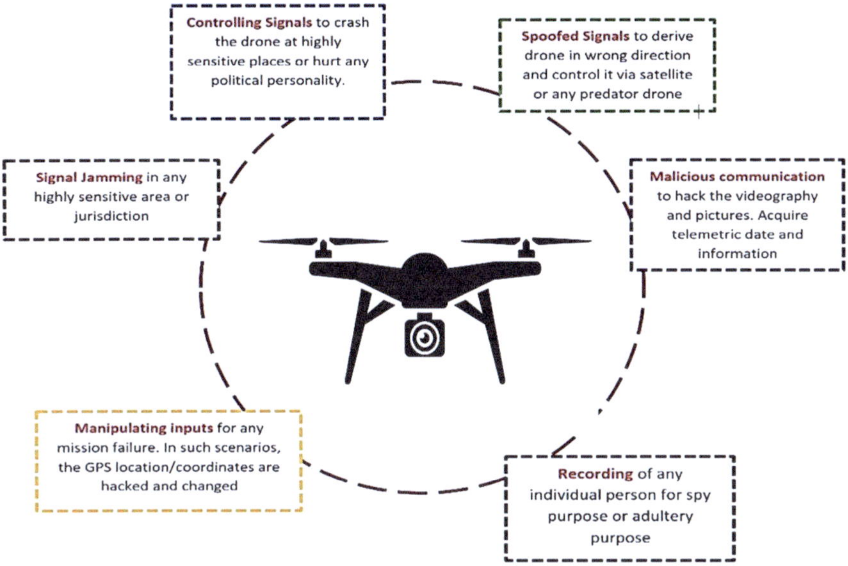

Fig. 1.5 Ethical and privacy concerns of UAVs [1]

main difficulties in controlling them. Predominantly within a crowded airspace close to airports, the use of drones raises the possibility of manned aircraft crashing, which creates a serious risk to life. For example, Gatwick airport was brought to a standstill for thirty hours in 2018, due to possible drones at the airport, this effected over one hundred and forty thousand people [53]. The effect on Gatwick alone shows the reason why governments need to enforce stricter rules and regulations on drone operators.

Drones being used in line with international law and especially within war operations is a critical component of drone control. Targeting within counter-terrorism operations in the middle east and other areas of conflict have caused deaths to terrorists or civilians, which has generated major ethical considerations. This was seen in Iraq where US soldiers were injured within a drone strike [13]. Additionally, in the near future with the use of AI will foresee further ethical considerations [65]. Critics have stated that drones break international law, which may cause civilian casualties, and appropriate control should be enforced. Lack of openness or acknowledgement about drone activities has created further discussion about legality and morality of employing drones for targeting.

Furthermore, not only legal systems need to be enforced but ethical rules are also required to control drone operations. Concerns within the decision-making procedure and the responsibility of AI being employed within autonomous drone systems have created a lot of questions [65]. For example, who will be accountable if an autonomous drone makes a mistake and causes civilian casualties? These issues have been brought to the surface, therefore ethical guidelines that autonomous drones ought to be brought in line with both human rights and international humanitarian

Fig. 1.6 Information governance [75]

law [39]. Humanity is in the beginning of an autonomous revolution and consideration of the use of autonomous drones within war cannot be delayed [21]. This shows the growth of technology and the importance of international law being up to date.

In addition, a Drone Information Governance (IG) framework should be installed at every level within every sector to ensure that drones are adhering to both laws and regulations. The information collected will be received, collected, and recorded correctly. The IG framework is an all-inclusive approach to managing information. The framework will implement roles, controls and a system of measurement that treats information as a treasured asset [81]. IG also provides an accountability framework to ensure the information is created and used correctly. An IG framework ensures that information is created, stored, used, shared, archived and then destroyed correctly [75]. The cycle of IG makes sure that all information is accounted for, and members of the organisation are held accountable (Fig. 1.6).

IG frameworks within a military environment are pertinent because when information isn't stored correctly the information could be lost or stolen, which can cause detrimental effects to organisations at the national level. The areas that military drones will be operating in or being used to effect can be highly sensitive. Thus, if the information was lost it could damage the country's reputation. This was seen when the Ministry of Defence (MOD) classified documents were left at a bus stop in 2021 [2], this could have caused serious issues at the national level and damaged relationships with partners. It is essential that all information within the public sector is stored correctly.

Legal and ethical issues that have been raised by the use of drones will also change as the technology within drones develops. International organisations, government and all industry players will have to evolve together [21]. All parties will have to ensure they communicate and cooperate continuously, to ensure that drones are utilised and implemented into society in a reasonable and ethical way. Information around drones needs to be governed correctly and all security vulnerabilities need to be secured to ensure the network is not compromised.

1.3 Research Methodology

A methodology has been developed to show a clear analysis of how the data is being collected. The methodology will show what the researcher has done, how the research was conducted and how the research was understood [71]. The methodology will then lead onto the data analysis by means of a security vulnerability assessment on the communication link between the GCS and drone when operating on the Wi-Fi, which will show the practical implementation of the data collected. The methodology will be broken down into the approach, data collection, data analysis and ethical considerations, followed by the methodical evaluations.

The research was conducted to create a greater understanding of how drones operate when using the Institute of Electrical and Electronics Engineers (IEEE) 802.11 [44], which is also known as Wi-Fi. The main objective was to find vulnerabilities between the GCS and the drone when operating on Wi-Fi. Once the research has been completed a practical assessment will be implemented on an FPV drone and an ethical hack on the vulnerabilities found will be conducted. Finally, once the ethical hack has been achieved an information governance framework will be written to create wider awareness on how to secure and manage all information when operating drones.

The fundamental concepts of a cause-and-effect relationship helps the reader understand the communication relationship between the GCS and the drone. The cause provides a factor or reason to why the communication link between the GCS and drone is unsecure, while the effect refers to the outcomes of the communication becoming compromised [103]. Causes and effects are linked because they are both interconnected; by understanding both of these it allows individuals to make precise decisions in order to solve the problem [91]. This enabled the researcher to critically analyse the IEEE 802.11 communication technique between the GCS and the Drone.

The data that has been collected by the researcher is qualitative data and this is non-numerical information that captures qualities of phenomena, characteristics, and a detailed descriptions [54], which will emphasise the vulnerabilities found for when drones are using communications through Wi-Fi [103]. The qualitative data has enabled the creation of a contextual and exploratory research question. This will also help the researcher define why the governance around drones is poor and why IG should be taken seriously when drones are being operated. Qualitative data will help both the research and reader gain an in depth understanding of how drones operate.

Furthermore, secondary data was used throughout the research, which provided a time-effective and cost-effective means of populating data in order to conduct the analysis on drones [76]. Secondary data provides information that has been published, organised or collected for other purposes other than the researchers specific research question [76]. This involved the use of pre-existing information to enable the researcher to find vulnerabilities within drone communications, when they were using IEEE 802.11 communication techniques and then conduct a practical assessment of how the vulnerabilities can be infiltrated. Utilising secondary data methods, types and practical implementation helped the researcher by pulling from a lot of sources and developed a further understanding on how drones are operated.

The data collection was conducted on a Hewlett-Packard (HP) ENVY x360 laptop and the ethical hack will also be demonstrated on the laptop within a virtual environment, which will be documented through a series of images. The data will be collected from different open-source databases, which enabled a wide area search of the information needed to facilitate the research problem. Utilising open-source data engines helped drive decision-making [57] and provide information that was needed to complete the research. The websites that are used are, science direct, Google scholar, Google, British Broadcasting Corporation (BBC) News, and all research is referenced accordingly. The literature review used search phrases such as drone communication vulnerabilities when using Wi-Fi or the use of drones within a modern environment in order to refine the search to ensure that the information is current and accurate [58]; search phrases were used on science direct, Google and Google scholar. BBC news was used to provide updated information on how drones are being used within modern conflict. Vulnerabilities will be researched using the same techniques and within a secure environment the vulnerabilities within Wi-Fi will then be tested.

A practical assessment of the vulnerabilities found within the communication link through Wi-Fi between the GCS and Drone, will be established using a virtual environment. VirtualBox will be used to ensure that ethical considerations are adhered to by using a Kali Linux virtual machine (VM). VirtualBox is an operating system software that operates as a hypervisor, by creating VMs and if required multiple VMs can be utilised. Additionally, an Alfa Network IEEE 802.11n long range universal serial bus (USB) adapter [4], will be attached to the VM. The device will establish a wireless network to monitor, detect and ethically hack communications that have been detected. This will be the reconnaissance stage in order to find and fix the connection between the GCS and drone.

Furthermore, once the reconnaissance stage has been finalised, the ethical hack can be conducted through a DoS or MitM attack. The ethical hack will break the communication link between the GCS and done, where the communication link will be manipulated. Figure 1.7 shows how a DoS attack can send the drone an overwhelming amount of information, which will cause the drone to lose communication with the GCS and is also known as jamming. The MitM attack can eavesdrop the drone's feed to see what the drone is looking at or inject malware to create false identities [85]. Furthermore, once the has been detected a KRACK or dictionary attack can be conducted [28], where the ethical hack can crack the drone's password and then gain control of the drone itself.

The drone used throughout the secure laboratory will be a DJI Ryze Tello, which has been voted one of the best drones for 2025 [34]. Additionally, the drone is not only handset capable but also FPV capable. The drone market initially started in

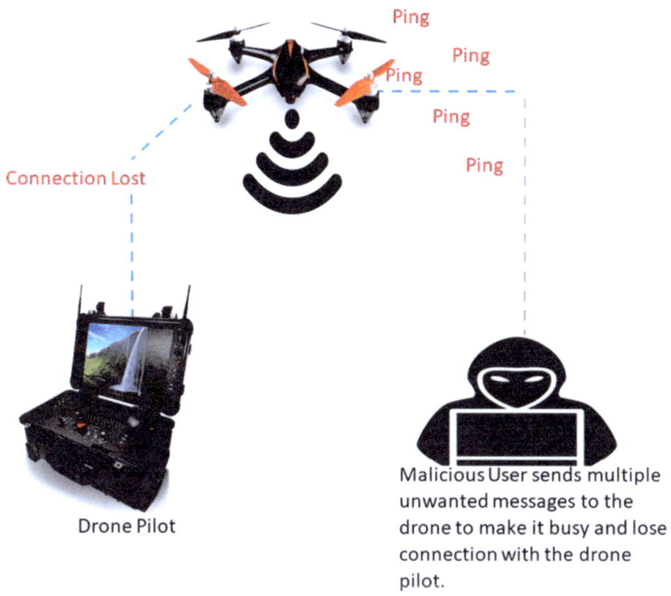

Fig. 1.7 DoS attack on a Wi-Fi enabled drone [55]

2010, which China was seen to dominate and pushing western competitors out of the market. In 2016 DJI have counted for two thirds of all consumers produced worldwide [41]. These systems are very capable at an affordable price.

China doesn't only manufacture the drone, but they also manufacture most of the component parts that are needed to assemble the drones [41]. DJI drones are currently being used within the Russian Ukrainian war, which shows the detrimental effects FPV drones can cause.

Once data analysis has been documented an Information Governance framework will be written to ensure that all the information held within the drones is handled correctly, which will state clearly defined roles and tasks for each member of an organisation [75, 81]. This will be clear for all drone operators and will protect the drones from further infiltration or penetration, ensuring that the drones remain protected and secure. The Information Governance framework will also protect the GCS to ensure that their data remains safe and doesn't get stolen or taken [79]. Additionally, this will help mitigate risk to life and enable more control over drones at the national level, where policies and permissions haven't caught up with the fast growing and evolving drone industry.

1.3.1 Methods of Analysis and Ethical Considerations

The methods of analysis used qualitative data which is used in non-numerical nature by expressing both the practical and literature part of the research in images, words or observations rather than numbers [54]. The research will also use subjectivity to reflect on personal experiences and perspectives. Additionally, the IG framework will be rich in context and will provide an in depth understanding of how the vulnerabilities are found, this will aid in mitigating the risk and create a wider awareness for how effective drones are used. Secondary data will be a key part in the framework in order to support the comparative study [76] by leveraging data from multiple sources, which will then compare the research and generalise the data to make the framework effective. Additionally, the availability of longitudinal data [76] will allow historical datasets to analyse trends over time for how drones are being operated currently.

It was essential throughout all research that ethics was followed at all times [18]. Informed consent was given to conduct the practical assessment within the researcher's workplace, and it was only the technology that was used. The drone will remain in the room to ensure that the assessment was conducted within a controlled environment. Confidentiality relates to an agreement between the researcher and the participant [17], that was granted by the researcher's workplace to conduct the assessments if there was no other participant involved. Respect for others was considered when operating the drone and ensuring that the drone wasn't operated within working hours and only operated within a secure room. The risk was mitigated within the research proposal with a risk assessment and then discussed with the supervisor to ensure that academic conduct has been followed at all times. Additionally, a dynamic risk assessment helps manage risk [74], this will be conducted throughout the assessment and discussed with the researcher's supervisor. Integrity was essential for ensuring that academic integrity is followed, and the work was only conducted by the researcher.

Furthermore, the Data Protection Act (2018) was followed throughout the research. All data that had been collected was used fairly, lawfully and transparently. The data was stored at the appropriate security on a two-factor authentication device and the file path was encrypted. This mitigated unauthorised processing, loss, access damage or destruction to the data. The virtual machine will use a type one hypervisor throughout the communication vulnerability assessment process which runs directly on the researcher's computer without using a separate operating system, this requires a separate machine to control the virtual environment. The highly secured virtualisation system only works if the operating system that is running within the hypervisor has direct access to the hardware [8], with nothing in-between that can be compromised in a cyberattack. Type one hypervisors are also capable of operating virtualised desktops, which the vulnerability assessment will be conducted through a Kali Linux VM, this can also be seen within Fig. 1.8. Therefore, this shows that the researcher has thought thoroughly about research ethics and data security.

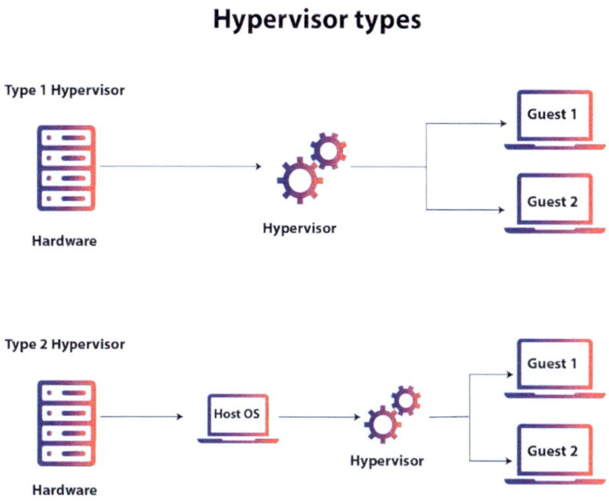

Fig. 1.8 Hypervisor types [8]

1.3.2 Methodological Choice and Evaluation

The methodology enables the research question to meet the research aim and produce a personalised awareness at every level for cyber-attacks on drones. This will be implemented by low-cost simple hacking techniques, which enables the data analysis to be set up for success and ensures that the research was able to outline some of the security vulnerabilities between the drone and GCS. Quantitative data was used, which helped express both the literature review and data analysis part of the assignment. The literature review found clear vulnerabilities when drones are talking to the GCS using the IEEE 802.11 standard. The vulnerabilities found that attack vectors like DoS, KRACK and a dictionary attack [28, 64] can be used to manipulate the communication or even control the drone. Therefore, this would set the data analysis up for success and enable the researcher to infiltrate the communication.

Secondary data enables the researcher to gather essential information in order for the research to be achieved, with the information pulled from multiple different search engines. Data collected was in accordance with the Data Protection Act (2018) and the research was conducted within an ethical manner. Although the cyber-attacks will be set up within a VM for ethical reasons, these types of attacks are being conducted within the real world and have the same effect. This has been seen in the Russian and Ukraine war, and it is essential that all drone communication vulnerabilities need to be secured.

1.4 Data Analysis and Critical Discussion

The data analysis will cover how the drone and GCS communicate through a wireless local area network (WLAN), which sits within the family of IEEE 802.11 protocols. This application has given the autonomy to work remotely from home, at the employees work place and now the operation of drones. WLAN was created in the 1990s, which originated from research within radioastronomy, which consisted of radio waves and complex equations within various different environments [33]. Wireless networks are said to be used by over five billion devices globally [33], this is the most modern way to establish a connection between two devices without the use of cables. Drones function as surveillance systems that can monitor radio frequencies and wireless fidelity [25]. However, there are detrimental security concerns focusing WLAN security for FPV drones.

The accessibility of WLAN has created a lot of opportunities for cybercriminals to find vulnerabilities within a wireless network. Cybercriminals have managed to find weaknesses within Wi-Fi protected access 2 (WPA2), service set identifier (SSID) and media access control (MAC) filtering, which WLAN security protocols operate off [96]. Drones are vulnerable within many different areas and various attack vectors can be used to infiltrate the connection between the GCS and drones. The attack vectors used within the ethical hack will be a MITM attack, which will include DoS attack, KRACK attack, dictionary attack and eavesdropping. The ethical hack will see the hacker situate themselves between the GCS and Drone, the data analysis can then be fabricated or modified [6]. The ethical hacker (EH) will initially conduct a DoS attack where the GCS will lose connection to the drone [27, 62], while the EH will establish a good two-way connection with the drone and then conduct a dictionary attack, using the brute force technique [90]. The EH can gain full control of the drone by manipulating the communication between the GCS and Drone, where the EH can eavesdrop the feed by seeing what the drone is looking into. This will show the research's aim and highlight the vulnerabilities between the drones that use Wi-Fi to communicate.

Additionally, once the security vulnerabilities have been documented the Drone Information Governance framework will be written to protect the communication link, drone, and GCS to ensure that organisations or individuals can be held accountable for their actions. Information governance is a super discipline that has evolved with new legislation that has been tightened, in order to meet legal demands, privacy concerns and to govern business or organisations [86]. Information governance frameworks were written due to external pressures such as data breaches, hacking and there are multiple policies that need to be followed or obeyed due to law [86]. Furthermore, Information governance is not just a project but an ongoing cycle that always needs to be updated for when new or updated legislations permits. Information governance is constantly evolving and should be updated once new risks have come to the surface. The communication between the GCS and drone is pertinent, which shows the importance of implementing an information governance framework to ensure that all data remains secure. The data analysis will now be conducted on the FPV drone through a DoS attack, Krack attack and dictionary attack in order to show the vulnerabilities within the communication link.

1.4.1 A Denial-of-Service Attack with a Dictionary Attack on an FPV Done

The first stage of the practical part of the research was ensuring that the DJI Ryze Tello drone was configured correctly by ensuring that the communication link between the GCS and drone are secured. Additionally, it was pertinent that the laboratory was established within a virtual and secure environment for ethical reasons [18], which will ensure that only the researcher will have access to the network. Wireless networks are used within everyday life which is assessed that one in three households use the default network, leaving the network name and password completely open [60]. Additionally, over 40% of homes use a password that is considered weak, and 15% of homes utilise an extremely outdated wired equivalent privacy (WEP) [60]. Figures 1.9 and 1.10 will talk through the wireless configuration. Wireless communication links must be secured correctly for the above reason and especially in a conflict situation because this could put lives at risk.

The FPV drone was configured using the Wi-Fi name of CONNORS-DRONE and the password was set as connor77 for research purposes only. A secure password should have at least twelve characters, contain a symbol, not be a previous password used, a combination of uppercase and lowercase letters [72]. A memorable phrase like "6MonkeysRLooking^" [72] is considered a strong password. However, to create awareness a commonly used password encryption has been used to show that using a weak password encryption can cause detrimental effects and then be infiltrated [60].

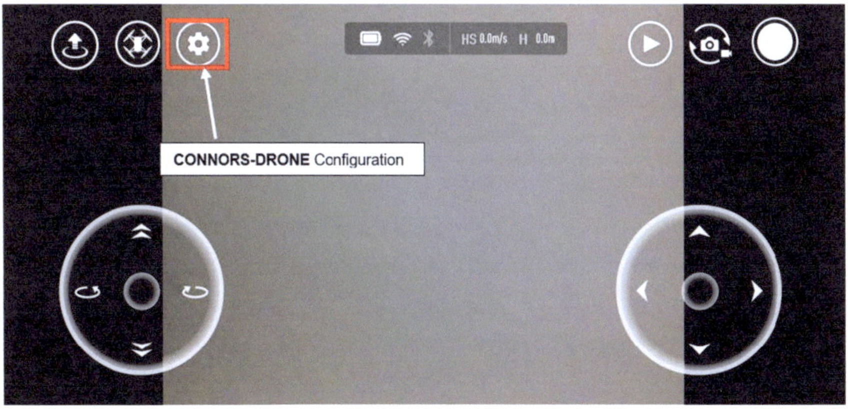

Fig. 1.9 Connors drone configuration

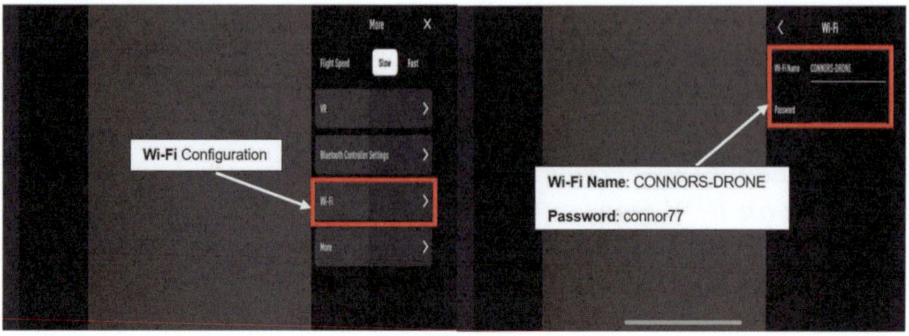

Fig. 1.10 FPV drones Wi-Fi name and password

Figure 1.11 shows that the drone has connected to the ground controlling station using the password connor77, which is now shown as connected on the furthest right-hand image of the picture. GCS often comes up in one of three categories, stationary, shipboard and mobile [102]. A stationary GCS is a permanent command post and is usually within a trailer or container, which can be transported to a battlefield and be ready for use. Shipboard GCS is located on a Navy vessel or any type of boat if a drone is required. Mobile GCSs are transportable and can be operated through any device that enables wireless communication and are often used within close quarters [102]. A mobile GCS was used for connecting to the FPV drone.

Once the configuration had been made between the FPV drone and GCS. The laboratory was created using VirtualBox [97], it was essential throughout the scenario that the Computer Misuse Act [30] was followed. Additionally, it was pertinent that all ethical hacking procedures implemented did not affect the internet and the ethical hack had been authorised with all risks considered. However, within cyber security ethical hacks are essential to test vulnerabilities [48, 56, 77]. The VM used was a 64-bit Kali Linux operating system (OS) and can be seen within Fig. 1.12.

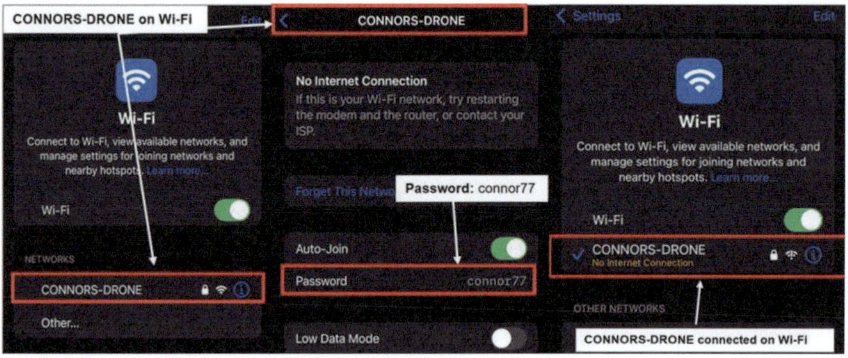

Fig. 1.11 CONNORS-DRONE connected to the GCS

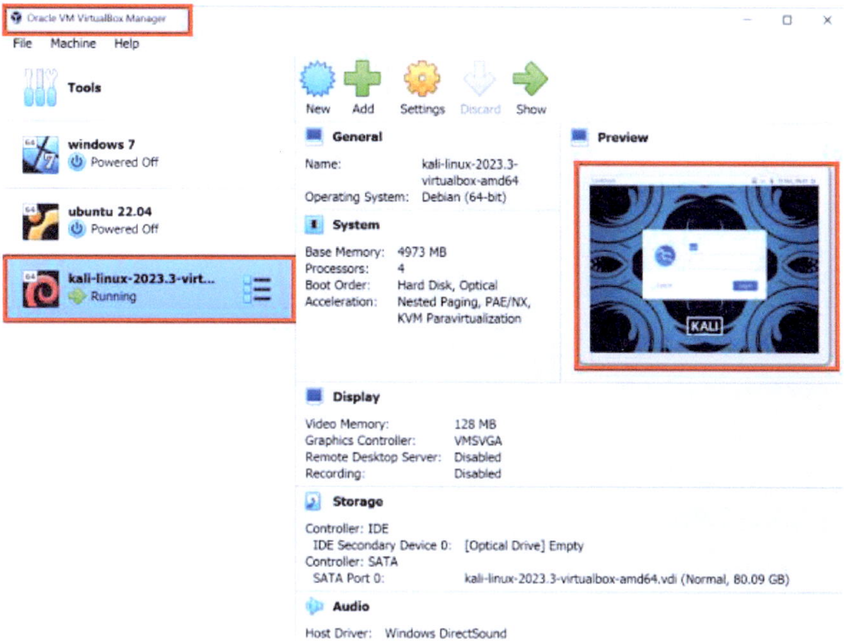

Fig. 1.12 VirtialBox and Kali Linux VM

The first WLAN standard was created in 1997, which was called IEEE 802.11 [19]. This has now evolved and there are many different variants coming under the IEEE 802.11 umbrella. The most recent standard is IEEE 802.11ax (Wi-Fi 6) which was approved in 2019 [26]. However, this is not available to everyone as the standards are being updated consistently. All standards have flaws and can be infiltrated by cybercriminals much like the DoS attack vector. A DoS attack on the drone will cause the connection between the GCS and drone to fail [105], this will be initiated through a bandwidth and connectivity attack. A bandwidth attack depletes all the network's resources, which causes the drone operator unable to pass requests. Connectivity attacks flood the network with too much information, which will cause the network to be exhausted and unable to answer any requests [105].

The laboratory below will demonstrate how a DoS attack works by means of deauthentication attack on the communication link between the drone and GCS. This will flood the bandwidth, and the GCS will be unable to gain access to the FPV drone. The first command that can be seen within Fig. 1.13 is "#airmon-ng check kill" to kill any process that could trouble the adapter. Followed by "# airmon- ng start wlan0" to start the adapter, which was then checked to ensure that the adapter was working and in monitor mode by the "#airmon-ng" and "#iwconfig" commands. This enables the ethical hacker to detect any networks that are operating within the adapters range. Additionally, the "# ip addr" command was given to find out the ethical hacker's internet protocol (IP) address which allows the ethical hacker to manipulate and view

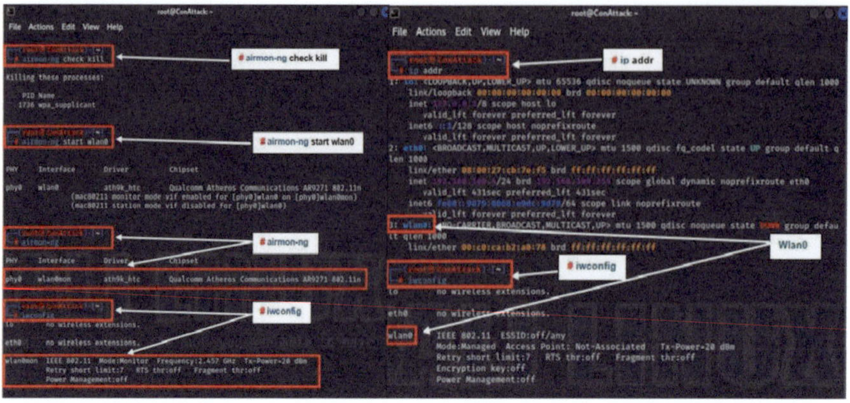

Fig. 1.13 Network adapter configuration

the devices network interfaces [23]. This is crucial information because it provided the IP addresses, netmasks and broadcast address [23]. In addition, the "#iwconfig" command shows that the network adapter is in monitor mode. The next stage of the laboratory will show how the ethical hacker will detect the communication link between the GCS and drone.

The ethical hacker conducts a wide area search by the "airdump-ng wlan0mon", this was used to search all wireless network devices within range. The ethical hacker is mainly looking for CONNORS-DRONE Basic Server Set Identifier (BSSID). This is the media access controls (MAC) physical address of either the access point or wireless router that is used to connect to Wi-Fi [9]. The search that was conducted by the ethical hacker found that CONNORS-DRONE BSSID is 48:1C:B9:E9:99:9B, which is the communication link between the GCS and drone. Also, the drone is operating on channel five (Fig. 1.14).

The network of the drone will be investigated by the "#airdump-ng -w connor2 -c 5 –bssid 48:1C:B9:E9:99:9B wlan0mon" command, which will tell the adapter to search all the devices that are using the network for the operation of the drone and GCS, which will capture anyone using the drone. The ethical hacker will now have everything needed to target the communication link between the drone and GCS [28]. Additionally, a file was created within the close area search command "-w connor2", which enabled a file to be created in order to capture the initial communication link between the GCS and drone, this can be seen within Fig. 1.15 with the EAPOL capture.

The communication link between the GCS and drone is now compromised because the ethical hacker now has everything needed to launch the deauthentication and dictionary attack vectors. The deauthentication attack vector was then initiated by using the command "#airplay-ng –deauth 0 -a 48:1C:B9:E9:99:9B wlan0mon". This broke the link between the GCS and the drone. The attack vector command is broken down as so, the "– deauth 0" will enable the deauthentication attack, which won't stop unless commanded by the ethical hacker. The "-a 48:1C:B9:E9:99:9B" was what

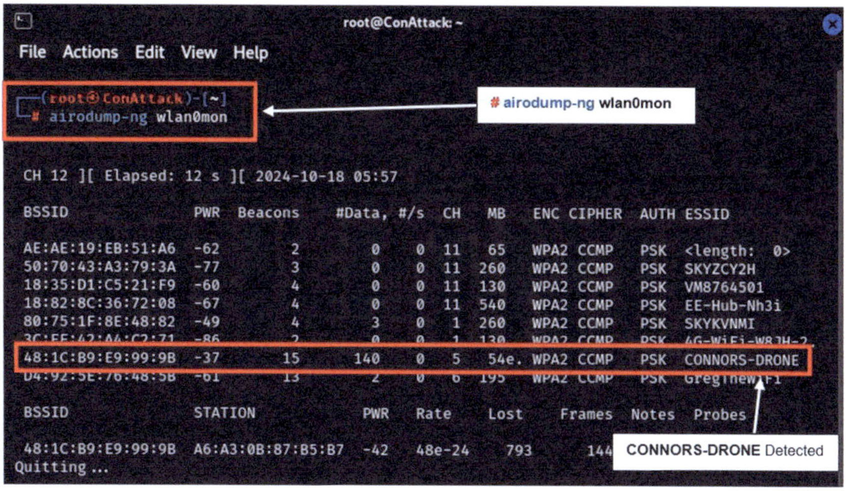

Fig. 1.14 Wireless wide area search

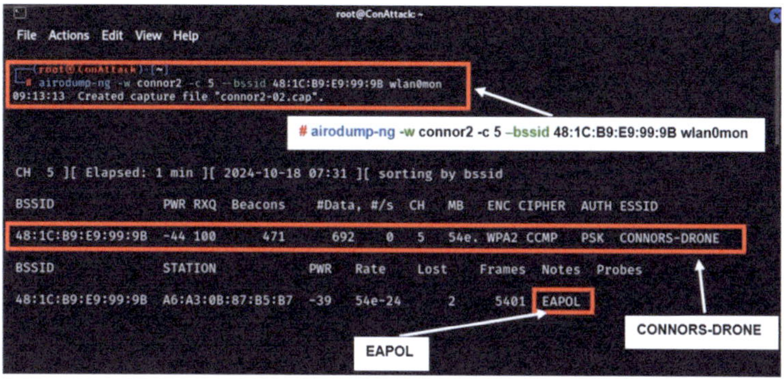

Fig. 1.15 Wireless close area search

targeted the communication link between the GCS and drone through the network adapter, "wlan0mon". This targeted the wireless networks BSSID that was communicating from the GCS to the drone, which is shown within Fig. 1.16 where the bandwidth attack is overwhelming the drone [105], and the GCS is unable to connect to the drone. The DoS shows that it is critical to secure the communication link between all controlling stations and drones because if the communication link is compromised by the wrong organisation it could cause severe risk to life, especially within a modern-day conflict [64]. Therefore, at the national level a framework needs to be pushed out with policies and permissions to ensure drone operators know how to operate the systems correctly.

Figure 1.17 shows that the ground controlling station was unable to connect to the drone due to the deauthentication attack. Simple hacking techniques have had

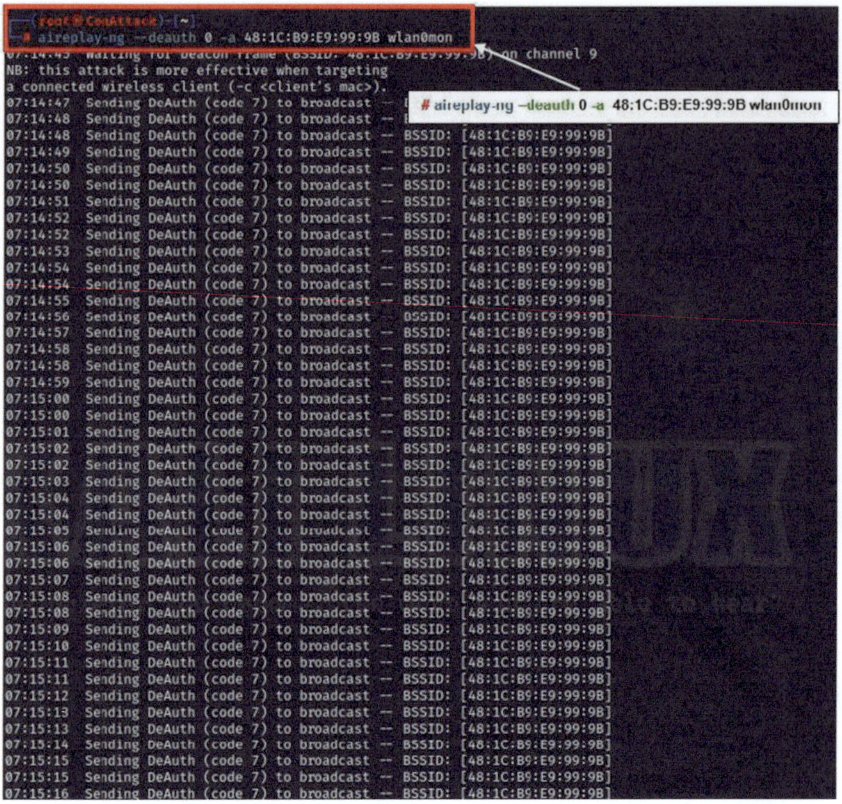

Fig. 1.16 Deauthentication attack vector

great success on the majority of systems that use the IEEE 802.11 standard. This now shows that even FPV drones that use Wi-Fi are not secure enough.

The ethical hacker can now conduct the KRACK or dictionary attack [28, 63], as the GCS doesn't have control of the drone. The command "wireshark connor2-01.cap" was given, which would see the four-way handshake. This was captured during the WPA2 EAPOL authentication process. This can be seen within Fig. 1.18 at the bottom of the image. Wireshark captures the initial communication link between the GCS and Drone, which was captured during the authentication process and captures the four-way handshake.

Furthermore, when the EAPOL was captured a password list was created to speed up the process. However, password lists can be downloaded and include every possible password for a specific system [63]. This enables any cybercriminal to have password lists when required to conduct a brute force dictionary attack. Additionally, this shows that the communication link between the GCS and drone is not secure. Therefore, there needs to be policies or frameworks put into place at the highest level (Fig. 1.19).

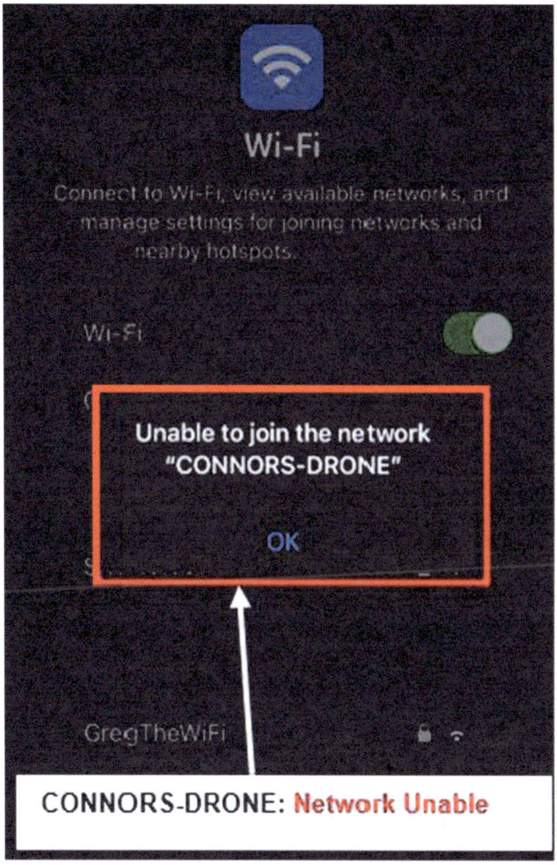

Fig. 1.17 Network unavailable

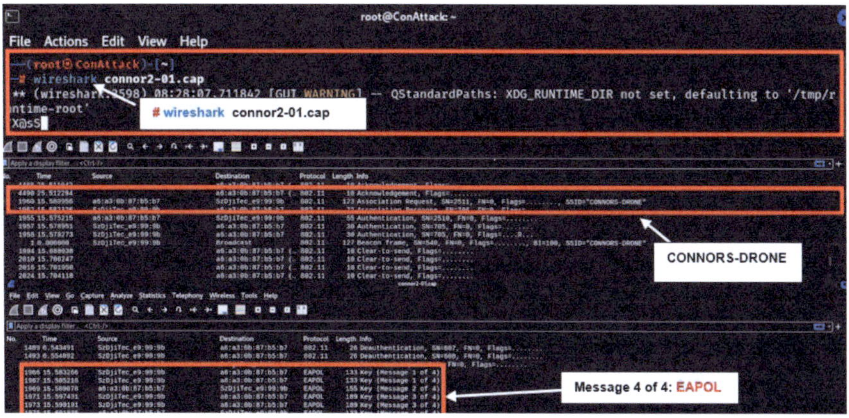

Fig. 1.18 Wireshark EAPOL capture

Fig. 1.19 Password list

The brute force attack is now ready to be conducted [90], which is initiated by the "# aircrack-ng connor-01.cap -w /home/kali/Desktop/password" command. The initial command of "aircrack-ng" orders the device to crack the password of CONNORS-DRONE. This will then be done through the file that has been captured from the GCS and the drones four-way handshake, which is known as EAPOL. The command also tells the device to use the password list that was saved to the desktop. The ethical hacker can now gain control of the drone and operate as the new GCS. Additionally, this type of attack has been around for a number of years [63, 90], but is still being operated to gain access to networks (Fig. 1.20).

The password of the drone is now shown within the Wi-Fi settings "connor77", and the ethical hacker can now take full control of the drone. The ethical hacker can now control the drone and operate as the new ground controlling station. The ethical hacker will then change the password and manipulate the drone for their own use. This has documented the vulnerabilities on drones that use the IEEE 802.11 standard as a communication technique (Fig. 1.21).

The practical part of the research has been documented to create a greater awareness on drone security vulnerabilities that use IEEE 802.11 as a form of communication. These simple hacking techniques can be detrimental to militaries around the world [46, 103]. The use of artificial intelligence could process these types of attacks a lot quicker than an individual could, which then could be weaponised as a defensive or offensive mechanism [38, 83] at a larger scale. The majority of FPV drones that are operating within the Russian Ukrainian war are using wireless communication to communicate from the GCS to the Drone.

Furthermore, this could be used as a defensive tactic for when drones are operating within the area of operation [68] and the defending military could gain control of the drone, then use them for their own defensive posture. In an offensive context the drone could be taken control of and then returned to the sender, the FPV would most likely have an explosive warhead attached, which would cause a detrimental effect to the original ground controlling station. All of the above highlights the reasons that this matter should be taken seriously and that both governance and security protocols should be put in place. An information governance framework will now be written to best protect drone operators and create awareness for the vulnerabilities that can occur within the communication link.

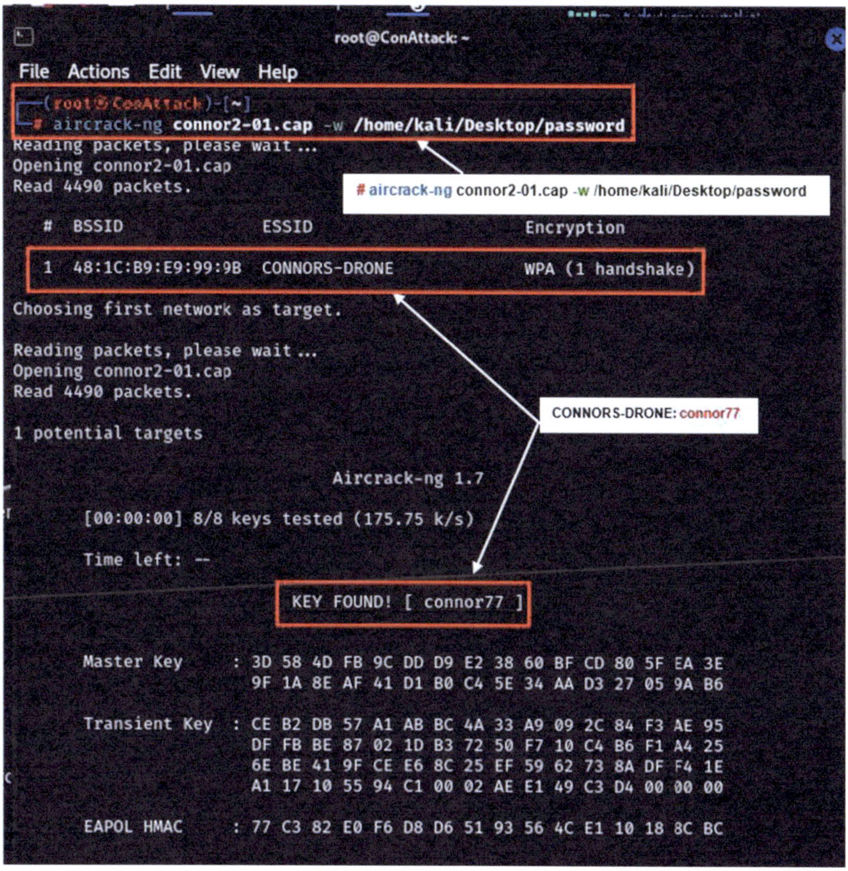

Fig. 1.20 Password crack

1.4.2 A Drone Information Governance Framework

The Drone Information Governance (DIG) framework will explore how all information is collected. The DIG will ensure that all information is recorded and stored correctly. Additionally, it will advise on how the information that drones collect should be stored, governed, and protected [98]. Furthermore, the DIG will explore different procedures when dealing with a wide range of information, which will include how sensitive information is accessed and authorised for use [95]. The DIG will also identify risks, protective measures and appropriate training, which will mitigate a breach of information occurring or information being used unlawfully.

The DIG will endeavour to incorporate all the relevant legal frameworks for how drone information is governed. It is essential that the DIG is in accordance with the UK General Data Protection Regulation [47]. It is pivotal that all drone operators handle information lawfully to ensure that the information is used purposely, fairly,

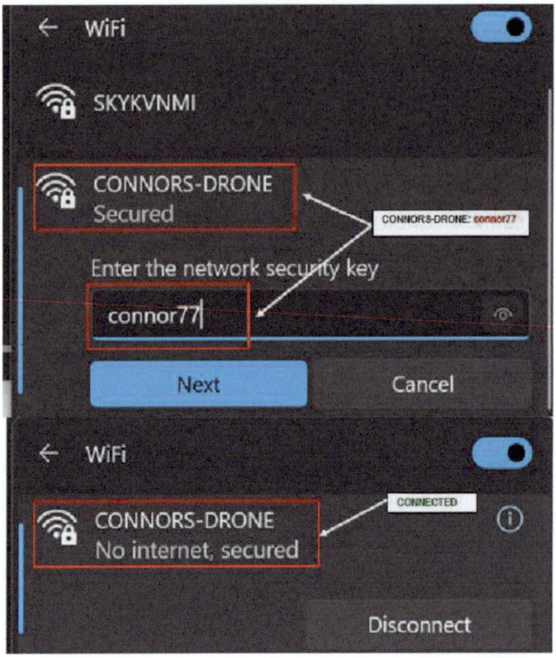

Fig. 1.21 Ethical hacker taking control of the FPV drone

safely and sensitively [95]. Furthermore, the DIG will identify how the government should identify the risk in which drones operate and identify that necessary training is needed prior to the operation of drones, which will ensure that all data is handled correctly.

The new COBIT framework has been created, for effective AI system governance [29]. The new COBIT creates a comprehensive and broad framework for new technologies. The new COBIT framework identifies how AI is becoming an essential driver for efficiency and innovation across all industries and organisations are facing increased pressure to ensure that these are governed correctly [29]. Therefore, with the growing evolution of drones it is essential that all governments, drone manufacturers and drone operators understand that they need to adhere to these laws. The DIG framework will now be broken down below.

1.4.2.1 Information Governance Framework on Drones

Drones are currently being operated within all environments. The drone information governance (DIG) provides security awareness to all drone ground controlling stations (GCS). The framework was created in 2025 to protect drone operators facing harm on an international level. DIG utilises a coordinated and comprehensive approach to the safeguarding of all information that drones hold or collect.

DIG provides a personal security awareness and should be implemented across all governments.

DIG creates a protective environment which should be known as 'the drone controller'. DIG offers a strategic framework to ensure the best available service is offered for the protection of all drone operators. The awareness must ensure that all drone operations are regulated and in accordance with the General Data Protection Regulation (GDPR). This will ensure that the DIG framework is fit for practice by ensuring legal and ethical regulations are complied with.

The purpose of this framework is to illustrate the way in which Drone Information Governance is implemented. The framework will ensure drone operators, drone manufacturers and drone sellers understand all drone procedures and regulations including the handling of all information. This will enable DIG to create awareness to all areas in which drones can be operated in order to have a unified understanding of DIGs goals, structures and purposes. This will mitigate the risk of all drone operators being misunderstood and drones being used unsuitably.

This DIG has created a legal framework and awareness, to ensure that all information in subject is managed, protected and used safely. The purpose of this is to mitigate the risk of all drone GCS and drone operators facing harm.

Without any exception, all drone operators are all within the scope of this framework. As drones are evolving in time, the framework will need to change. It is expected all drone operators will adhere to framework modifications.

All drone operators must sign a legal contract and comply with this framework. Any expectations must be clearly written with the legal agreement between all drone users, manufacturers and organisations of subject.

The objective here is to:

- To guarantee that DIG creates and follows all mandated frameworks, which will include supporting policies. Clear precise procedures and governances are put in place to implement legal and best practices to ensure that all drone operators remain safe.
- To reduce risk to drone operators, the policy will ensure that all data integrity is held at the highest standard. A risk management and disaster plan will be completed. This will enable a strategic and immediate disaster recovery plan to be written if a disaster within drone operations was to arise.
- To provide a secure and controlled environment for all drone operators. This environment will give the drone operators confidence and trust that their information is protected entirely.

Government creates drone related policies to ensure the correct security standards are in place with the correct licences and directs drone manufacture to enforce training prior to the purchase of drones.

Drone Manufacturers ensure that drones cannot be purchased unless the correct licences and credentials have been granted. Additionally, training on the drone must be conducted prior to use.

Chief Information officer (CIO) governs and approves the framework being implemented in DIG. The CIO is the chief authoritative figure for IG and is responsible for authorising the overall governance of all information. The CIOs written consent must be documented before any changes in policy are put into practice. The CIO is also head of information governance and the board.

Senior Information Risk Owner (SIRO) is the head of risk management within DIG. The SIRO must ensure an effective disaster plan and risk management is completed within the DIG. The drone operators are subject to a written agreement. SIRO will be held accountable if objective 4.2 is breached. The SIRO will be a member of the board who will authorise the regulations and procedures when managing the risk within all drone operations.

Information Asset Owners (IAO) will monitor the operation of information held on drones. IAO must grant access for information being shared and make sure that it is compliant with the framework. IAO is responsible for upholding objective. They will lead in the security, protection and classification. IAO will also be a member of the board.

Team Leaders (TL) will provide both governance and guidance to drone operations. TL will ensure that the drone Information governance framework is implemented within their organisations, who are responsible for managing team training. TL must ensure Security Operating Procedures (SyOPs) are protected and signed in their retrospective teams. TL must report to the SIRO and IAO every fortnight on team goals, objectives and progress.

Drone Operators (DO) must adhere to both laws and the framework. They must maintain high levels of professionalism and follow the DIG regulations. All drone operators must complete up to date mandatory training and comply with the government's guidance and instructions. All drone operators will sign SyOPs, prior to drone operations to show that they have understood and agree to the DIG.

1.4.2.2 Policy Framework

DIG has developed an effective policy framework. This is to provide guidance in a Drone Information Governance framework and how it will be implemented into drone operations. The framework will also ensure all drone operators and manufacturers understand the framework produced.

DIG will create a learning culture. This will enable all drone operators to maintain and develop drone operations. This will be achieved by ensuring that all drone operators undertake mandatory training, and the government will invest in drone operator's development, skill set and satisfaction.

DIG will create a security-based environment. This will be achieved by all drone operators signing a confidentiality agreement and making sure the framework is adhered to. This will ensure all data and information is only available to authorised parties. A security first environment will also be upheld through ensuring all drone operators understand their roles and responsibilities.

DIG will have effective internal and external communication. Internally all drone operators will achieve effective team working and management. Every drone operator will conduct integrated exercises, where they will share ideas, knowledge and experience. Externally they will create a secure network, where both parties can talk effectively and freely to identify lessons learnt.

DIG adhere to related legal Acts and policies. This is to ensure that both policies and permission procedures are in line with all legal frameworks. Associated Policies and Acts are identified in the table below:

Policies	Description
The Official Secrets Act (OSA)	Provides the main legal act that provides protection to the United Kingdom (UK), which protects it against espionage and the unlawful leak of information. This will be followed when drones are being operated by the UK military
The Public Record Act	The legislation governing all public records. It establishes a cohesive regulatory framework for public records, which will be followed by DIG
The Data Protection Act	Having the right to find out the information that all drone operators store or hold about individuals to ensure data is protected
The Freedom of Information Act	This provides a general right to access all types of information held by public authorities
Information Sharing Policy	DIG will ensure that drone operators have the appropriate protection when sharing information. This will endeavour to give equal and fair access to the information as required
Documents and Records Management Policy	It will promote effective management when drone operators are handling information
Information Security Policy	This will protect all drone operators information assets. Security measures is within the policy to protect the drone and information

1.4.2.3 Monitoring Measurement and Review Mechanisms

The framework will be monitored annually by the board. In the event of new risks or threats arising to drone operators, the framework will be reviewed and updated. This framework will be updated to analyse risk that could cause contextual harm; this will monitor drone operations continuity and ensure disaster recovery is in place. This will include adding mandatory training to all drone operators and ensure procedures are in place.

Drone operators' goals, tasks and outcomes will be reviewed on a monthly basis by the board. Team Leaders will be expected to provide updates on drone operations, weaknesses, strengths and vulnerabilities must be addressed.

If drone operator breaches the framework regulations, standards and rules drone operators will have their drone licence taken. It should then be decided if the drone operator's licence should be sanctioned and a fine should be given. The board will decide if the drone operator is fit to operate after remedial training has been conducted.

Approval of this framework stands with the CIO. This is because the CIO is the chief authoritative figure for the Drone Information Governance framework.

1.5 Conclusion and Future Work

In conclusion the report has conducted a thorough literature review on drones, which includes how they are currently being operated within modern warfare, the use of drones by non-state actors, the possibilities of how drones will evolve, with the drone communication methods and how drones should be governed. The methodology ensured that the data analysis was set up for success by using secondary and qualitative data. Furthermore, the use of a virtual machine allowed the data analysis to be ethically sound. This enabled the data analysis to conduct a security analysis of the vulnerabilities of drones that operate using the IEEE 802.11 (Wi-Fi) standard, by using simple hacking techniques which then created a DIG framework. The DIG framework was created to secure all information around drones. Although the data analysis was set up within a virtual environment, the effect would be the same in environments that are not secure. The research has critically identified flaws within drone process and implementation, which has created an awareness on the security vulnerabilities. The data analysis has identified that simple hacking techniques could be weaponized by integrating artificial intelligence, for offensive or defensive weapons, this could have detrimental effects to militaries worldwide. It is pivotal to have a critical awareness that drones are becoming more advanced and used within different environments. However, this report has demonstrated the use of drones have both positive and adverse effects.

Future work will see the simple hacking techniques integrated with artificial intelligence to either take control of drones for an offensive or defensive purpose. The DIG framework ought to be strategically used at national level to ensure that all drone operators have necessary training in order to operate a drone. This will positively enable drone operators to conduct a number of tests to ensure they correctly implement updated policies and laws to operate drones correctly. Furthermore, this will endeavour to enable drone operators to require a licence prior to purchasing a drone. Thus, mitigate risks to life and ensure that drones are controlled at the national level, to help the government manage the use of drones. This report has added to the breadth of literature around drone security vulnerabilities and created a DIG framework. This research urges a DIG framework to be implemented at the earliest opportunity. Nevertheless, this research does not reject the fact that future work is needed to ensure the framework is implemented and adjusted to remain fit for purpose.

References

1. Abro GEM et al (2022) Comprehensive review of UAV detection, security, and communication advancements to prevent threats. Drones. Multidisciplinary Digital Publishing Institute (MDPI). https://doi.org/10.3390/drones6100284. Accessed 20 Jan 2024
2. Adams P (2021) Classified Ministry of Defence documents found at bus stop. https://www.bbc.co.uk/news/uk-57624942. Accessed 20 Jan 2024

3. Aerial (2024) The evolution of drone technology over the years. https://aerialspotter.com/the-evolution-of-drone-technology-over-the-years/. Accessed 31 Jan 2025
4. ALFA (2024) AWUS036NHA. https://alfa-network.eu/awus036nha. Accessed 22 Dec 2024
5. Alphaswift (2023) The evolution of drone technology: past, present, and future. https://alphaswift.com/blog/the-evolution-of-drone-technology. Accessed 20 Jan 2025
6. Amoordon A et al (2022) A single supervised learning model to detect fake access points, frequency sweeping jamming and deauthentication attacks in IEEE 802.11 networks. Mach Learn Appl 10:100389. https://doi.org/10.1016/J.MLWA.2022.100389. Accessed 20 Jan 2024
7. Andersson JJ, Simon S (2024) Minding the drone gap drone warfare and the EU. https://www.iss.europa.eu/sites/default/files/EUISSFiles/Brief_2024-17_Drones%20%281%29.pdf. Accessed 12 Nov 2024
8. appviewx (2025) Hypervisor. https://www.appviewx.com/education-center/hypervisor/. Accessed 10 Jan 2025
9. Atera (2024) Computer terms unwrapped: what is BSSID? https://www.atera.com/blog/computer-terms-unwrapped-what-is-bssid/. Accessed 11 Jan 2025
10. Baghai C (2024) The MQ-9 reaper: the pinnacle of unmanned aerial superiority. https://christianbaghai.medium.com/the-mq-9-reaper-the-pinnacle-of-unmanned-aerial-superiority-ef11bcf875b0. Accessed 11 Dec 2024
11. Barnes J (2025) A weapon resembling a toy telephone is changing the face of warfare on the Ukrainian battlefield. The Telegraph. https://www.telegraph.co.uk/world-news/2025/01/10/russia-fibre-optic-drone-ukraine-war/?msockid=10b2bada09e3685304dda98f08946932. Accessed 11 Jan 2025
12. BBC (2019) Drones monitor south coast of England for migrant boats. BBC News. https://www.bbc.co.uk/news/uk-england-kent-50673241. Accessed 7 Nov 2024
13. Berg R (2023) US hits Iran-backed militias in Iraq after attack. BBC News. https://www.bbc.co.uk/news/world-middle-east-67819124. Accessed 12 Nov 2024
14. BOEING (2022) MQ-25. https://www.boeing.com/defense/mq25#news. Accessed 7 Dec 2024
15. BOEING (2024) MQ-28. Boeing. https://www.boeing.com/defense/mq28#gallery. Accessed 5 June 2024
16. Borsari F, Gordan B (2023) An urgent matter of drones. CEPA. https://cepa.org/comprehensive-reports/an-urgent-matter-of-drones/. Accessed 3 June 2024
17. Bos J (2020) Confidentiality. In: Research ethics for students in the social sciences. Springer International Publishing, Cham, pp 149–173. https://doi.org/10.1007/978-3-030-48415-6_7. Accessed 20 January 2024
18. Boyle G (2023) Addressing ethical issues in your research proposal. https://www.open.edu/openlearn/education-development/postgraduate-studies/addressing-ethical-issues-your-research-proposal. Accessed 10 Jan 2025
19. Bradley M (2023) 802.11 standards explained: 802.11ax, 802.11ac, 802.11b/g/n, 802.11a, Lifewire. https://www.lifewire.com/wireless-standards-802-11a-802-11b-g-n-and-802-11ac-816553#toc-80211aj. Accessed 15 Dec 2023
20. Brooke-Holland L (2015) Overview of military drones used by the UK armed forces. www.parliament.uk/commons-library|intranet.parliament.uk/commons-library|papers@parliament.uk|@commonslibrary. Accessed 6 Oct 2024
21. Brown A (2023) Ethics, autonomy, and killer drones: can machines do right? Comp Strategy 42(6):731–746. https://doi.org/10.1080/01495933.2023.2263333. Accessed 20 Jan 2024
22. Caballero-Martin D et al (2024) Artificial intelligence applied to drone control: a state of the art. Drones. Multidisciplinary Digital Publishing Institute (MDPI). https://doi.org/10.3390/drones8070296. Accessed 20 Jan 2024
23. Cao D (2024) 2 ways to use Linux ip addr command. https://www.howtouselinux.com/post/linux-ip-addr-command. Accessed 10 Jan 2025
24. Cheetham J (2023) Sea drones: what are they and how much do they cost? https://www.bbc.co.uk/news/world-europe-66373052. Accessed 10 NoV 2024

25. Chen HH (2024) Developing a custom communication protocol for UAVs: ground control station and architecture design. Internet Things 27:101319. https://doi.org/10.1016/J.IOT.2024.101319. Accessed 20 Jan 2024
26. CISCO (2025) What Is 802.11ax? https://www.cisco.com/c/en/us/products/wireless/what-is-802-11ax.html. Accessed 11 Jan 2025
27. CISCO Press (2015) Wireless LAN implications, problems, and solutions. https://www.ciscopress.com/articles/article.asp?p=2351131. Accessed 5 Dec 2024
28. CLOUDFLARE (2025) What is a KRACK attack? | How to protect against KRACK attacks. https://www.cloudflare.com/en-gb/learning/security/what-is-a-krack-attack/. Accessed 10 Jan 2025
29. COBIT (2025) Leveraging COBIT for effective AI system governance. https://www.isaca.org/resources/white-papers/2025/leveraging-cobit-for-effective-ai-system-governance. Accessed 1 Feb 2025
30. Computer Misuse Act 1990 (1990) Computer misuse act. https://www.legislation.gov.uk/ukpga/1990/18/pdfs/ukpga_19900018_en.pdf. Accessed 6 Aug 2023
31. Corera G (2023) Ukraine war: cyber-teams fight a high-tech war on front lines. BBC. https://www.bbc.co.uk/news/world-europe-66686584. Accessed 3 June 2024
32. Corera G, Mackintosh T (2024) Red Sea: US, UK and French destroy dozens of Houthi drones. BBC News. https://www.bbc.co.uk/news/world-middle-east-68524596. Accessed 9 Nov 2025
33. CSIRO (2020) Bringing Wi-Fi to the world. https://www.csiro.au/en/research/technology-space/it/wireless-lan. Accessed 5 Dec 2024
34. Daly D (2024) Best drones for 2025. Flying. https://www.flyingmag.com/guides/best-drones-for-2025/. Accessed 24 Dec 2024
35. Davies B (2024) Putin red-faced after his £11,500,000 stealth drone is shot down by own warplane. https://metro.co.uk/2024/10/05/putin-red-faced-11-500-000-stealth-drone-shot-warplane-21740515/. Accessed 11 Oct 2024
36. DHL (2024) DHL logistics trend Radar 7.0. https://www.dhl.com/content/dam/dhl/global/csi/documents/pdf/glo-csi-logistics-trend-radar-7-0.pdf. Accessed 7 Nov 2024
37. DJI (2021) The use of drones in agriculture today. DJI Enterprise. https://enterprise-insights.dji.com/blog/drones-in-agriculture#:~:text=Agriculture%20drones%20can%20be%20used%20to%20do%20anything,costs%2C%20improved%20crop%20quality%2C%20and%20increase d%20yield%20rate. Accessed 7 Nov 2024
38. Emadi A (2024) Aerial intelligence: how AI-powered drones can revolutionize industries. Forbes. https://www.forbes.com/councils/forbestechcouncil/2024/09/18/aerial-intelligence-how-ai-powered-drones-can-revolutionize-industries/. Accessed 11 Dec 2024
39. FCDO (2024) The UK and international humanitarian law. https://www.gov.uk/government/publications/international-humanitarian-law-and-the-uk-government/uk-and-international-humanitarian-law-2018. Accessed 20 Dec 2024
40. Fox (2017) Two plead guilty in first case of drug smuggling by drone on U.S.-Mexico border. Fox News. https://www.foxnews.com/world/two-plead-guilty-in-first-case-of-drug-smuggling-by-drone-on-u-s-mexico-border. Accessed 11 Nov 2024
41. Franke U (2025) Drones in Ukraine: four lessons for the West. https://ecfr.eu/article/drones-in-ukraine-four-lessons-for-the-west/. Accessed 11 Jan 2025
42. Fruhlinger J (2024) What is the CIA triad? A principled framework for defining infosec policies. https://www.csoonline.com/article/568917/the-cia-triad-definition-components-and-examples.html. Accessed 12 Dec 2024
43. Gambrell J (2024) Yemen's Houthi rebels launch drone boat that hits ship in Red Sea as missile strikes another. https://apnews.com/article/yemen-houthi-rebels-red-sea-attacks-israel-hamas-war-2faa48176d4f773003b6a4a3ad5ca73d. Accessed 8 Nov 2024
44. Gebrehiwet L, Negussie Y, Tesfaye E (2024) A review on drone ground control station, configurations, types and the communication systems. Orange Publ Int J Interdiscipl Sci Eng Appl IJISEA Int Peer-Rev J. www.ijisea.org. Accessed 22 Dec 2024

45. Gibbons-Neff T, Shyvala Y (2024) 'Jamming': how electronic warfare is reshaping Ukraine's Battlefields, Russia-Ukraine War. https://www.nytimes.com/2024/03/12/world/europe/ukraine-drone-russia-jamming.html. Accessed 3 June 2024
46. Goldenberg T (2024) Hezbollah's drones are a fierce and evasive threat to Israel. https://apnews.com/article/hezbollah-drones-iran-israel-war-hamas-iran-houthi-386ae3c8deeb4c8997e64c954c3670e5. Accessed 10 Nov 2024
47. GOV (2018) Data Protection Act. https://www.legislation.gov.uk/ukpga/2018/12/contents/enacted. Accessed 7 June 2024
48. GOV.UK (2021) Ethical hackers collaborate with defence to strengthen cyber security. https://www.gov.uk/government/news/ethical-hackers-collaborate-with-defence-to-strengthen-cyber-security. Accessed 29 Sept 2023
49. Greenall R (2024) Russia and Ukraine trade missile and drone strikes. BBC News. https://www.bbc.co.uk/news/articles/c79z0vje935o. Accessed 12 Dec 2024
50. Greenwood Aerospace (2023) ISR aviation: intelligence, surveillance, and reconnaissance in modern military. Greenwood Aerospace. https://www.governmentprocurement.com/news/isr-aviation. Accessed 11 Nov 2024
51. Gupta A, Giri DK, Tripathy T (2024) Detection and tracking of uncooperative unmanned aerial vehicle using a 2-degree of freedom robotic manipulator. IFAC-PapersOnLine 57:333–338. https://doi.org/10.1016/J.IFACOL.2024.05.057. Accessed 20 Jan 2024
52. Hambling D (2016) Cheap off-the-shelf drones are changing how so-called Islamic State and other groups are fighting—and the world's militaries are racing to catch up. BBC Future. https://www.bbc.com/future/article/20161208-how-is-is-using-consumer-drones. Accessed 11 Nov 2024
53. Hannett J, Bish A (2024) Gatwick drone warning ahead of Christmas break. https://www.bbc.co.uk/news/articles/c049w23y25no. Accessed 20 Dec 2024
54. Hassan M (2024) Qualitative data—types, methods and examples. https://researchmethod.net/qualitative-data/#:~:text=Qualitative%20data%20refers%20to%20non-numerical%20information%20that%20captures,experiences%2C%20and%20subjective%20insights%20rather%20than%20numerical%20measurement. Accessed 22 Dec 2024
55. Hassija V et al (2021) Fast, reliable, and secure drone communication: a comprehensive survey. https://doi.org/10.1109/COMST.2021.3097916. Accessed 20 Jan 2024
56. Heiding F, Katsikeas S, Lagerström R (2023) Research communities in cyber security vulnerability assessments: a comprehensive literature review. Comput Sci Rev 48:100551. https://doi.org/10.1016/J.COSREV.2023.100551. Accessed 20 Jan 2024
57. Hiller W (2023) 10 great places to find free datasets for your next project. https://careerfoundry.com/en/blog/data-analytics/where-to-find-free-datasets/. Accessed 22 Dec 2024
58. Hindy J (2023) 20 Google search tips to Google like a pro. https://www.lifehack.org/articles/technology/20-tips-use-google-search-efficiently.html. Accessed 22 Dec 2024
59. ICAO (2023) International Civil Aviation Organization. https://www.icao.int/safety/UA/UASToolkit. Accessed 20 Jan 2024
60. Islam R (2024) Securing your wireless network: an in-depth guide to Wi-Fi password best practices. https://www.gurusoftware.com/securing-your-wireless-network-an-in-depth-guide-to-wi-fi-password-best-practices. Accessed 11 Jan 2025
61. Ismael HM, Tariq Mustafa Al-Ta Z, Emails mordasshani A (2021) Authentication and encryption drone communication by using HIGHT lightweight algorithm. Turk J Comput Math Educ. Authentication and Encryption Drone Communication—ProQuest. Accessed 11 Jan 2025
62. Jiang P, Wu H, Xin C (2022) A channel state information based virtual MAC spoofing detector. High-Confid Comput 2(3):100067. https://doi.org/10.1016/J.HCC.2022.100067. Accessed 11 Jan 2025
63. Kenneth (2024) Download a collection of passwords and wordlists for Kali Linux (2025). https://kennyvn.com/wordlists-password-dictionaries-for-kali-linux/. Accessed 10 Jan 2025

64. Khan AK (2023) How to prevent your drones from DOS attack. Khan Drones Origin. https://khandronesorigin.com/2023/01/27/drones-under-attack-the-threat-of-denial-of-service-dos-attacks/#:~:text=A%20DoS%20attack%20on%20a%20drone%20happens%20when,or%20a%20distributed%20denial%20of%20service%20%28DoS%29%20attack. Accessed 5 June 2024
65. Kleinman Z (2024) US Air Force denies AI drone attacked operator in test. https://www.bbc.co.uk/news/technology-65789916. Accessed 20 Dec 2024
66. Ko Y et al (2021) Drone secure communication protocol for future sensitive applications in military zone. Sensors 21(6):2057. https://doi.org/10.3390/s21062057. Accessed 20 Dec 2024
67. Lagrone S (2023) MQ-25A stingray IOC pushed to 2026 following manufacturing delays. https://news.usni.org/2023/04/04/mq-25a-stingray-ioc-pushed-to-2026-following-manufacturing-delays. Accessed 11 Dec 2024
68. Loh M (2025) Ukraine's drone jammers are proving decisive amid a new push on Russian soil, pro-Kremlin milbloggers say. Business Insider. https://www.msn.com/en-us/news/world/ukraines-drone-jammers-are-killing-it-amid-a-new-push-into-russian-soil-pro-kremlin-milbloggers-say/ar-AA1x1Qjc?ocid=BingNewsVerp. Accessed 7 Jan 2025
69. Mackintosh T (2024) US warplane shot down in Red Sea 'friendly fire' incident. BBC. https://www.bbc.co.uk/news/articles/cj30zk1jnmno. Accessed 22 Dec 2024
70. McArther M (2024) B.C. company using drones to pressure wash windows and buildings. CBC News. https://www.cbc.ca/news/canada/british-columbia/drones-window-washing-bc-1.7319479. Accessed 20 Dec 2024
71. McCombes S (2022) What is a research methodology? | Steps & Tips. Scribbr. https://www.scribbr.co.uk/thesis-dissertation/methodology/. Accessed 3 June 2024
72. Microsoft (2025) Create and use strong passwords. https://support.microsoft.com/en-us/windows/create-and-use-strong-passwords-c5cebb49-8c53-4f5e-2bc4-fe357ca048eb. Accessed 11 Jan 2025
73. Milasauskas T, Jaškūnas L (2024) FPV drones in Ukraine are changing modern warfare. https://www.atlanticcouncil.org/blogs/ukrainealert/fpv-drones-in-ukraine-are-changing-modern-warfare/. Accessed 10 Nov 2024
74. Minett A (2020) The purpose of risk assessments explained. https://www.chas.co.uk/blog/purpose-of-risk-assessments/. Accessed 6 June 2024
75. MRC (2021) Information governance services. http://mrcit.com/mrcit/information-lifecycle-governance-consulting-services/. Accessed 20 Dec 2024
76. Muhammad H (2024) Secondary data—types, methods and examples. https://researchmethod.net/secondary-data/. Accessed 22 Dec 2024
77. NCSC (2023) Joint report on publicly available hacking tools. file:///C:/Users/conno/OneDrive/Desktop/Joint%20report%20on%20publicly%20available%20hacking%20tools%20(NCSC).pdf. Accessed 20 Sept 2023
78. Panasovskiy M (2022) U.S. national guard tests MQ-9 Reaper drone with modified AGM-114 Hellfire R-4 extended-range missile. https://gagadget.com/en/175493-us-national-guard-tests-mq-9-reaper-drone-with-modified-agm-114-hellfire-r-4-extended-range-missile/. Accessed 12 Dec 2024
79. Pantelimon G et al (2019) Survey of multi-agent communication strategies for information exchange and mission control of drone deployments. J Intell Robot Syst Theory Appl 95(3–4):779–788. https://doi.org/10.1007/s10846-018-0812-x. Accessed 20 Dec 2024
80. Read B (2017) Life-saving drones. https://www.aerosociety.com/news/life-saving-drones/. Accessed 20 Dec 2024
81. Robinson S, Cole B (2024) Information governance. TechTarget. https://www.techtarget.com/searchcio/definition/information-governance. Accessed 20 Dec 2024
82. Russel L (2024) What's going on with drones spotted over air bases in the UK? https://news.sky.com/story/whats-going-on-with-drones-spotted-over-us-air-bases-in-uk-13261593. Accessed 30 Nov 2024

83. Ryan R (2025) Military drones and AI regulation: a UK drone lawyer's perspective. https://www.suasnews.com/2025/01/military-drones-and-ai-regulation-a-uk-drone-lawyers-perspective/. Accessed 7 Jan 2025
84. Sharma A et al (2020) Communication and networking technologies for UAVs: a survey. J Netw Comput Appl 168:102739. https://doi.org/10.1016/J.JNCA.2020.102739. Accessed 20 Dec 2024
85. Sivasankari N, Kamalakkannan S (2022) Detection and prevention of man-in-the-middle attack in IoT network using regression modelling. Adv Eng Softw 169:103126. https://doi.org/10.1016/J.ADVENGSOFT.2022.103126. Accessed 20 Dec 2024
86. Smallwood R (2019) Information governance. https://books.google.co.uk/books?hl=en&lr=&id=z3nADwAAQBAJ&oi=fnd&pg=PT11&dq=smallwood+information+governance&ots=71nJpzx2Re&sig=v2gqqkw_AR Tbxenb4ztan-4fLqY#v=onepage&q&f=false. Accessed 10 Jan 2025
87. Sommerville Q (2024) Ukraine thrown into war's bleak future as drones open new battlefront. BBC News. https://www.bbc.co.uk/news/articles/cne4vl9gy2wo. Accessed 9 Nov 2024
88. Sussman B (2022) The drone cyberattack that breached a corporate network, BlackBerry. https://blogs.blackberry.com/en/2022/10/the-drone-cyberattack-that-breached-a-corporate-network. Accessed 20 Dec 2024
89. Swaminathan S, Raajan NR (2022) Improved spectral allocation in optical wireless communication based on quadrature amplitude modulation using metaheuristic wavelength division multiplexing. Optik 271:170121. https://doi.org/10.1016/J.IJLEO.2022.170121. Accessed 20 Dec 2024
90. Swinhoe D (2020) What is a dictionary attack? And how you can easily stop them. https://www.csoonline.com/article/569677/what-is-a-dictionary-attack-and-how-you-can-easily-stop-them.html2. Accessed 10 Jan 2025
91. This vs. That (2023) Cause vs. effect. https://thisvsthat.io/cause-vs-effect. Accessed 22 Dec 2024
92. Troianovski A (2022) Putin announces a "military operation" in Ukraine as the U.N. Security Council pleads with him to pull back. The New York Times, 23 February. https://www.nytimes.com/2022/02/23/world/europe/putin-announces-a-military-operation-in-ukraine-as-the-un-security-council-pleads-with-him-to-pull-back.html. Accessed 9 Nov 2024
93. UK Parliament (2014) The situation in Iraq and Syria. https://committees.parliament.uk/writtenevidence/54858/pdf/. Accessed 22 Oct 2014
94. US Air Force (2024) MQ-9 reaper. https://www.af.mil/About-Us/Fact-Sheets/Display/Article/104470/mq-9-reaper/. Accessed 10 Dec 2024
95. van Dijk N, Casiraghi S, Gutwirth S (2021) The "Ethification" of ICT governance. Artificial intelligence and data protection in the European Union. Comput Law Secur Rev 43:105597. https://doi.org/10.1016/J.CLSR.2021.105597. Accessed 20 Jan 2024
96. Virgin Media (2024) How to enhance your Wi-Fi security. https://www.virginmedia.com/help/broadband/enhance-your-wifi-security. Accessed 5 Dec 2024
97. VirtualBox (2023) Welcome to VirtualBox.org! https://www.virtualbox.org/. Accessed 17 Sept 2023
98. Weill P, Ross J (2005) A matrixed approach to designing IT governance. MIT Sloan Manag Rev 46:26–34. https://www.researchgate.net/publication/236972814_A_Matrixed_Approach_to_Designing_IT_Governance. Accessed 10 Jan 2025
99. WFP (2024) Using drones to deliver critical humanitarian aid. World Food Programme. https://drones.wfp.org/updates/using-drones-deliver-critical-humanitarian-aid. Accessed 8 Dec 2024
100. Whittingham S (2024) Prison 'like an airport' as drones fly in drugs. BBC. https://www.bbc.co.uk/news/articles/c2l9kw511nwo. Accessed 10 Nov 2024
101. Wild Hornets (2024). https://dykishershni.com/en.html. Accessed 7 Nov 2024
102. Xingkai T (2025) RC video ground control stations: link between UAV and ground. https://xingkaitech.com/ground-control-station/. Accessed 11 Jan 2025

103. Yaacoub JP et al (2020) Security analysis of drone's systems: attacks, limitations, and recommendations. Internet of Things 11:100218. https://doi.org/10.1016/J.IOT.2020.100218. Accessed 20 Dec 2024
104. Yousaf J et al (2022) Drone and controller detection and localization: trends and challenges. Appl Sci (Switzerland) 12(24). https://doi.org/10.3390/app122412612. Accessed 20 Dec 2024
105. Zhang CY (2011) DOS attack analysis and study of new measures to prevent. In: Proceedings—2011 international conference on intelligence science and information engineering, ISIE 2011, pp 426–429. https://doi.org/10.1109/ISIE.2011.66. Accessed 20 Dec 2024

Chapter 2
Effect of Open-Source Tools on Commercial Off The Shelf (COTS) Small Unmanned Aerial Systems (sUAS) That Use WiFi 802.11 Transmission System

Nick Costello and Hamid Jahankhani

Abstract Unmanned Aerial Systems (UAS) and their associated technology is a rapidly growing and developing sector with employment across national infrastructure, conflict areas and commercial entities. As more governments, businesses and consumers acknowledge the benefits offered from intelligent, smart devices such as UAS, the risk of device compromise through cyber-attack could not only cause significant data breach, but also present a major risk to safety. However, their sought-after capabilities to provide real-time data often takes priority over the need for robust cyber security measures. Governmental procured Commercial Off The Shelf (COTS) UAS are required to operate with more strict security implementations including WiFi Protected Access 2 (WPA2), encryption and 802.11w implementation. This research investigates the current UAS cyber security landscape and demonstrates vulnerabilities within COTS UAS that can be exploited. Three stages of attacks were designed and conducted on three WiFi enabled COTS UAS using low cost equipment and open source programmes. The communication links were investigated and targeted for cyber attacks including Denial of Service (DoS), deauthentication, Password Recovery and Internal Access. An open source packet sniffing and injection suite of tools, named Aircrack-ng, and a USB WiFi antenna were used to monitor, inject and disrupt the UAS connections. The successful methodology and associated commands are outlined for the attacks and stages within the research. The extensive results documented vulnerabilities within the UAS including successful deauthentication and 4 way handshake capture against Protected Management Frame (PMF) capable systems. Finally, the research addresses a selection of the current UAS security challenges and provides mitigation for the issues identified, to enhance cyber security controls and management.

N. Costello · H. Jahankhani (✉)
Northumbria University London, London, UK
e-mail: Hamid.jahankhani@northumbria.ac.uk

U. B. Chaudhry and H. Jahankhani, *Autonomous Revolution: Strategies, Threats and Challenges*, Advanced Sciences and Technologies for Security Applications,
https://doi.org/10.1007/978-3-031-96782-5_2

Keywords Unmanned aerial systems · Cybersecurity · Hacking · WiFi communication · COTS

2.1 Introduction

The primary motivation for choosing and conducting this research was to explore how COTS sUAS could be impacted by low resource cyber attacks. Research performed for this project sought to be original and multifaceted, it was hoped that the results, recommendations and methodology would provide useful information to UAS vendors and other researchers.

The overall aim of the research was to assess the effectiveness of cyber-attacks on COTS sUAS that implement WiFi 802.11 transmission type by combining low-cost devices and open-source tools. Monitor UAS wireless communications and conduct network scanning to identify and exploit potential vulnerabilities that could compromise the system. Enhance security by recommending mitigations or implementations against similar attack types.

sUAS have become widespread within commercial and non-commercial markets due to their rapidly increasing technology, capabilities and multi-purpose applications. This demand is largely due to the view they provide, easy deployability and significantly reduced cost compared to traditional avionic assets [61]. The invasion of Ukraine in 2022 highlighted the application of conventional military UAS into conflict, however it also documented the growing incorporation and importance of small commercially available UAS on a modern battlefield for both kinetic and non-kinetic usage, demonstrating significant success.

With the increased lethality, effectiveness and utilisation of UAS, Counter-UAS (C-UAS) is a priority for many nations, directing significant financial and personnel resources to create a viable solution but documenting mixed results [61]. Additionally, the size, cost and technical expertise of employing a competent C-UAS system makes it unlikely to defend less strategically important objectives.

Most high-end commercial UAS, currently employed in conflict areas for detailed intelligence, surveillance and reconnaissance (ISR) taskings are controlled via WiFi communication [48]. The commercial UAS's reliance on wireless communication presents opportunities for an effective C-UAS solution through interrogation and manipulation of wireless transmissions [46]. Therefore a low-cost network adaptor and open source software could provide an option to reduce, or mitigate, the success of UAS effects by targeting the WiFi communication network.

2.2 Literature Review

The background for this research focusses on two main areas of importance within sUAS landscape and applies them specifically to this project.

The first area provides a brief overview of UAS variations, types and their various use cases through their rapid technological development and subsequent growth within conflicts. Secondly, it discusses the UAS architecture with regards to hardware, networking capabilities and communication transmissions.

The second area examines the UAS cyber security requirements, potential impacts and risks associated with poor implementation, and vulnerabilities discovered within UAS. This area concludes with related research, based on vulnerabilities identified in devices with similar architecture, that could be customised to influence COTS UAS.

2.2.1 UAS Types, Proliferation and Usage

2.2.1.1 UAS

A UAS incorporates the Unmanned Aerial Vehicle (UAV), the communication data links through WiFi or radio, GCS and satellites to transmit Global Positioning System (GPS) signals [64] (Fig. 2.1). A UAV, or drone, is an aircraft which operates as a pilot-less aerial vehicle, controlled remotely by an operator on the ground or autonomously by an onboard computer [56]. UAVs are operated using the communication link between the GCS and the UAV, either via the GPS or manually via the pilot through Line of Sight (LOS) [53].

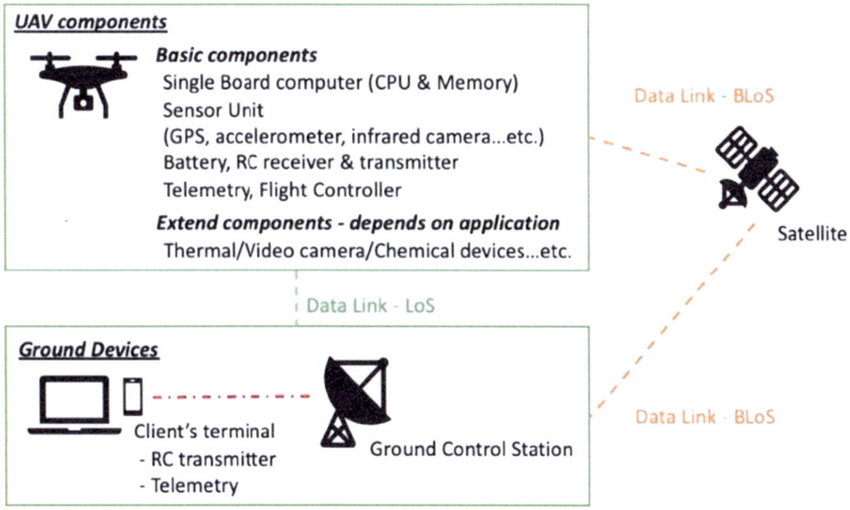

Fig. 2.1 UAS components [56]

2.2.1.2 Categories and Types

UAS are categorised into recreation, commercial and military categories [61]. Recreational UAVs are designed for personal enjoyment or photography and video recording [39]. Commercial UAS applications include agriculture, search and rescue, environmental monitoring, products delivery, communication relays, building site monitoring and traffic monitoring [56]. Military UAS include combat operations, intelligence gathering, reconnaissance and more [53]. Current UAS deployments have blurred the lines between Commercial and Military UAS utilisation.

UAVs are classified into 3 types: multi-rotor, fixed-wing and hybrid. Multi-rotor UAVs use the concept of vertical take-off and landing principle to provide continuous coverage for a particular area [15]. Multi-rotor UAVs require the least resources and are the easiest to construct [51]. The actual power consumption of a UAV depends on various factors such as system performance, sensors, payloads, data processing and loads external or internal [53]. The Civil Aviation Authority (CAA) defines where UAVs can operate based on their weight (Fig. 2.3) and capabilities [56]. The research conducted in this project will be focussed on multi-rotor sUAS Class 1 in categories <2–25 kg.

2.2.1.3 Growth and Conflict Application

The UAS global market growth is estimated to increase from £15.8bn to over £30.2bn between 2020–2025 due to significant software improvements [56]. UAV development has created more than 600,000 jobs and provide net cost savings of £16bn in the UK [56]. Within China, research predicts that the market for sUAS will rise to 75bn yuan [64]. The market to inspect utility transmission and distribution infrastructure is predicted to reach $4.1bn annually by the end of 2024 with the US Drone Energy Alliance suggesting that sUAS under 25 kg will be the platform of choice due to ease of use and cost [33].

The invasion of Ukraine highlighted the incorporation and importance of small commercial UAVs on a modern battlefield for lethal and non-lethal effects [45]. These UAVs, able to be purchased quickly without long procurement and testing cycles, have provided critical capabilities for tactical level units through enemy surveillance and directed artillery fire, significantly reducing the time from detection of forces to an accurate strike within 3 min. This, combined with flight and transport of packages, make UAVs a critical asset [61]. Additionally, the recent technological advancements have provided the ability for flight controls through smartphone applications instead of dedicated remote controls [61] creating ease of use, accessibility and vast ad-hoc, interconnected networks [40]. UAS conducting ISR taskings are weather tolerant assets deployed continuously to an area with sophisticated payloads that provide high-definition imagery. Their associated high costs and capabilities deters these types from lethal, disposable utilisation [35].

Weaponised commercial UAVs through First Person View (FPV), provide highly effective one-way attack and loitering munitions [3], capable of hitting fast moving targets at 12 km ranges. Additionally, commercial UAVs are repurposed and adapted to incorporate military requirements such grenade dropping abilities to provide additional effectiveness. The general public's perception and view on UAS changing from large, expensive, high tech military RPAS to small commercial UAS utilised for surveillance, information operations and explosive delivery will impact countries' future acquisitions.

2.2.2 *UAS Architecture*

UAS architecture including components, communications and capabilities is displayed at Fig. 2.2.

2.2.2.1 UAS Structure

UAV Component

The UAV airframe is the light, pilotless aircraft utilised for flying. The UAV element comprises of a power system, sensors, control system and communication module

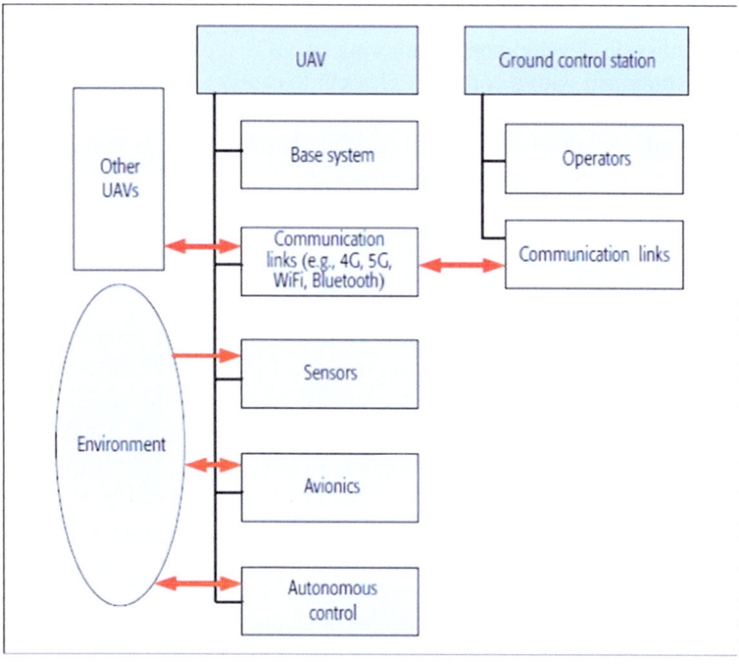

Fig. 2.2 UAS architecture [3]

Table 2.1 Common UAV OS

Flight controller	Open source	Operating system
ArduPilot	Yes	ChibiOS [22]/NuttX [23]/Linux
PX4	Yes	NuttX/Linux
Paparazzi	Yes	ChibiOS/Linux
DJI	No	Custom

[3]. The flight computer consists of critical system flight sensors including pressure sensors, accelerometers, GPS and magnetometer [24]. Payloads are sensors, including Infra-Red (IR), electro-optical (EO), radar and others to collect, process and transmit information for analysis [4]. The base system module forms the Operating System (OS) (see Fig. 2.5) and the foundations of the UAV, linking modules to support communication. The mission controller is the onboard computer and is used to determine the flight plan, mission assignment and control payload sensors [57]. The avionic module converts received commands into commands for flight apparatus from the controller [22] (Table 2.1).

GCS Component

The most important part of the UAS is the base station, which is a ground computer system [57]. The base station serves as a platform for UAV operations, payload operations and mission development monitoring. A GCS typically consists of a wireless router combined with a computer element to capture, process and display data received from the UAV [56]. It should facilitate open architecture and be compatible with various platforms to control multiple UAVs, payloads and communicate effectively over wireless means effectively [52].

Data Link Component

Within UAV-GCS centralised architecture, every UAV must directly connect to the GCS to obtain data [63]. UAV operations will dictate the adopted communication infrastructure as this will have a critical impact on the range [53]. These mission sets can be split into LOS, the transmission of signals happens through direct radio waves [4], or BVLOS (Beyond Visual Line of Site) where the transmission of signals is conducted through satellite communications or relay UAVs [53]. The communication functions incorporate various wireless standards including 4/5G, WiFi, WiFi Direct, and Bluetooth [41], providing significant capability in support systems through multimedia streaming [12]. UAVs optimised with internet use over 4G/LTE can improve efficiency between components, especially over long distances [41]. However, UAS reliance on wireless communication by its nature presents opportunities and vulnerabilities for the system [30].

2.2.2.2 Wireless Communications

IEEE 802.11 standards are globally integrated and used for wireless local area networks (WLAN) and ad hoc networking for a variety of devices, including UAVs [63]. IEEE 802.11 protocol networks provide mobility, flexibility and expansion of communication infrastructure across 2.4 and 5 GHz bands [5].

IEEE 802.11i-2004 standard, known as WPA2, introduced the concept of a security association into IEEE 802.11 to provide link-layer protection for WiFi communications [54]. It defines protected WiFi networks with enhanced security solutions, known as RSN [54]. WPA2 utilises AES and Counter Mode Cipher Block Chaining-Message Authentication Code (CCMP) to provide strong data confidentiality. This protocol is employed within 2.3 billion devices, including the majority of UAS, but has known significant vulnerabilities [8].

In 802.11i standard, when a station connects to an Access Point (AP), it must complete four phases known as the 802.11 State Machine including Network Discovery, Authentication and Association, and 4 Way Handshake (Fig. 2.3).

802.11 Network Discovery

Beacon frames containing the Service Set Identifier (SSID), MAC address, channel and capabilities are periodically transmitted by the AP at a consistent interval to advertise their presence and enable active or passive scanning stations to join that network (Fig. 2.7) [26]. The SSID is the name of the wireless network broadcasted

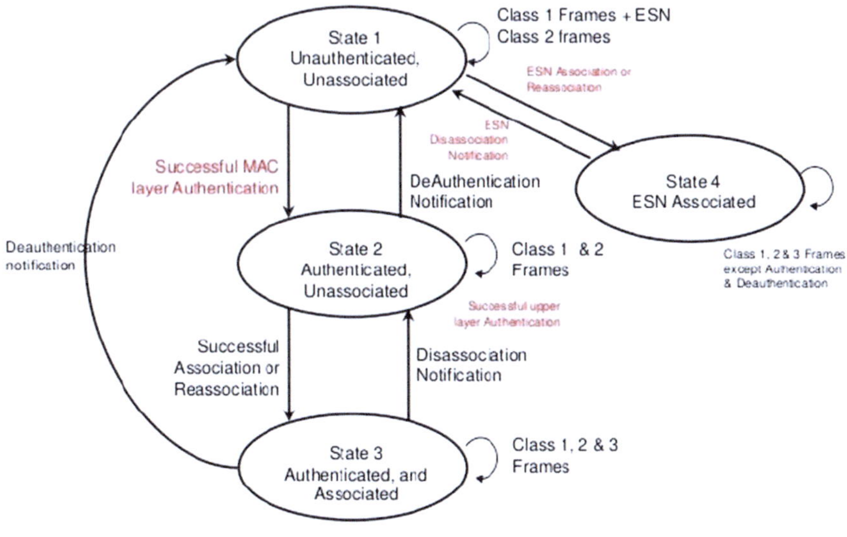

Fig. 2.3 802.11i state machine process [26]

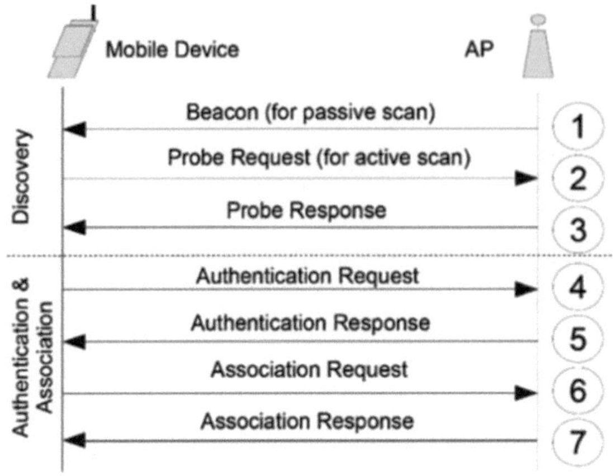

Fig. 2.4 Network discovery, authentication and association process [8]

by the router or AP [38]. The MAC address is unique and assigned to an individual device [6].

The station sends a Probe Request frame to the BSSID selected from the information within the Beacon frame to verify network availability [26]. The AP identified returns a Probe Response with BSSID capabilities, requirements and station's MAC address [6] (Fig. 2.4).

802.11 Authentication and Association Process

To join a wireless network, an 802.11 authentication and authorisation process must be completed. In WPA2 Pre-Shared Key (PSK), the station transmits an Authentication Request Frame and the AP responds back with an Acknowledgement Frame. If the response frame is "successful" then the station has been authenticated by the AP [8]. In WPA3 protocol, the Dragonfly handshake and Simultaneous Authentication of Equals (SAE) is implemented by exchanging 4 authentication frames.

This entire process described in stages 4–7 (Fig. 2.7) is conducted without encryption and incorporates the Open System authentication to join the network [26]. Any security implementation is enforced in the next stage during the 4-way Handshake Phase [21]. At the end of this process, the Pairwise Master Key (PMK) is derived from the PSK and cached on both devices and State 3 has been achieved.

4-Way Handshake

The phase involves the exchange of 4 Extensible Authentication Protocol (EAPOL) messages (Fig. 2.9), generating encryption keys to encrypt the data transmitted [8]. The PSK is used to permit admission to the network with an additional unique new key generated for data encryption. At the end of authentication, the client has its

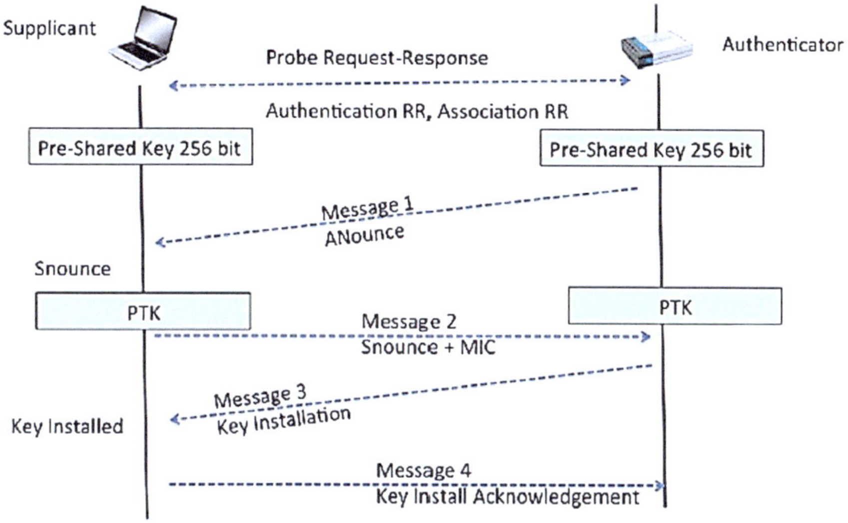

Fig. 2.5 4-way handshake process

PMK, random number (SNonce), own MAC address and AP's MAC address through previous stages. The AP has its PMK, random number (ANonce) and MAC addresses (Fig. 2.5).

Once devices have installed security keys, the pairwise data communication between AP and client now will be encrypted at the link layer by the session key using the ciphers negotiated [54]. Although this implementation protects all data frames, all management frames remain unprotected as these phases are conducted prior to security key negotiation [26].

Protected Management Frames (PMF)

IEEE 802.11w amendment standard (PMF) was introduced into the 802.11i standard in 2014 and made mandatory from 2018 in both WPA2 and WPA3 devices. PMF encrypts data frames and improves the resiliency of the network to prevent eavesdropping and forging to establish confidentiality and replay protection [36]. Management frames include beacons, probes, association, disassociation, authentication and deauthentication used to connect and manage client connections.

PMF achieves enhanced security by adding a Message Integrity Code (MIC) element to the frames transmitted by the AP [56]. The MIC ensures that management frames received have not been injected or altered by a third party and discarded if failed [51], enabling the clients to remain connected to the desired AP [26]. The Security Association Teardown Protection prevents spoofing attacks by conducting an SA Query procedure to verify the authenticity of disassociation requests [54]. PMF has 3 configurations:

- Disable: Disables PMF for this network.
- Capable: Default option for encrypted networks utilising WPA2 and negotiate the use of PMF if capable.
- Mandatory: Only PMF capable clients can connect.

If both AP and client possess capability and implemented, PMF is automatically invoked during association and further communication is encrypted [54].

2.2.3 UAS Cyber Security

2.2.3.1 Importance of Protection

UAS's wide employability and increased autonomy, warrants strong cyber security measures [14]. UAS function on complex software, wireless communication systems and data exchange protocols making them highly vulnerable to malicious actors attempting to gain access to the UAV [59]. UAV exploitation could cause flight control manipulation, compromise on-board sensors and payload interception, leading to severe consequences including property damage, loss of control, potential endangerment of human lives and public safety [50].

To prevent risks to UAV's functionality, integrity and safety, robust measures must be implemented to mitigate potential cyberattacks, maintain public and regulatory trust in their secure operation, and protect sensitive data [50]. This cyber resilience is crucial for maintaining commitments to critical national infrastructure, government services and sensitive taskings [50].

2.2.3.2 Hardware and Software Issues

The exponential demand for UAS has reduced the vendors abilities and motivation to mitigate common security vulnerabilities within devices [22]. UAS are often designed with lightweight, energy efficient hardware which can cause issues for implementing robust, high resource, industry standard security measures [37]. The additional hardware and computing resources required for these implementations challenges the balance for maintaining optimal flight performance and capabilities [50].

Key capabilities such as video transmission encryption require significant resources to maintain real time connection which will impact performance [22], presenting a unique challenge in UAS cyber security [37]. Outdated software, unpatched vulnerabilities and poor coding practices all compound the issue and create large attack surfaces for malicious actors [50].

2.2.3.3 Communication Vector

Wireless communications vulnerabilities are a critical concern as these are targeted by malicious actors for eavesdropping, jamming and/or packet injection [45]. Compromise of communications links affects the confidentiality and integrity of data transmitted, providing the potential for control manipulation and data theft [50].

2.2.3.4 Standard Cyber Security Requirements

Security features are derived based on the perspective of stakeholders including national security agencies, aviation authorities, manufacturers and users with the majority of UAS operators not deeming cyber security as a key aspect of the system. However, minimum UAS security should consist of system access to only authorised operators, guaranteed availability during operational period, confidentiality and integrity through encryption processes, and non-repudiation protocols to ensure operator responsibility [37]. Common UAS platforms document poor UAV protection with weak authentication schemes, limited or null cryptographic suites and limited frequency range and hopping abilities [61].

2.2.3.5 COTS Requirements

Designed and Defence constructed military UAS, deployed with the associated enhanced cyber security capabilities, are well documented within government organisations but their high cost and specialist use prohibits widespread application and reduced mission compatibility. COTS UAS bridge the gap between bespoke military UAS and commercially prevalent UAS. The systems provide the same airborne capabilities as widely available commercial UAS but are adapted aftermarket to the requirements of the purchasing organisation by the manufacturer, instead of the organisation initiating complex custom-made or bespoke solutions. This method enables high reliability by using proven parts and quick progression in development and employment, which are usually limited by budget and time resources [23].

COTS UAS procured, incorporated and utilised within Government organisations have strict security requirements prior to purchase that are not required in standard civilian commercial systems due to nature of operation. For example, these systems require all transmissions between the UAV and GCS to be encrypted to industry standards i.e. AES256 and the system must have robust means to authenticate the user i.e. WPA2.

2.2.4 *Vulnerability Research*

2.2.4.1 Categories and Types of Attacks

Small commercial UAS are vulnerable to various attack vectors [45], demonstrated multiple times through research and displayed in Tables 2.3, 2.4 and 2.5 [58]. Weak security by design processes, poor policies for adoption of authorisation and authentication, partial application testing for UAV control and limited implementation of isolation mechanisms all contribute to significant limitations within UAS security [1].

2.2.4.2 UAS Vulnerabilities

UAV Component

Successful UAV hardware attacks have demonstrated various system vulnerabilities and attack vectors [63]. Researchers identified a vulnerability in an older Parrot AR Drone that enabled installation of compromised firmware due to no signature validation, subsequently causing the UAV to shut down and crash [44].

GPS Jamming and Spoofing attacks prevent the UAV from obtaining legitimate GPS signals by sending commands or data, as if they were transmitted from the control station [14]. GPS jamming causes the UAV receiver to malfunction through attacker's GPS signal, interfering with receiving legitimate GPS signals.

GPS spoofing is a more advanced method since the attacker transmits false messages that may modify the UAV path and direction, rather than produce distractive signals [41]. GPS spoofing attacks using Software Defined Radios (SDR) exploited a lack of authentication and encryption with spoofing no-fly areas and causing the UAV to land immediately. He et al. [22] used similar methodology on any coordinate during autonomous navigation to cause the UAV to fly to unintended locations [22].

Communication Process

Wireless devices by their nature are subjected to multiple attacks due to accessibility (Fig. 2.12). This issue is compounded by lack of secure protocol implementation and the prevalent usage of open Wi-Fi networks presents a fundamental vulnerability that has been exploited numerous times [58].

DoS attacks target the wireless communication process by overloading the system with requests or packets to compromise memory or processing capacity, preventing system availability [45]. Another significant attack is deauthentication, where management frames are sent to two communicating devices to disconnect them. A deauthentication attack is used against a UAV to inhibit communication and prevent further connectivity [16]. This vector can be utilised to capture the 4 way

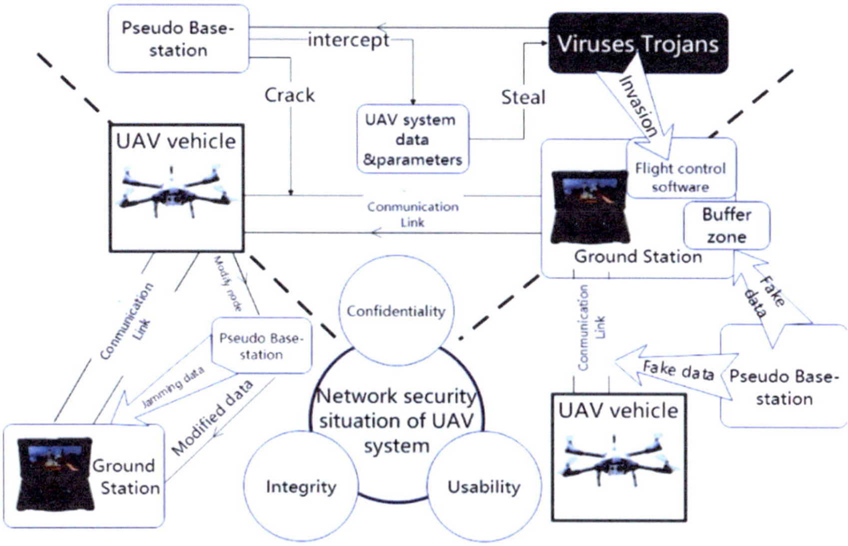

Fig. 2.6 Topography of wireless communication attacks

handshake when users reconnect, and launch subsequent dictionary, brute force and mathematical attacks to access the system LAN [41].

Researchers [19] conducted deauthentication attacks on commercial UAVs controlled by a smartphone application to compromise the device and attain access. In 2019, Westerlund et al. compromised 2 recreational UAVs through 5 attack variations, demonstrating vital security and privacy concerns [29]. Hooper et al. [25] used basic security attacks and equipment to exploit multiple vulnerabilities in commercial Parrot Bebop UAS by disconnecting the primary user, and DoS attacks [25]. Kamkar et al. transmitted deauthentication frames to disconnect the operator from the network and enabling an attacker to hijack the UAV. Tian et al. used a DoS intrusion tool for UAV communication, interrupted the connection through deauthentication and then filled the UAV memory, preventing the controller reconnecting to the system [55]. Karmakar et al. [29] used open-source scripts to change attacker's device MAC address and conduct unauthorised device pairing.

MITM attacks enables an attacker to monitor communication between two parties and attain sensitive data, which they may modify without the victims' knowledge or permission if required [63]. This could be passive where the attacker gains information to the UAS's positioning, or active where information is modified [16]. These Authors demonstrated a side-channel attack on a UAV's Wi-Fi network, demonstrating the potential to determine whether a specific point of interest is being video streamed by a UAV. Research has identified a protocol implemented in professional UAVs and employed by police departments, that lacked the required authentication and encryption. Subsequently, these vulnerabilities were exploited through MITM, FPV analysis and UAV remote control through protocol messages (Fig. 2.6).

Open ports on UAS enable device configuration change through Telnet, FTP and HTTP. This allowed attacker to read files from the FTP folder and provided the ability to upload malicious files if required. Authors identified two vulnerabilities in UAV's OS including Telnet protocol which enabled an attacker to send commands to the UAV, killing the UAV's main process and crashing. Additionally, there was a lack of authentication between the GCS and UAV, enabling an unauthenticated attacker to send protocol messages to take photographs, disable video feed and control the UAV. A similar vulnerability was also discovered and exploited to disrupt the system and manipulate important files [9].

A review was conducted by Warnakulasooriya and Segev [58] on recreational UAS attacks that utilised WiFi communications, demonstrating further the vast number of vulnerabilities and attacks, summarised in Tables 2.3, 2.4 and 2.5.

2.2.4.3 Related Research

Majority of research identified and discussed has been conducted on recreational/commercial UAS with inadequate, restricted or no security implementations. Therefore, the vulnerabilities identified within studies affecting recreational devices does not indicate that similar issues will be evident in COTS UAS that should be technologically advanced, operating with specific robust hardware and software e.g. WPA2, PMF, secure protocols, AES256 encryption and resilient password policies to enhance security. Very limited research on specific COTS UAS or military UAS vulnerability testing has been discovered. Therefore, literature and research discussing the aforementioned security implementations and potential vulnerabilities should be explored to demonstrate the feasibility of this project.

Colter et al. [16] conducted various attacks on a COTS DJI Tello EDU UAV including MITM, spoofed command transmission and DoS. The spoofed commands attack was successful, causing a critical failing of the system [16]. Researchers [32] performed deauthentication attacks against three types of UAV, including one COTS and two recreational types that implemented the IEEE 802.11 protocol to interrupt and deny communications between UAV and GCS. Two recreational UAVs were successfully hijacked with low skill scripts due to poor security procedures but attacks on the commercial UAV was unsuccessful due to device hardening. However, researchers acknowledged with more time, similar results could have been achieved. Researchers [47] discovered vulnerabilities with the Enhanced WiFi protocol (encrypted but no authentication) within commercial DJI UAVs, enabling a successful attack and UAV control by cracking the WEP. This methodology can be applied to WPA2 protocol implementation, although presenting a much stronger challenge.

Zhi et al. [64] conducted a deauthentication and dictionary attack on a WPA2 DJI WiFi platform that enabled the researchers to compromise the system [64]. This attack is similar to various research conducted on WPA2 AP routers and handshake capture using deauthentication to exploit protocol. Nozomi Networks [31] conducted vulnerability analysis against commercial DJI Mavic 3 UAV utilising WiFi communication protocol. The researchers achieved Handshake capture through deauthentication and used a cloud based GPU to crack the WiFi password, enabling connection to the UAS. However, both studies discussed were conducted on communication links without PMF enabled.

With the introduction of PMF on WPA2 and WPA3, deauthentication and disassociation attacks should not be effective. However, PMF vulnerabilities have been identified via successful DoS attacks against WPA3 protected networks using script manipulation, documenting PMF's susceptibility to DoS attacks [13]. Researchers using WPA3Fuzz found 3 vulnerabilities through DoS attacks against several devices implementing WPA3 [42]. Research has demonstrated the effectiveness and success of a DoS attack has against an enabled device when targeting PMF enabled routers [28].

Therefore, based on the known technology implemented and prior research conducted on similar areas, the aim of the research should be achievable in COTS UAS and address an area with limited research previously conducted.

2.3 Research Methodology

2.3.1 Introduction

The section provide the background of the study, applicable technology and relevant research. The aim of the subsequent three are to present the research conducted on the three UAS including the Research Methodology, Results and Analysis. This section focuses on the methodology process, hardware and software used to collect the data.

The primary aim of this testing is to conduct an overview and assess the impact of various attacks on three UAS, previously identified within the research proposal to determine whether COTS UAS are secure in their stated configurations. The UAS will be subjected to the same types of attacks initially and further investigated if required based off their capabilities or identified vulnerabilities. The UAS will be subjected to reconnaissance, DoS attacks, Deauthentication attacks and attempted unauthorised access to the internal system. Each stage will gather the results and data from the identified attack types.

2.3.2 Motivation

The prevalence, ingenuity and adaptation of UAS within conflicts has documented the exponential use of commercial UAS, providing opportunities to identify vulnerabilities and exploits. A low-cost device and open source tools could provide the opportunity to reduce or mitigate the success of UAV effects by targeting the UAS communication network.

2.3.3 Experiment Setup

The experimental setup consists of several hardware and software components described below. The UAS were selected based off their designation as COTS UAS, system availability and their use limited to non-operational commitments. All UAS comprised of a GCS, a UAV and utilised WiFi communication links.

2.3.3.1 Hardware

The hardware utilised for this research comprised of:

1. Parrot Anafi Thermal UAS
2. Parrot Anafi USA UAS
3. SkyRanger R60 UAS
4. WiFi Alfa Network Adaptor AWUS036NHA
5. HP Laptop running a Kali Linux Operating System

Parrot Anafi Thermal UAS

Parrot UAS are Linux-based quad-coptors which are smart device pilotable, utilising 802.11 wireless communication to provide on screen control and real time video link [49].

The onboard 802.11 WiFi modules enable a client controller through a smartphone, tablet or dedicated SkyController to control the aircraft [25]. The devices bind to the aircraft or SkyController directly by connecting to an AP via an application that runs on Android and iOS platforms. Once both powered up, the application initiates this discovery process with a TCP handshake between the controller and UAV to establish communication channels and data is sent between connected devices through JSON records via UDP [25].

Parrot Anafi Thermal has been developed and built to meet the requirements of professionals ranging from the construction industry and energy sector, to civil security agencies. It utilises WiFi 802.11a/b/g/n transmission systems with an operating frequency 2.4–5.8 GHz through WPA2 using AES128 encryption key. It consists of a dual-core CPU and quad-core GPU embedded system running a Linux-based OS with a built in Wi-Fi AP, GPS and camera. This UAV can be controlled by a smart device connected to the SkyController 3 (Fig. 2.7).

Parrot Anafi USA UAS

Parrot Anafi USA is an Anafi Thermal-based UAV with additional enhanced cyber security configurations and increased robustness to hardware to meet the demands of crisis response, security forces, military and pan-government agencies.

Fig. 2.7 Parrot ANAFI Thermal

Fig. 2.8 Parrot ANAFI USA

The network communication between UAS and controller are authenticated with WPA2 AES CCMP cyphering and each UAV/controller pair has a unique encryption key generated. With regards to disassociation attacks, Parrot claim "The protection of the 802.1x management framework is activated to prevent all known disassociation attacks, which could cause service interruption" (Fig. 2.8).

SkyRanger R60 UAS

SkyRanger R60 is designed for secure, aerial intelligence for disaster response, overwatch and security. SkyRanger uses WiFi and Secure Network Pairing with AES256 encryption. The SkyRanger UAS includes a quadcopter and is controlled remotely by a pilot using a flight controller containing a small screen to input commands and

Fig. 2.9 Aeryon SkyRanger R60

to receive information. R60s are deployed with over 20 militaries and in use with public safety and industrial operators in over 30 countries (Fig. 2.9).

WiFi Alfa Network Adaptor AWUS036NHA

The Alfa Network AWUS036NHA is a popular wireless WiFi USB antenna that supports monitor mode. It is routinely used for wireless penetration testing due to its high sensitivity enabling it to detect weaker signals, compatibility with open source software, and supports Monitor Mode and Packet Injection.

2.3.3.2 Software

Kali Linux Setup

Oracle VM VirtualBox (v7.0.12) installed and a Kali Linux (v2023.3) VM loaded onto the laptop. Kali has many pentesting appropriate software suites preinstalled and all software was running at the newest release version through sudo apt upgrade. Cog was the username utilised within the profile.

Open Source Software

A variety of software (all open source and free) was used to facilitate the attack types.

1. Monitoring software
    a. Wireshark is a packet analyser tool used to inspect network traffic and conduct packet or protocol analysis on communication links. This was used to collect information and to monitor the deauthentication attack.
    b. Nmap is a network scanner used to discover hosts and services on a network by transmitting packets and analysing the response.

   c. Kismet is a remote packet sniffer used for WiFi, Bluetooth, RF etc. Information is provided through a GUI.

2. Attack software

   a. airodump-ng. Utilised for packet capture of raw 802.11 frames and to monitor the communications.
   b. aireplay-ng. This tool allows frame injections and was used to conduct a DoS and Deauthentication attack on the UAVs.
   c. Metasploit Framework. A project that provides information about security vulnerabilities and exploits.
   d. Hashcat. This is a password recovery tool that offers multiple modes for obtaining the information required effectively.

2.3.4 *Stages and Attack Types*

The methodology and attack vectors were applied to all three UAS. The attacks were separated into 3 stages; DoS, Deauthentication and Unauthorised Internal Access.

2.3.4.1 Stage 1—DoS Attacks

The first attacks are variations of the DoS attacks, which compromises the Availability aspect of the CIA triad [32]. This type of attack prevents legitimate users from accessing the system by overloading the network with information [32].

Aireplay-ng will launch DoS attack through Beacon Frame flood and Deauth requests to assess the effects on the system. Beacon Frames are not protected by PMF [54] and therefore selected to test device parameters. In both types, packets will be sent at an exceptionally rapid rate, to overwhelm the target and subsequently become inoperable [58]. Due to the security implementations such as PMF and Frequency Hopping within COTS UAS, these attacks are theorised to be unsuccessful. However, manipulation of flags, including reason codes within the software, types and packet rate can be altered and tested to potentially produce an effect (Fig. 2.10).

2.3.4.2 Stage 2—Deauthentication Attack

The second attack vector is a deauthentication attack, where the researcher aims force the operator to disconnect from the UAV, making the system inoperable [32]. Additionally, when the operator reconnects, the 4-way handshake may be captured within the WPA2 implementation [14]. Research documenting this attack is widely published, however with PMF enabled across devices, it is theorised that only a DoS

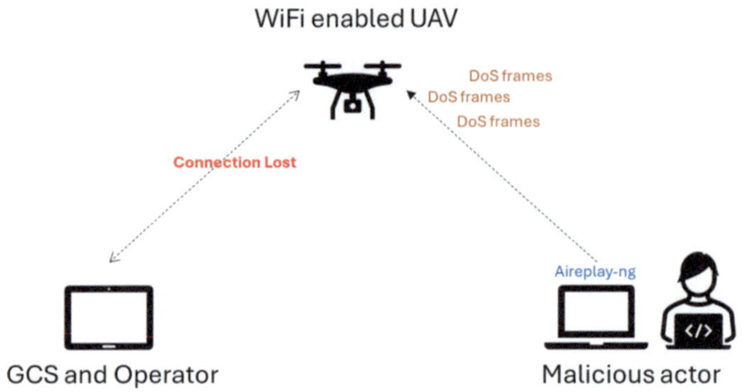

Fig. 2.10 Denial of service attack

Table 2.2 Deauth-rc code

Code	Description
6	Class 2 frame received from non-authenticated STA
7	Class 3 frame received from non-associated STA (Aireplay-ng default)
46	In a disassociation frame: disassociated because authorized access limit reached
47	In a disassociation frame: disassociated due to external service requirements

effect through resource exhaustion in computational limited UAS may be achieved [54].

Aireplay-ng delivers the attack by specifying the UAV's MAC address, the controller's MAC address, setting the appropriate flags to have the desired effect, and delivering packets to both. Number of packets will be altered in line with research e.g. short burst vs prolonged attack [54] to affect potential PMF vulnerabilities. Airodump-ng is utilised to monitor communications and notify the attacker if the WPA2 handshake has been captured.

Reason codes are flags within the deauthentication frames that provide a reason for the deauthentication to occur. These can be changed based on the scenario to manipulate the systems. The codes selected below were based on their potential for success, due to variation between deauthentication or disassociation frame, and to test the previous claims from Parrot Anafi USA (Table 2.2; Fig. 2.11).

2.3.4.3 Stage 3—Unauthorised Internal Access

Unauthorised Internal Access attack is based on the attacker successfully connecting the network directly and accessing the resources within the UAS. This connection can be achieved through password recovery, connecting during deauthentication attack or through legitimate credentials.

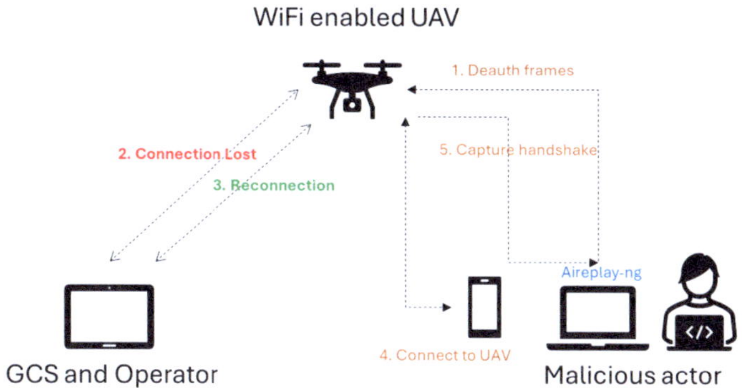

Fig. 2.11 Deauthentication attack

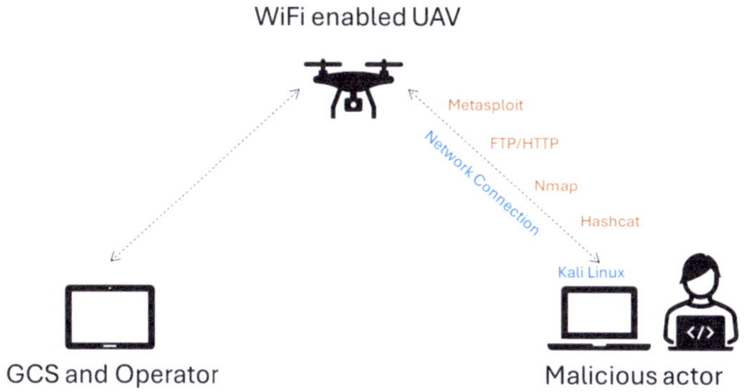

Fig. 2.12 Wireless unauthorised internal access attack

Once accessed, nmap will return a list of open ports, services running and device's OS, that could lead to vulnerability identification. If open access is available on protocols such as FTP and Telnet, then internal access could potentially be achieved quickly. Metasploit can be used to exploit known vulnerabilities derived from information through nmap scans (Fig. 2.12).

2.3.5 Experimental Methodology

The testing was conducted in a controlled environment and external networks not affected.

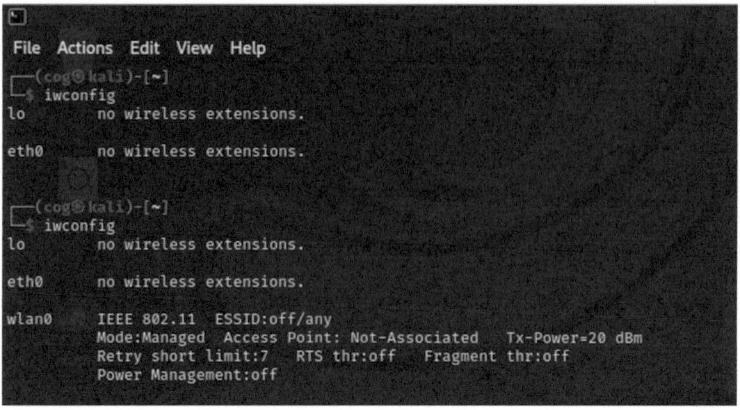

Fig. 2.13 iwconfig environmental check

2.3.5.1 Preliminaries

These preliminaries are required for attack Stages 1 and 2 on all UAS.

1. Environment check: Current wireless interfaces and configurations through iwconfig (Fig. 2.13).
2. Tool Installation: Aircrack-ng suite is pre-installed within Kali Linux. Command sudo apt install aircrack-ng would install software if not.
3. Switch to monitor mode: Network interface switch through sudo airmon-ng start wlan0. Wlan0 is now designated as wlan0mon. Use iwconfig to verify interface has transitioned to wlan0mon (Fig. 2.14).
4. Network monitor: A network scan initiated with sudo airodump-ng wlan0mon to identify active, local wireless networks and connected clients (Figs. 2.15 and 2.16).
5. Targeted Packet Capture: This process requires the BSSID and channel number to isolate the devices and network. The command sudo airodump-ng -d (BSSID) -c (channel) -w (name of packet capture file) wlan0mon to acquire packets from only target's BBSID and channel for future analysis within Wireshark. This terminal remains open to assess the effects of the Stage 1 and 2 attacks including potential handshake capture (Fig. 2.17).

2.3.5.2 Stage 1—DoS Testing

1. Information required for testing and device capabilities was gathered through Kismet, Wireshark packet analysis and airodump-ng, including the AP's and client's MAC.

    a. Wireshark opened with wireshark filename.pcap file captured during the airodump-ng command.

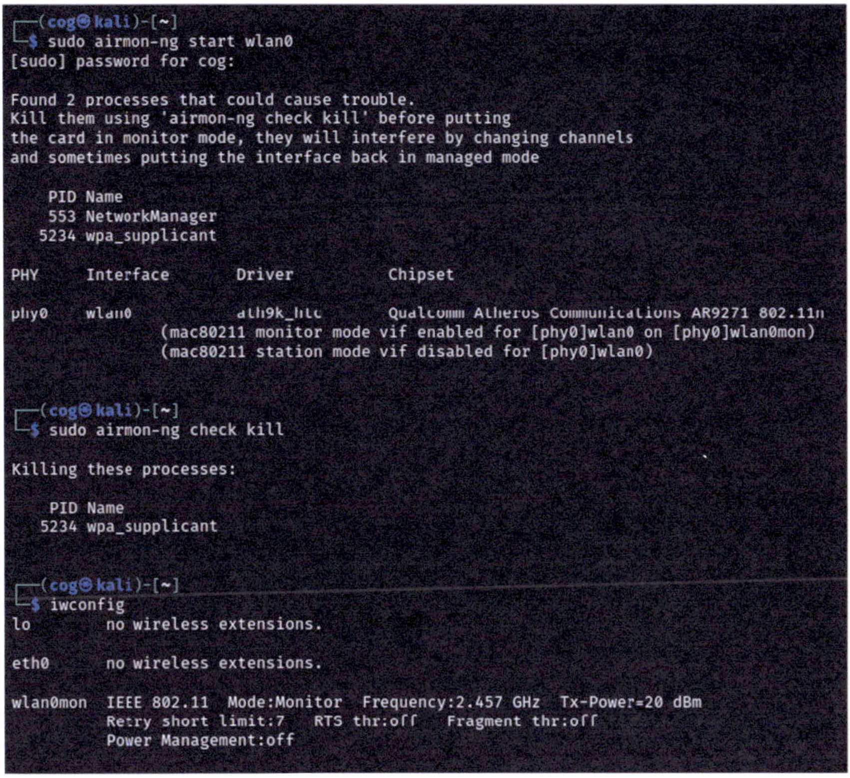

Fig. 2.14 wlan0 to wlan0mon process and commands

   b. Kismet is initiated through the CLI via Kismet command and viewed in the Firefox Browser.

2. To attack the specific network, a new terminal is opened with Kali and the default aireplay-ng deauthentication attack is launched using command sudo aireplay-ng --deauth (number of packages sent) -a (AP MAC Address targeted) -c (client MAC Address) wlan0mon. The –deauth 0 flag ensures the attack continues without limitation parameter.
3. A similar version of the attack is conducted but the default reason code (code 7) (deauth-rc) within aireplay-ng is amended to include codes 6, 7, 46, and/or 47 with sudo aireplay-ng --deauth 0 -a -c wlan0mon --deauth-rc.
4. Injecting management frames via beacon frame flood attack using the command sudo aireplay-ng -2 -v 8 -u 0 -w 0 -h -a wlan0mon. -2 flag selects the aireplay-ng interactive replay attack, -v 8 signifies beacon frames and remainder are for no WEP encryption.

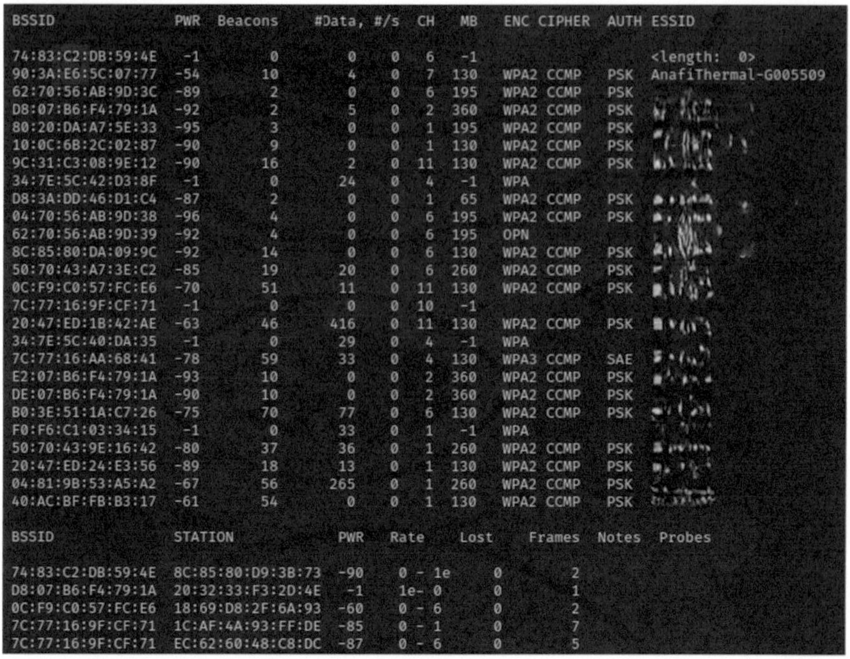

Fig. 2.15 airodump-ng network scan example results

Fig. 2.16 Identified target network and information

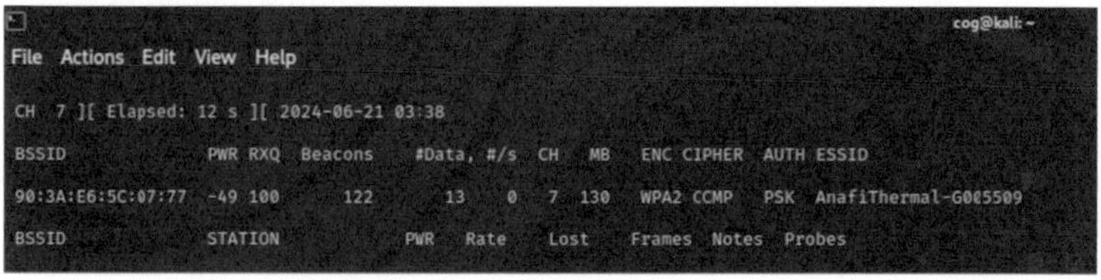

Fig. 2.17 Target network packet capture. Remains running whilst conducting attacks

2.3.5.3 Stage 2—Deauthentication Attack Testing

1. The same default aireplay-ng Deauth attack to be conducted as Stage 1.
2. The deauthentication reason code (deauth-rc) change attack as Stage 1 but with increased combinations and run concurrently.
3. Any WPA handshake captures are to be analysed through Wireshark for EAPOL messages confirmation.

2.3.5.4 Stage 3—Internal Access Attack Testing

1. Connect to UAS WiFi network through legitimate credentials or password recovery (if handshake captured) via Kali Linux. Tool hcxpcapngtool to convert the.pcap file with the captured handshake into a hashcat readable hash file. This hash file can then be run with hashcat command to ascertain the hashed password.
2. Once on the network, iwconfig confirms connection and ip address show ascertains the IP range for the network. Scan network with nmap scan using -sC, -O or -sV (or similar) and identify open ports, services and versions through enumeration. Nmap –script vuln is then used to identify potential vulnerabilities with the UAS, providing the CVE and potential method to exploit.
3. Analyse ports and protocols and exploit vulnerabilities through Metasploit Framework or similar open source tools.

2.3.6 Limitations

Research will only be conducted on three UAS with vulnerabilities and exploits limited to 802.11 spectrum. UAVs will not be airborne when conducting assessments and only currently non-operational devices are used. UAS have been granted permission to be researched and utilised by product owner. MAC addresses of the UAV and controller were identified and specified to increase effectiveness and to reduce the risk of unintended disruptions.

This will not be incorporated with additional sensors to identify type or movement of UAV. Other researchers could combine with additional antennas to increase range or incorporate with SDN for increased attack surfaces.

2.3.7 Ethical, Legal and Professional Issues

It is important to identify the scope of data and information collected, with clear documentation of what it will be used for to product owners and stakeholders prior to collection. The data can only be used for the purposes of this project and will be destroyed after a defined reasonable period has passed. Care must be applied to resources and connectivity parameters employed throughout the project. Data and systems utilised will be maintained within a protected area and network. Any data collected will be subject to encryption at rest and in transit. On project conclusion, any and all raw or processed data will be destroyed, and any sensitive data exported out will be appropriately anonymised as stipulated within DPA 2018 and UK GDPR.

All research will be conducted on site and no sensitive information will be identified within the report. Any sensitive imagery will be removed if discovered deliberately or accidentally. Any significant vulnerabilities discovered will be reported to the duty holder, product owners and vendors if required. All research will be conducted ethically and within the legal regulations instructed in the Computer Misuse Act 1990. The researcher cannot foresee any professional issue that may occur due to this research. If these research findings can be extrapolated to other systems, then this may benefit the organisation.

2.4 Research Results

2.4.1 Experiment Results

The results are initially segmented by UAS and then by stages of testing. The attack vectors are divided into type of attack, commands used, effect on UAS and subsequent evidence including terminal screenshots and UAS imagery.

Due to R60's communication link between GCS and controller operating on 5 GHz band, no DoS or Deauthentication attacks could be conducted with researcher's equipment. However the WiFi communication link was set to WPA2-Personal and therefore theoretically should have a similar effect. Additionally, the volatility of the system, through age and extended usage, created difficult conditions for internal access and therefore only limited information was ascertained before the system shutdown but enough nmap scans were conducted prior to failure.

2.4.2 Stage 1—DoS Testing

2.4.2.1 Parrot Anafi Thermal

Information

The AP's MAC 90:3A:E6:5C:07:77 (UAV) and client's MAC A0:14:3D:C0:01:5F (SkyController 3) were ascertained and RSN flags within Wireshark captures identified that the system was PMF capable with flag set to True (Figs. 2.18 and 2.19).

Default DoS Deauth

The default aireplay-ng deauthentication attack was launched, set initially with packet number of 8 which had an effect, as seen in previous research on PMF but unfortunately was not documented. Subsequently, this attack with -0 flag had no visual or noticeable effect on the network, controllability or communications after 2 and 4 min (Fig. 2.20).

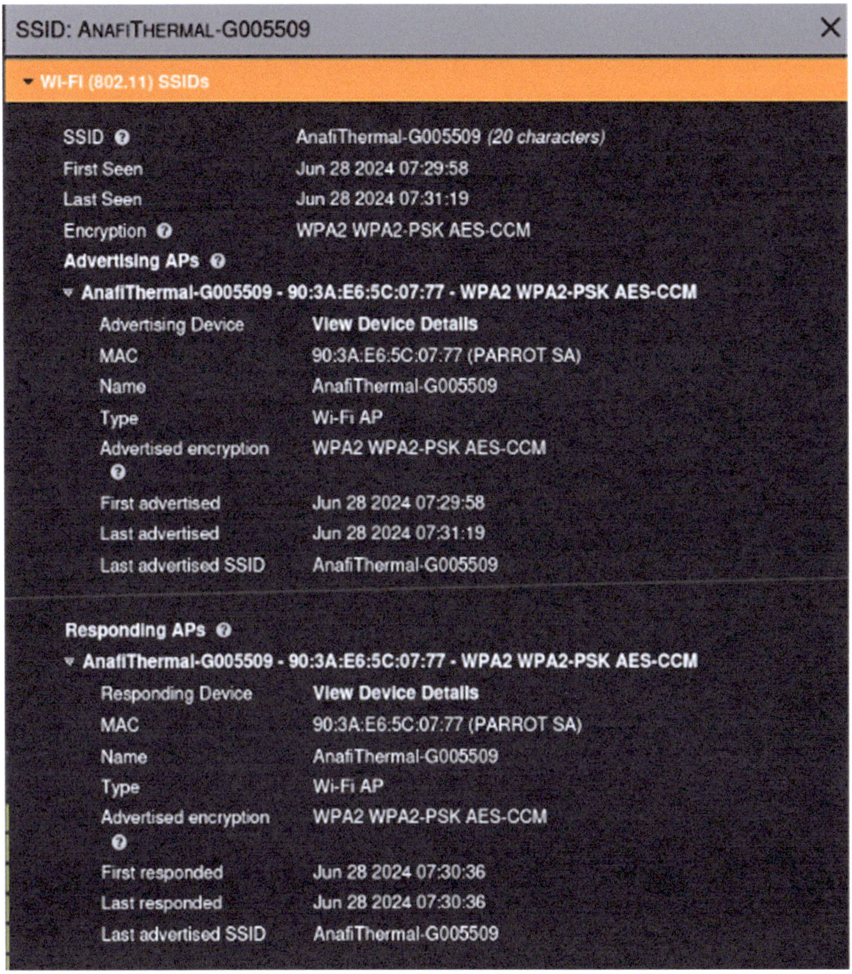

Fig. 2.18 Kismet information from Anafi Thermal UAS

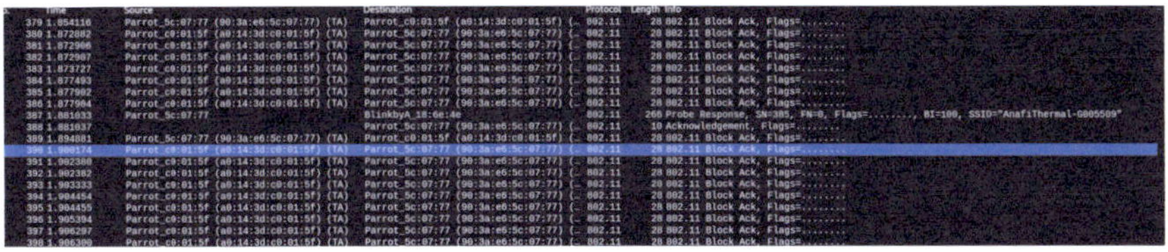

Fig. 2.19 Wireshark information from packet capture including RSN and MAC addresses

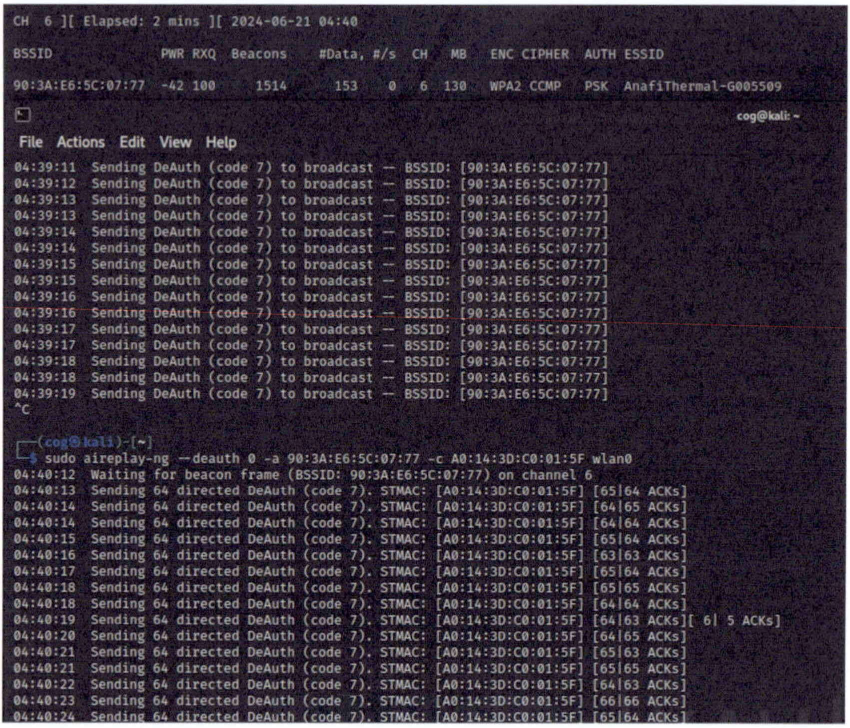

Fig. 2.20 Airodump-ng target network and default aireplay-ng attack running

Fig. 2.21 Anafi Thermal connection pre-DoS default deauth attack and during

Deauth-rc DOS

The deauth-rc within aireplay-ng was altered to codes 6 and 7 to provide an additional vector. Two deauth requests were sent simultaneously to assess effect on computation resources. This attack had a noticeable effect on the network and device, initially causing latency followed by lost connection and inoperability. The devices did re-establish communication links after the attack was concluded (Fig. 2.21).

Fig. 2.22 Anafi Thermal connection pre-DoS Beacon Frame flood attack and during

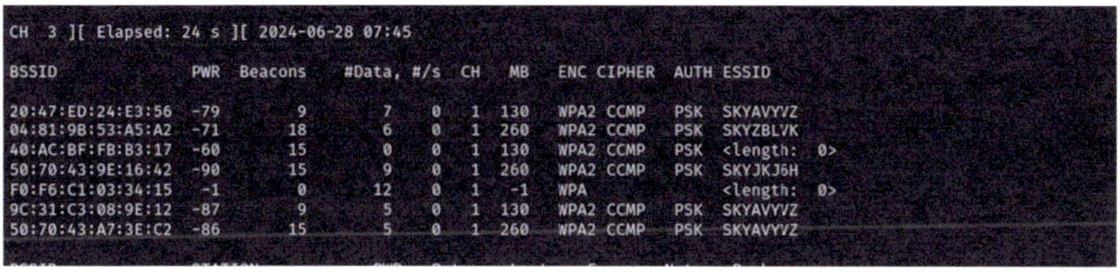

Fig. 2.23 No network advertisement for Anafi Thermal during subsequent scan

Beacon Flood

Beacon Frame flood was conducted to assess the attack's effects on the network and was replicated on 4 terminals, causing a significant effect on the network including latency, poor controllability and decreased image resolution. Once the attack was complete, functionality resumed, however the AP stopped transmitting Beacon Frames up to 2 min post-attack, impacting on the ability for external devices to connect and on network discovery protocols (Figs. 2.22 and 2.23).

2.4.2.2 Parrot Anafi USA

Information

The AP MAC 90:3A:E6:BC:33:36 (UAV), client MAC A0:14:3D:91:52:2E (Controller) and RSN capabilities, including PMF, was obtained (Figs. 2.24 and 2.25).

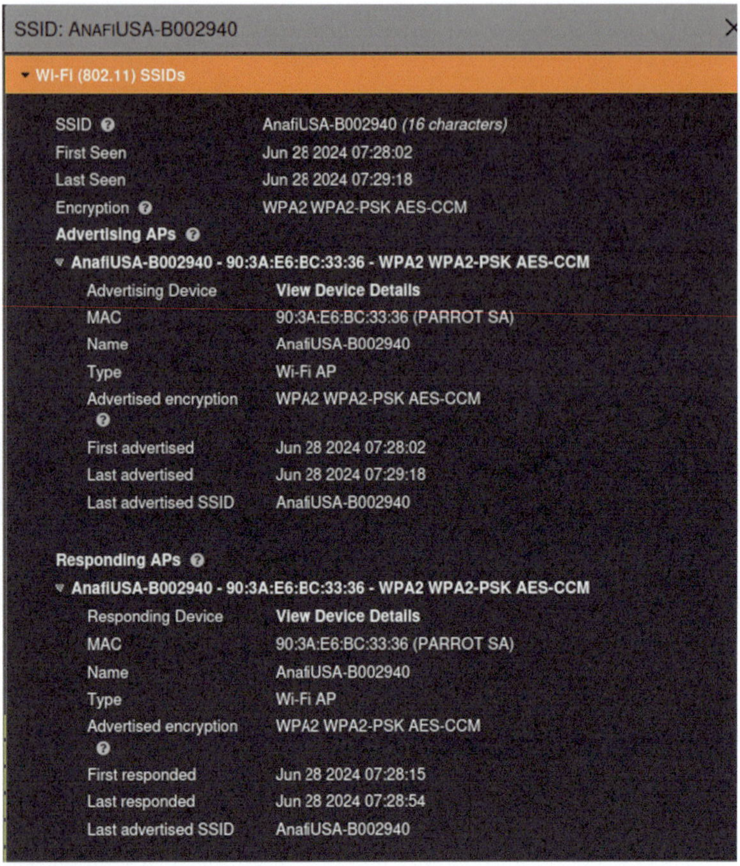

Fig. 2.24 Kismet information from Anafi USA UAS

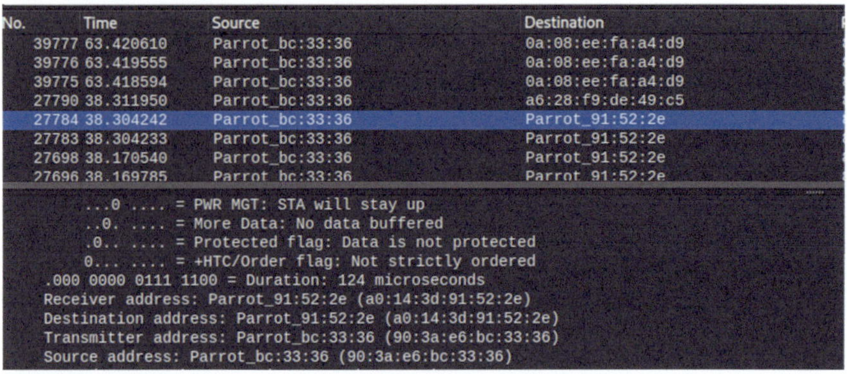

Fig. 2.25 Wireshark information from packet capture including RSN and MAC addresses

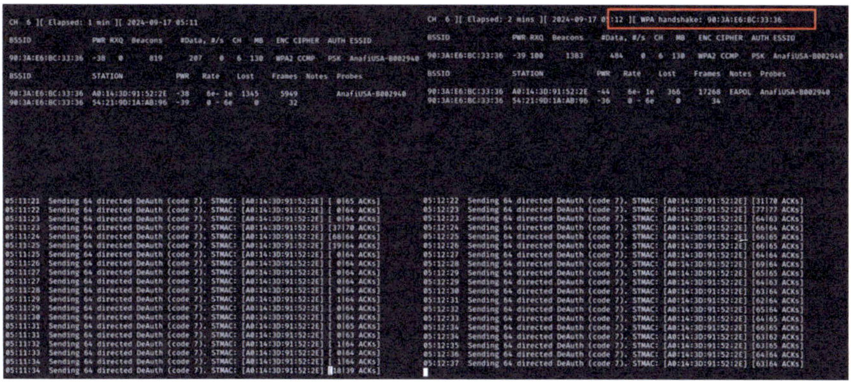

Fig. 2.26 Airodump-ng target network and default aireplay-ng attack running. Capture after 2 min (Test 1)

Fig. 2.27 Pre and during DoS attack

Default DOS Deauth

Although against the initial thesis and previous experimentation, this default Deauth attack disconnected the UAV and controller within 1 min. The WPA handshake was captured after 2 min and Wireshark analysis documented four EAPOL messages were captured. This test was conducted twice to check validity (Figs. 2.26, 2.27, 2.28 and 2.29).

Deauth-rc DOS

The deauth-rcs were changed to 6 and 7 with two deauth requests sent simultaneously, causing the devices to disconnect and be inoperable after 30 s. However, no handshake

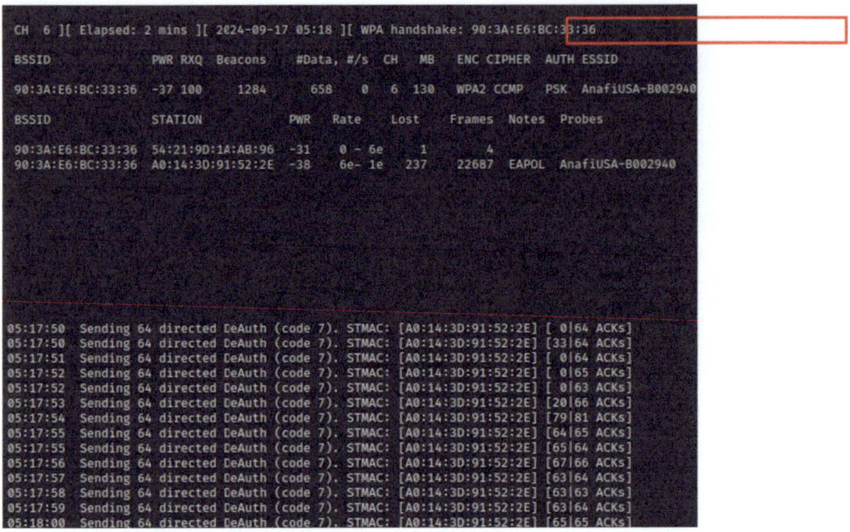

Fig. 2.28 Displaying handshake capture in Test 2

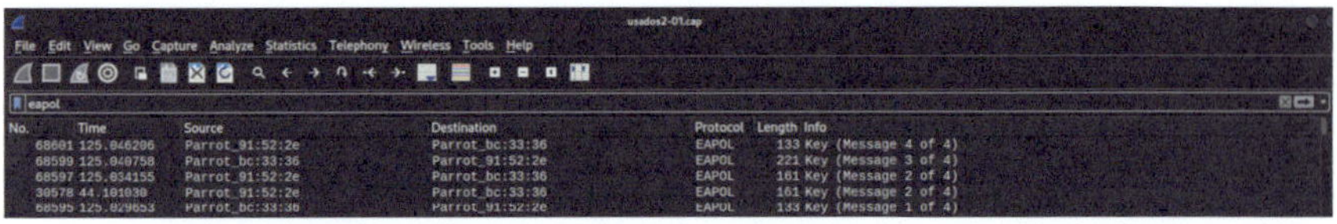

Fig. 2.29 Wireshark EAPOL message 1–4 capture confirmation

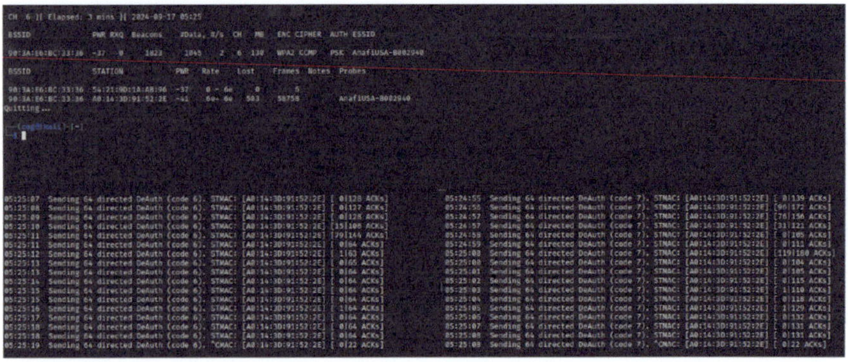

Fig. 2.30 Aireplay-ng DoS attack with deauth-rc 6 and 7

was captured during this test and Wireshark analysis confirmed no EAPOL messages were documented (Fig. 2.30).

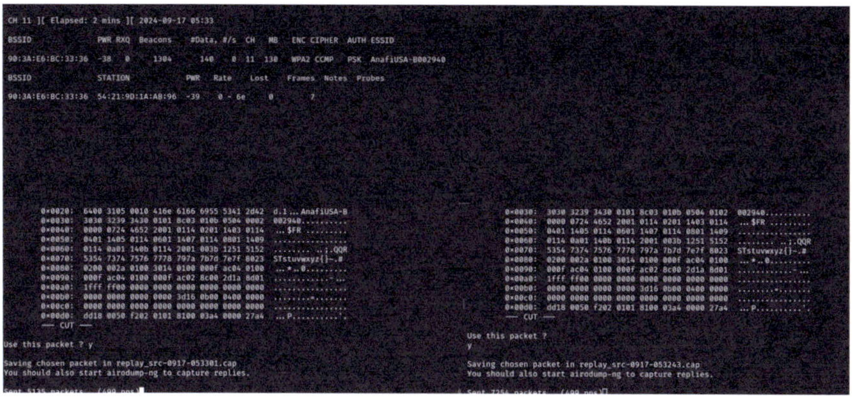

Fig. 2.31 Beacon Flood attack

Fig. 2.32 DOS effect on connectivity (red connection)

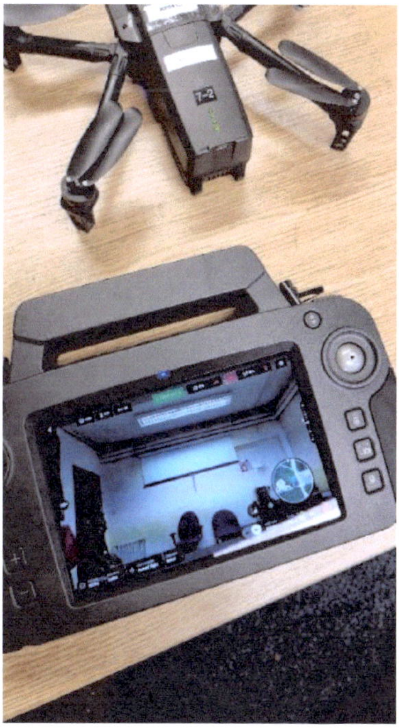

Beacon Flood

This attack caused latency on the network initially, however the attack was ineffective after 30 s due to channel switch protocol within the Anafi USA system. The AP also stopped sending out Beacon Frames post-attack for up to 2 min (Figs. 2.31 and 2.32).

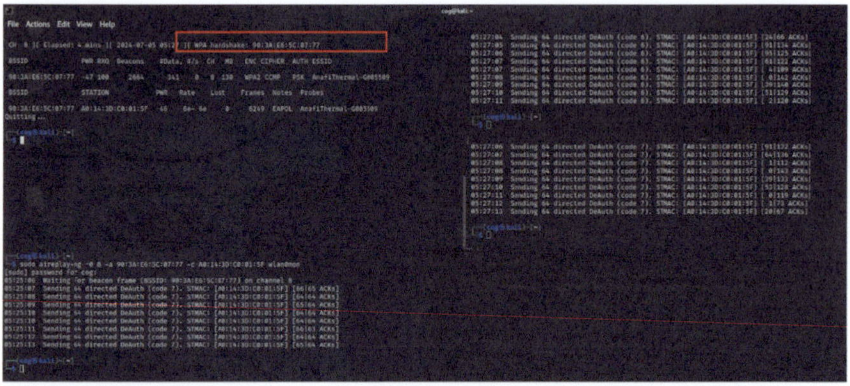

Fig. 2.33 WPA handshake capture during deauth attack using codes 6 and 7

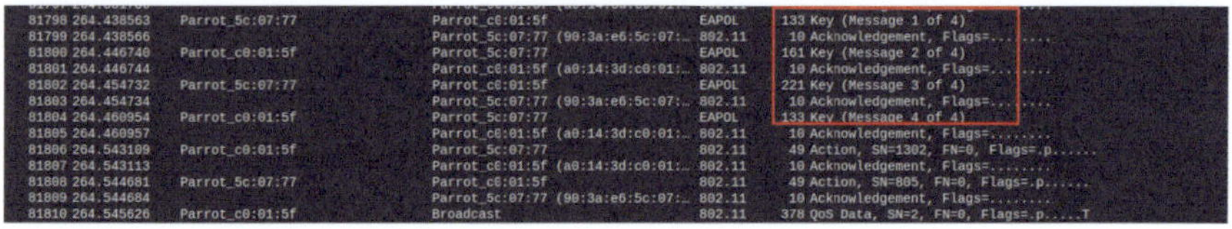

Fig. 2.34 EAPOL messages 1–4 captured

2.4.3 Stage 2—Deauthentication Attack Testing

2.4.3.1 Parrot Anafi Thermal

Default

The default aireplay-ng Deauth attack was conducted as Stage 1 and produced no impact on the UAS.

Deauth-rc

The deauth-rcs were set to 6 and 7, with two deauth requests sent concurrently creating a noticeable effect on the network and causing the device to lose connection. After 2 min the WPA handshake was captured through this attack and devices re-established communication links once the attack was concluded. This process was repeated 2 more times to check for consistency, documenting the same results (Figs. 2.33 and 2.34).

Fig. 2.35 Conducting deauthentication attack on Anafi Thermal with deauth codes 6, 7, 46 and 47

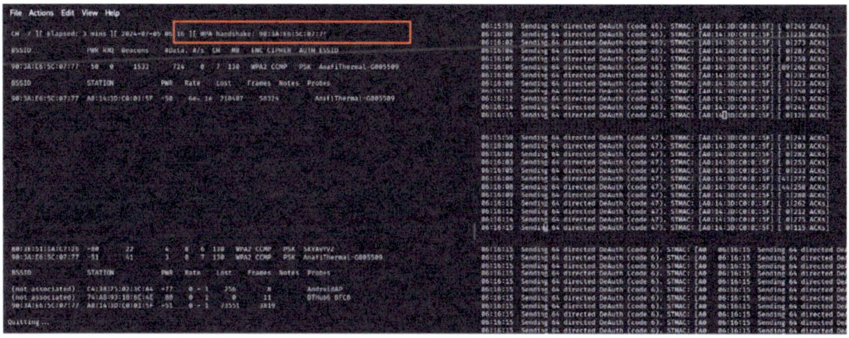

Fig. 2.36 Handshake capture during 6, 7, 46 and 47 deauth attack

4.4.1.3 Deauth-rc Enhanced

This attack was then enhanced to include deauth codes 6, 7, 46 and 47 causing a lost connection between UAV and GCS, providing no controllability. After 3 min, this attack captured the WPA handshake and devices regained connection once the attack concluded. This process was repeated 2 more times for consistency with the same results (Figs. 2.35, 2.36 and 2.37).

Fig. 2.37 Anafi Thermal during attack

2.4.3.2 Parrot Anafi USA

Default

The default aireplay-ng Deauth attack was conducted on Stage 1 and had already produced the desired effect and handshake required.

Deauth-rc

The deauth-rcs were combined and initially set to 6 and 7. These commands were demonstrated in Stage 1 with a detrimental effect on the UAS but with no handshake capture.

Deauth-rc Enhanced

This attack was then increased to include deauth codes 6, 7, 46 and 47 causing a lost connection between UAV and GCS, providing no controllability within 25 s. After 2 min, this attack captured the WPA handshake, with messages 1–3 captured and would enable password recovery due to only Messages 1–3 being required. This process was repeated 2 more times for consistency with the same results (Figs. 2.38, 2.39 and 2.40).

Deauth-rc was then set to 6, due to the higher success rates, and 4× deauthentication requests were sent concurrently to attempt to attain all four messages. This attack had a significant effect on the network causing the device to lose connection and inoperable and after 1 min, the WPA handshake was captured. The devices did re-establish communication links after the attack was complete, however this took a significantly longer than in previous tests (Figs. 2.41 and 2.42).

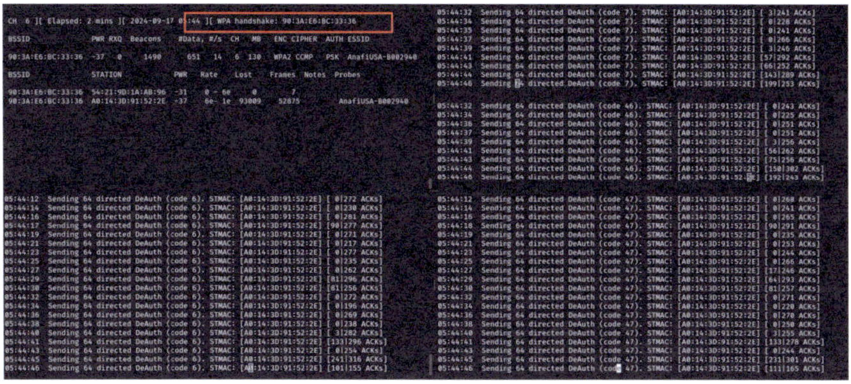

Fig. 2.38 Handshake capture on Anafi USA

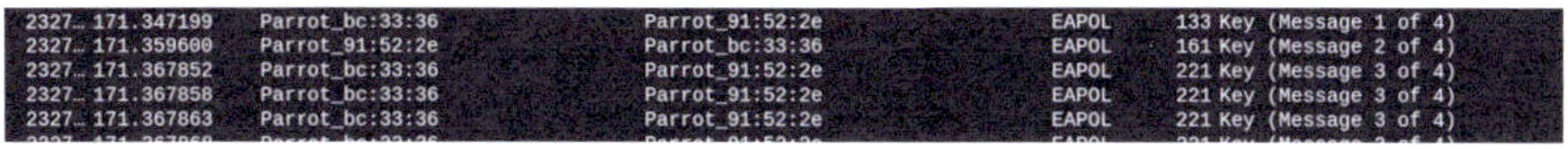

Fig. 2.39 EAPOL Messages 1–3 captured

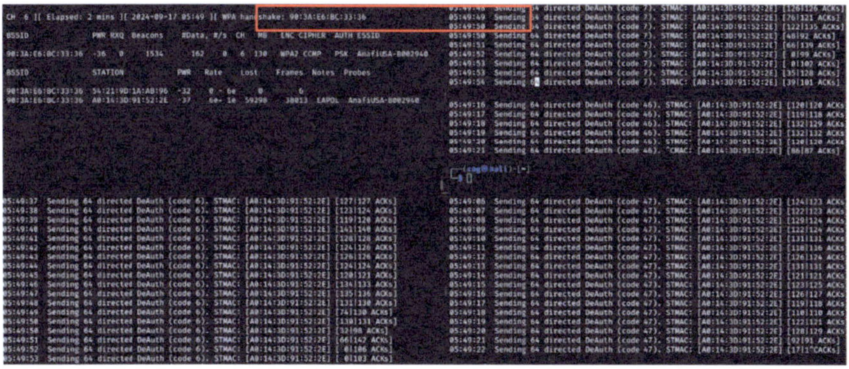

Fig. 2.40 Deauth attack using codes 6, 7, 46 and 47 on Anafi USA (Test 2)

2.4.4 Stage 3—Internal Access Attack Testing

2.4.4.1 Parrot Anafi Thermal

Network Access

The Kali Linux machine connected to the Anafi Thermal network using the credentials provided through hashcat. Hcxpcapngtool was used to convert the.pcap files to

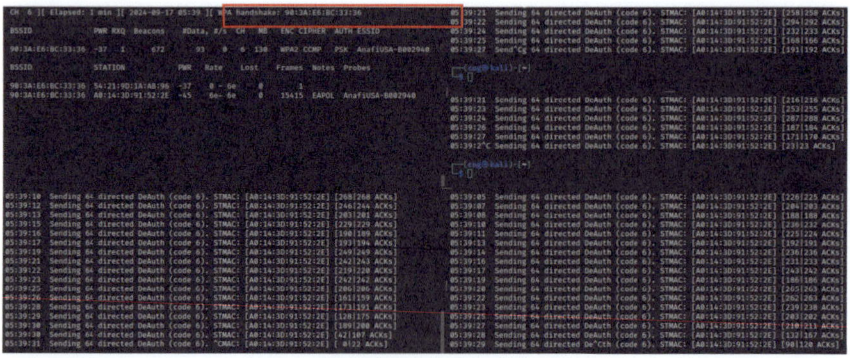

Fig. 2.41 WPA handshake capture during deauth attack using code 6 four times

Fig. 2.42 Wireshark EAPOL confirmation

hashcat readable hc22000 files. This.hc22000 file was combined with hashcat hash-cracking software with a suitable mask to identify the password. For research, 6 of the 12 characters were inputted as known to increase the speed of cracking to less than 60 min. However, using online GPUs with the full password mask, this could be conducted in 2–4 h due to low password complexity seen (Fig. 2.43).

Environment

Once connected, the IP range for the network was ascertained as 192.168.42.99/24 (Fig. 2.44).

Nmap

Scans were performed with various flags to enumerate the system for large amounts of information. Initially the devices on the network were identified through nmap 192.168.42.99/24, providing insight into the services open on each device. Scans were conducted on each identified host using sudo nmap 192.168.42.28 (the controller) or 192.168.42.1 (the UAV) -O -sV to identify OS, ports, services, versions and states combined with –script vuln.

Open ports and services were identified in the scans including FTP, used to transfer files unencrypted from a server to a client and Hyper Text Transfer Protocol

Fig. 2.43 File conversion through hcxpcapngtool and hash crack through hashcat

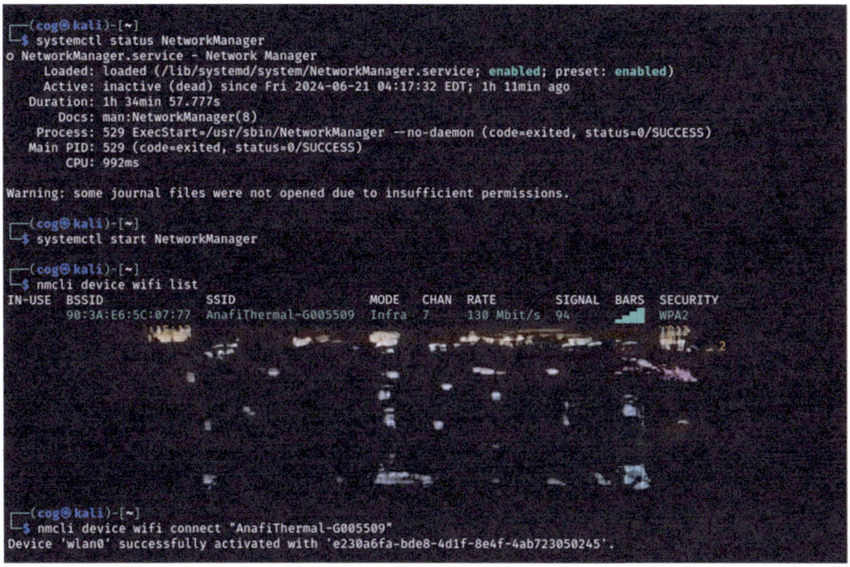

Fig. 2.44 Adaptor setup and wireless AP connection

(HTTP) for unencrypted communication between web clients and servers. Real Time Streaming Port (RTSP) was open on port 554, which is based on HTTP to stream media and monkeycom on port 9898, used for videophone/file transfer programs. Initial findings display that FTP is open and anonymous login is available, and the

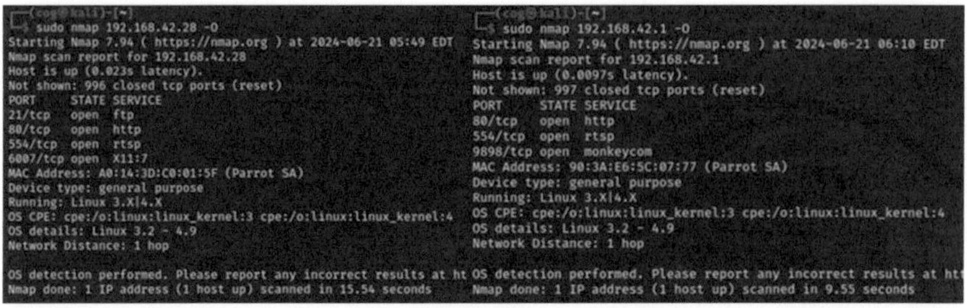

Fig. 2.45 Services and OS on 192.168.42.28 and 0.1

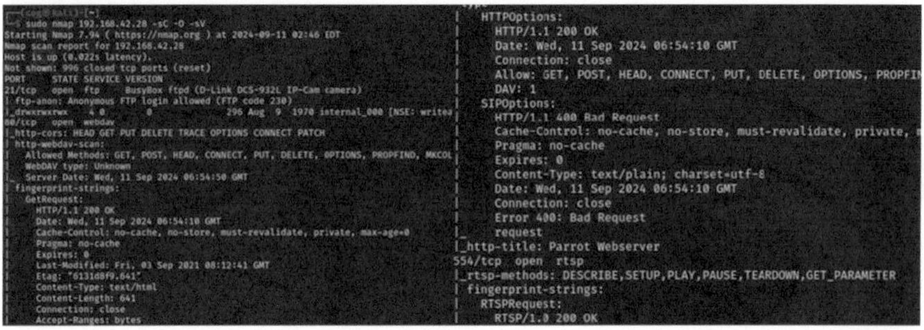

Fig. 2.46 Scan results for 192.168.42.28

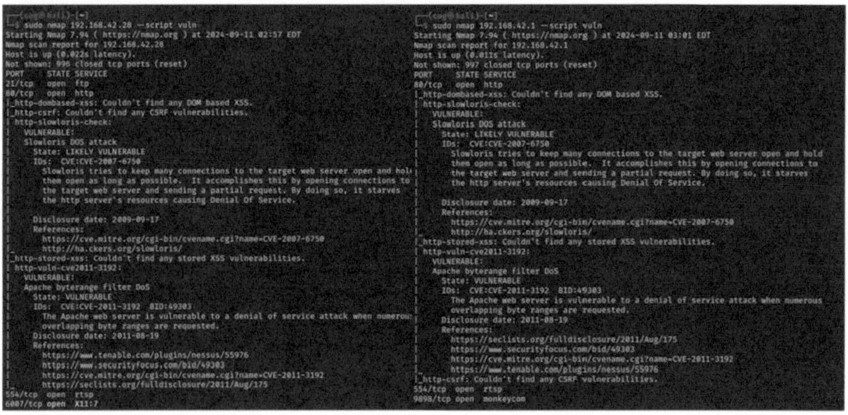

Fig. 2.47 Nmap script vuln scan on 192.168.42.28 and 0.1 results

HTTP server is susceptible to slowloris and Apache DOS attack (Figs. 2.45, 2.46 and 2.47).

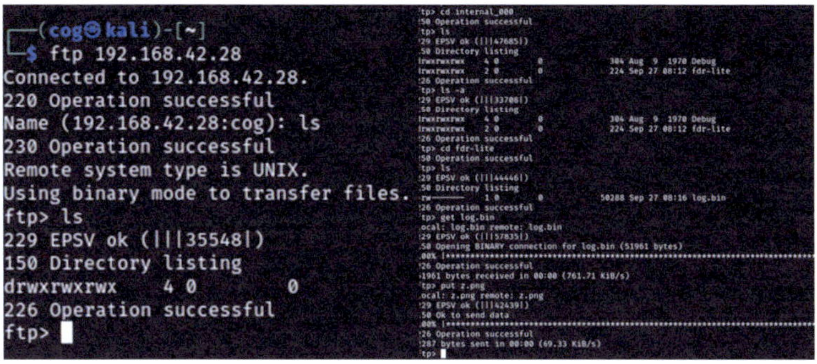

Fig. 2.48 FTP access and processes

FTP

Anonymous login enabled the Kali machine to connect and identified the directory as fully read, write and executable. Files were able to be up and downloaded via get and put commands with directories created through mkdir.

Sensitive information could be ascertained via files availability. Additionally, shell files were able to be uploaded that could be executed to create a reverse shell. The ability to gain root privileges may have been available through mount processes documented in previous studies (Fig. 2.48).

HTTP

Opening Firefox browser and inputting 192.168.42.1, accessed the Parrot Webserver and provided all the images and videos that had been recorded on the UAV without additional authentication. Inputting 192.168.42.28 provided the SkyController webserver and both these services allowed files to be uploaded onto the server.

Metasploit Exploitation

RTSP: An exploit was identified within Metasploit to target Linux RTSP devices through exploit/linux/misc/Hikvision_rtsp_bof. This targets a buffer overflow in the request parsing code of appliances and was theorised that it would apply in some form to the RTSP within the Parrot devices. Although it did not create the desired reverse tcp meterpreter shell, it did cause the video feed to stop (although devices remained connected) and would only resume once all devices were reset (Fig. 2.49).

Linux: Linux OS version was identified in the nmap scans but no effective exploit was identified within current Metasploit Framework.

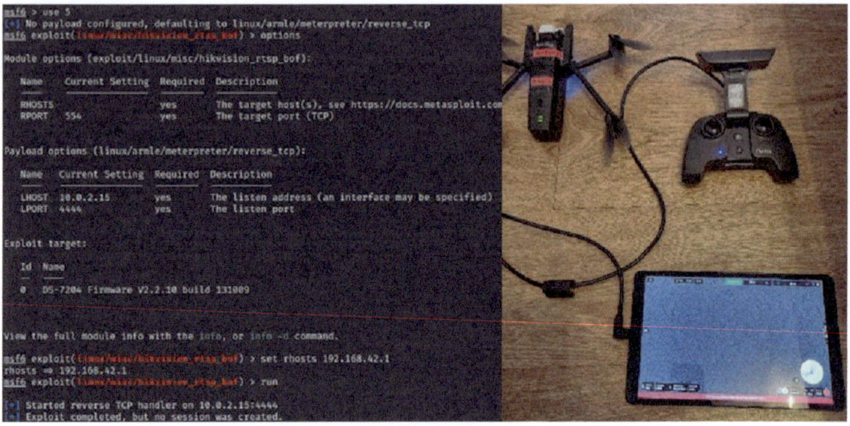

Fig. 2.49 RTSP Metasploit exploit and subsequent connection but not video link

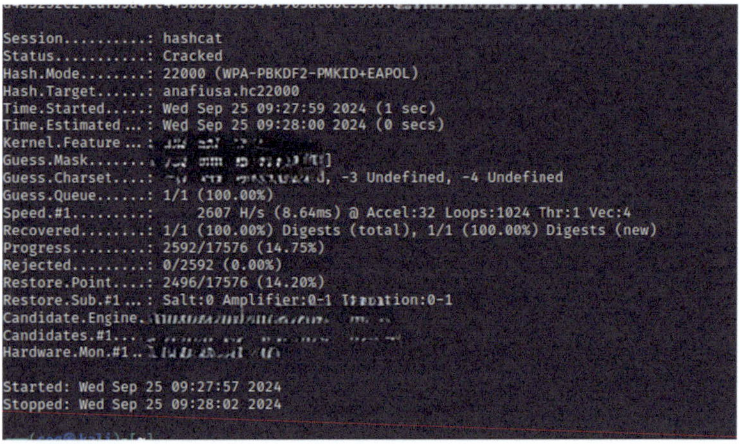

Fig. 2.50 Anafi USA password capture using hashcat

2.4.4.2 Parrot Anafi USA

Network Access

The Kali Linux machine connected to the Anafi USA network using the credentials ascertained through hashcat, using the same hash cracking process as the Anafi Thermal. The password was recovered in a similar timeframe using the same technique (Fig. 2.50).

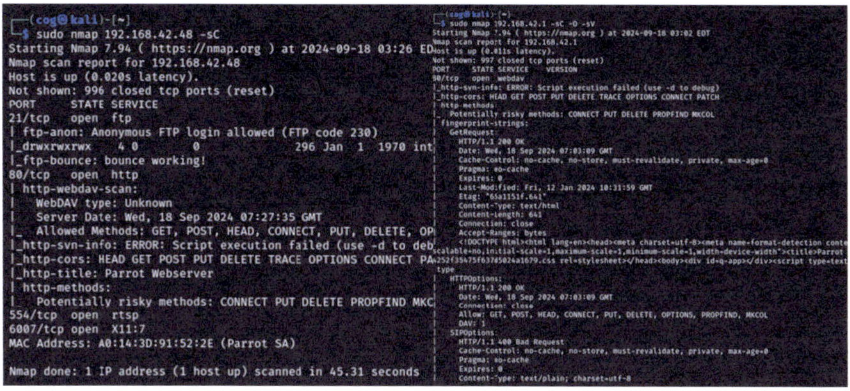

Fig. 2.51 Scan results for 0.1 and 0.48

Environment

Once connected, ip address show was run to ascertain the IP range for the network as 192.168.42.88/24.

Nmap

Scans were run to enumerate the system with hosts identified through nmap 192.168.42.88/24. Scans were conducted on each identified device using sudo nmap 192.168.42.48 (the controller) or 192.168.42.1 (the UAV) -O, -sC and -sV to identify OS, ports, services and states.

Ports and services were similar to those identified within Anafi Thermal with FTP, HTTP, RTSP, monkeycom and X11:7. Initial findings display the same vulnerabilities as the Anafi Thermal (Figs. 2.51 and 2.52).

FTP

Anonymous log in enabled the Kali machine to connect and noted that the directory is fully read, write and executable again. The same configurations and information were displayed and available in the USA version, presenting potentially the same vulnerabilities.

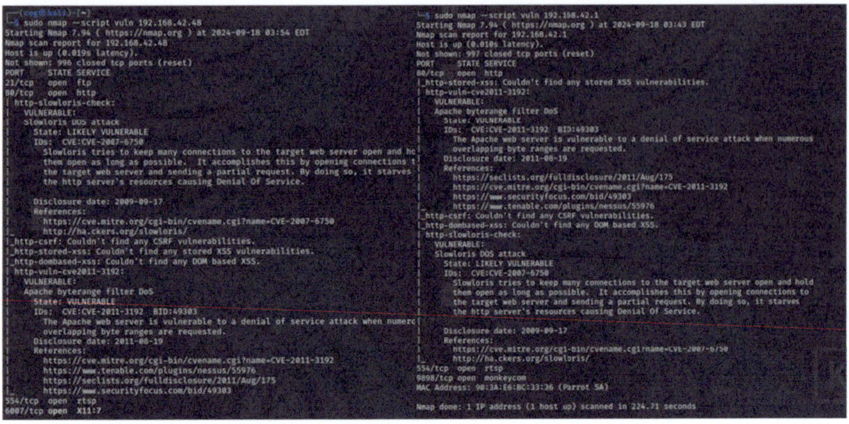

Fig. 2.52 Nmap script vuln scans for 0.1 and 0.48

Fig. 2.53 Anafi USA pre and post RTSP exploit attack

HTTP

The Parrot Webserver search provided all the images and videos that had been recorded on the UAS without additional authentication with both these services allowing files to be uploaded onto the server.

Metasploit Exploitation

RTSP: The same vulnerability identified within Anafi Thermal, was adjusted to target the Anafi USA system. The same results occurred, causing the video feed to stop and video connection only reestablished once all devices were reset (Fig. 2.53).

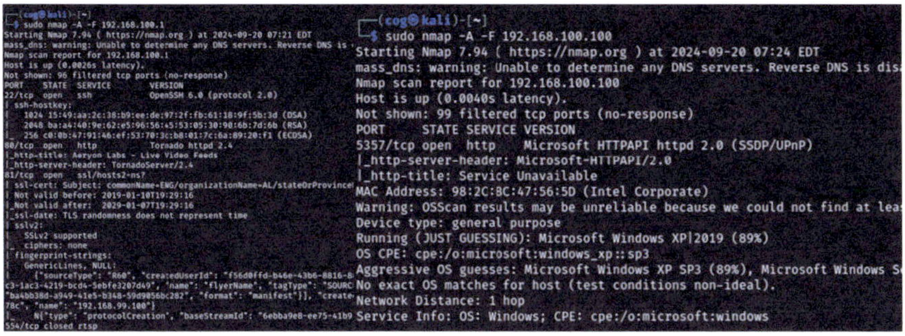

Fig. 2.54 Nmap -A scan for 192.168.100.1 and 0.100

2.4.4.3 Aeryon SkyRanger R60

Network Access

The Kali Linux machine connected to the Aeryon SkyRanger network using the legitimate credentials provided due to inability to impact WiFi.

Environment

The IP range for the network was identified as 192.168.100.1/24.

Nmap

Initially the devices on the network were identified through nmap 192.168.100.* showing the hosts. Scans were conducted on each identified host using sudo nmap 192.168.100.1 (the router), 192.168.100.100 (the GCS) and 192.168.100.101 (the UAV) with -A to identify all OS, ports, services, states and versions.

Open ports and services were found in the scans. Secure Shell (SSH) is a method for sending commands to a computer securely over an insecure network, and HTTP and RTSP on 80 and 554 respectively were identified. Web Services on Devices API (WSDAPI) used to develop client applications was identified on port 5357. The Window XP OS on 192.168.100.101 is deprecated and vulnerable to multiple attacks. Initial findings display that HTTP may have potential CSRF vulnerabilities and port 81 is vulnerable to SSL POODLE attack (Figs. 2.54 and 2.55).

Metasploit Exploitation

Methods and payloads to enumerate and exploit the system were located within Metasploit Framework Console including Windows XP vulnerabilities and POODLE

```
┌──(cog㉿kali)-[~]
└─$ sudo nmap --script vuln 192.168.100.1
Starting Nmap 7.94 ( https://nmap.org ) at 2024-09-20 07:27 EDT
Pre-scan script results:
| broadcast-avahi-dos:
|   Discovered hosts:
|     224.0.0.251
|   After NULL UDP avahi packet DoS (CVE-2011-1002).
|_  Hosts are all up (not vulnerable).
mass_dns: warning: Unable to determine any DNS servers. Reverse DNS is disabled. Try using -
Nmap scan report for 192.168.100.1
Host is up (0.0026s latency).
Not shown: 995 filtered tcp ports (no-response)
PORT     STATE  SERVICE
22/tcp   open   ssh
80/tcp   open   http
| http-csrf:
| Spidering limited to: maxdepth=3; maxpagecount=20; withinhost=192.168.100.1
|   Found the following possible CSRF vulnerabilities:
|
|     Path: http://192.168.100.1:80/
|     Form id:
|     Form action: r2uAdd
|
|     Path: http://192.168.100.1:80/
|     Form id:
|_    Form action: r2uAdd
|_http-stored-xss: Couldn't find any stored XSS vulnerabilities.
|_http-dombased-xss: Couldn't find any DOM based XSS.
81/tcp   open   hosts2-ns
| ssl-poodle:
|   VULNERABLE:
|   SSL POODLE information leak
|     State: VULNERABLE
|     IDs:  CVE:CVE-2014-3566  BID:70574
|           The SSL protocol 3.0, as used in OpenSSL through 1.0.1i and other
|           products, uses nondeterministic CBC padding, which makes it easier
|           for man-in-the-middle attackers to obtain cleartext data via a
|           padding-oracle attack, aka the "POODLE" issue.
|     Disclosure date: 2014-10-14
|     Check results:
|       TLS_RSA_WITH_AES_128_CBC_SHA
|     References:
|       https://www.securityfocus.com/bid/70574
|       https://cve.mitre.org/cgi-bin/cvename.cgi?name=CVE-2014-3566
|       https://www.imperialviolet.org/2014/10/14/poodle.html
|_      https://www.openssl.org/~bodo/ssl-poodle.pdf
554/tcp  closed rtsp
8088/tcp open   radan-http
```

Fig. 2.55 --script vuln scan on 192.168.100.1

attack. However, these were unable to be conducted due to system issues (Tables 2.3, 2.4 and 2.5).

Table 2.3 Stage 1 summary of events

Stage	Attack vector	Tools	Type	Deauth-rc	Desired effect		
					Thermal	USA	R60
1—DoS	Information	Kismet/ws/ Airodump	Monitor	N/A	Yes	Yes	N/A
	Default DoS Deauth	Aireplay-ng	Deauth	7	No	Yes	N/A
	Deauth-rc DoS	Aireplay-ng	Deauth	6 and 7	Yes	Yes	N/A
	Beacon Flood	Aireplay-ng	Injection	N/A	Yes	No	N/A

Table 2.4 Stage 2 summary of events

Stage	Attack vector	Tools	Type	Deauth-rc	Desired effect		
					Thermal	USA	R60
2—Deauth	Default Deauth	Aireplay	Deauth	7	No	Yes	N/A
	Deauth-rc Deauth	Aireplay	Deauth	6 and 7	Yes	No	N/A
	Deauth-rc Enhanced	Aireplay	Deauth	6, 7, 46, 47	Yes	Yes	N/A

Table 2.5 Stage 3 summary of events

Stage	Attack vector	Tools	Type	Services	Desired effect		
					Thermal	USA	R60
3—Internal	Hashcrack	Hashcat/ hcxpcapngtool	Crack	Password	Yes	Yes	N/A
	Vulnerabilities found	Nmap	Recce	Open ports	Yes	Yes	Yes
	Impact on system	Metasploit	Exploit	RTSP	Yes	Yes	N/A

2.5 Analysis of Research Results

2.5.1 Security Architecture

Security architecture's goal is the identification of various security risks and subsequent implementation of a robust network [18]. A single method will not guarantee device protection and therefore a combination of capabilities and an overall integrated approach is required to reach high levels of security [18]. COTS UAS are likely to be the target for threat actors, ranging in sophistication, to exploit the data within the UAS. Therefore, the cyber security implementation must reflect the threat.

The most practical security framework to apply is the defence in depth principles, to mitigate issues identified within this research and within. This involves a multi-layered approach to network security appropriate to securing UAS.

With the 2 distinct areas tested and analysed, the recommendations will be split into the Wireless Communication components and the Internal Device issues.

2.5.2 Wireless Communication Components

The attacker can influence the UAS whenever they are within proximity of the UAV's WiFi signal and this was demonstrated without requiring extensive skillsets or expensive equipment. Tailored Beacon Flood and Deauthentication attacks rendered the UAV inoperable, potentially leading to a catastrophic incident or significant impact on UAS tasking, and could have been combined with connecting a malicious controller to hijack UAV device [10]. Higher security measures and more secure protocols should be implemented to prevent this occurrence. The sections to follow will aim to provide mitigation to these attacks through various levels of sophistication.

2.5.2.1 Wireless Sniffing

Critical information was ascertained from the UAS including MAC, BSSID, RSN and encryption, providing the elements necessary to conduct a targeted attack. Monitoring and analysis of wireless signals is almost impossible to detect if communications are being monitored, logged or information extracted. Disabling the SSID broadcast prevents beacons transmitted with the network's name, providing a low-level element in the defence in depth principle [8].

Novel research has shown that signals can be protected by other nodes within the UAS network transmitting artificial noise to confuse the eavesdropper and mask confidential information transmitted [65]. Covert UAS communication techniques could be implemented to exchange information securely and discreetly to ensure sensitive information remains inaccessible and concealed to unauthorised parties [34]. This is particularly applicable to military or governmental COTS UAS where discretion is a significant issue. This can be configured using an upper and lower limit of the UAV's transmitting power, combined with optimal height data [62].

Additionally, wireless physical layer data can be protected through properties such as superposition and diffusion to provide secrecy in transmissions [2]. This layer is critical for communication and security controls of the UAS. Hadi et al. [20] and is achieved by altering the attributes of cellular channels including interference, fading and noise boosting for the receiver, whilst reducing the signal quality for the eavesdropper. This can be combined with cryptographic protocols to add an additional level of security [20].

2.5.2.2 Deauthentication

PMF/MFP is implemented in defence against DoS attacks such as dissociation and deauthentication. The UAS tested had PMF capabilities in their security architecture, identified through the RSN information transmitted and captured in Wireshark. The MFPC flags were set to 1, signifying devices' ability to implement MFP. This means all devices on the network are to be PMF capable for PMF to be enforced, whereas the most secure option would be to ensure all clients and AP are PMF Required which would have prevented communication without it [54]. This would have been displayed by the flags on MFP Required set to 0.

The PMF implementation was ineffective against basic deauthentication requests, short bursts of requests or more complex, continuous multi layered requests. The deauthentication reason codes should not have affected the wireless communication's integrity or availability, but this was documented multiple times and especially effective with reason code 6. Although theoretically PMF should prevent these attacks, it faces multiple challenges due to difficulties with implementation [54]. Issues identified within the research could be due to the issues below and should be explored further:

- Large resource intensive cryptographic requirements.
- Miscommunication or misconfiguration during 802.11 State Machine process.
- Inability to prevent deauthentication when utilising short bursts [54].
- Insufficient support from the OS and WiFi chip's driver.
- PMF's inability to defend DoS attacks when utilising WiFi jamming [54].

2.5.2.3 Flooding

PMF does not protect pre-authenticated management frames such as beacon frames [54]. The beacon flood attack caused severe latency in the Anafi Thermal system by overloading and compromising its operating ability. These large number of false messages affected the systems processing power and buffers creating an inability to transmit the correct data between UAV and GCS [34]. The channel switching through Frequency Hopping Spread Spectrum (FHSS) parameters within the Anafi USA mitigated this attack within 30 s and UASs should incorporate similar security.

2.5.2.4 Wireless Jamming

Signal jamming on network links within UAS caused DoS on the targeted system. Consequences vary based on the link and type of UAS from compromising in-flight input through GCS, to loss of C2 link and video link, which could be disastrous.

Mitigation techniques either rely on detection and countermeasure deployment during jamming or mitigate the effects of the jamming. These vary including FHSS (as seen in the Anafi USA), which although successful in this research, would not be effective against smart jammers which are far more sophisticated that those utilised in this study [2]. With COTS UAS, the security configuration should mirror the potential threat and implement appropriate measures.

Research into Intelligent Reflective Surface Multiple-Input Multiple-Output coupling has significantly lowered hardware and power usage resources, creating an appropriate system for UAS [60]. Detection techniques have evolved from noise monitoring to learning based approaches using directional signal parameters to effectively detect, locate and identify jammers. However, these methods result in a temporary loss of UAS connection which may be unacceptable for UAV operations [2]. Machine Learning (ML) using convolution neural network can be implemented to calculate jamming signals and prevent UAS communication issues [20].

2.5.2.5 WPA2

Modern WiFi-enabled UAS with correctly configured WPA2 security protocol and a robust random key, will protect most systems against a low-resource attack [2]. This protocol contains known issues and is implemented within COTS UAS with minimal consideration to the area of operations. This was fundamentally ineffective as deauthentication was achieved consistently and 4-way handshake captured with relative ease, providing credentials to access the network by offline bruteforce.

WPA3 should be implemented as the Dragonfly handshake increases the complexity of PMK and ensures enhanced authentication through robust ECC key exchange and AES-GCMP encryption. This protocol prevents offline attacks as the PMK is independent of the PSK [54]. PMF is also mandatory within WPA3 preventing potential compatibility issues documented.

2.5.2.6 Password Complexity

Password complexity for the UAS was low due to character selection and length, which contributed to the hash cracking achieved in a reasonable time. This could have been completed much faster using a more powerful computer, GPUs, or establishing a Dictionary Dataset and applying these to reduce the keyset [11]. Therefore password complexity should be increased significantly to include special characters etc. to enhance security.

The COTS UAS tested would be utilised frequently for a variety of non-kinetic taskings. Therefore, password recovery would enable a persistent threat to continually access the UAS network, creating a significant risk.

2.5.3 Internal Device Issues

When discussing these vulnerabilities in relation to Governmental COTS UAS usage, the impact of internal access is significant. Access could provide attackers the ability

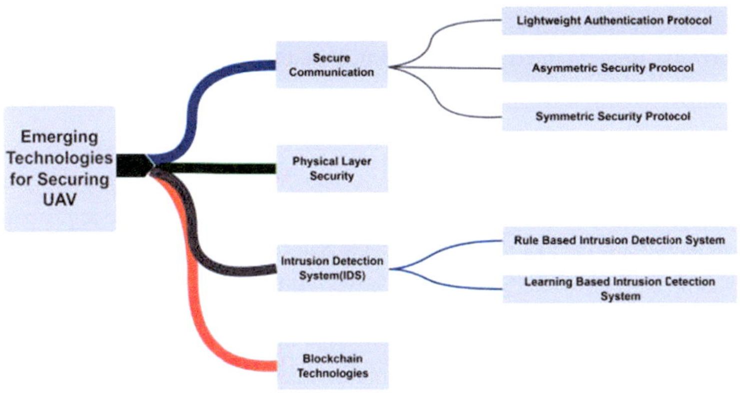

Fig. 2.56 Emerging technologies for UAS security [20]

to reverse engineer imagery to determine operator's location, access metadata and GPS information or similar sensitive information, leading to high-risk situations.

Implementing appropriate secure protocols, cryptographic techniques, robust key management practices and MFA, ensures the CIA of UAS communications and the overall security of UAS operations, through both established and emerging technologies [20] (Fig. 2.56).

2.5.3.1 MAC Filtering

MAC Address Filtering is a primitive access control method where the MAC address, which is assigned to an individual device, is allocated to a whitelist enabling access to the network [27]. MAC filtering only accepts connections from the previous list of devices [17] and although easily circumvented, it provides another layer within the defence in depth principles [6].

2.5.3.2 Secure Communication

Authentication is a crucial component of UAS communication for trusted and secure transmission between GCS and UAVs [34]. Due to size and processing power, cryptographic overheads are to be minimised to reduce time lag and process expenditures in UAS [34]. Lightweight Cryptography (LWC) symbolises the schemes that have a smaller memory consumption and computation complexity to extend cryptography use to computationally and resource constrained devices such as UAS [20]. One such system is an upgrade Temporal Credential-Based Anonymous Lightweight Authentication Scheme (TCALAS) for IOD networks which provides security for attacks including availability and traceability attacks, and successfully achieves authentication process within 2.29 ms [2].

Blockchain technology is evolving into a solution for mitigating problems within UAS network, implemented for transparency, auditability and traceability [20].

2.5.3.3 Communication Spoofing

Spoofing is where the attacker effectively assumes the identity of a legitimate user, device or network node by identifying as a GCS or UAV [2]. Network access provides the potential to conduct these attacks where communication protocols within UAS e.g. MAVlink are vulnerable. Although not performed, this could have been exploited to take full control of the UAV by the researcher via MITM or code injection [2].

Spoofing can be countered by strong authentication utilising LWC primitives to authenticate all devices. Additionally, a smart Intrusion Detection System (IDS) can effectively detect spoofing attempts by monitoring from a central overseeing position [7]. This IDS is most successful when integrated in ad-hoc networks, based on malicious packet identification and combined with ML techniques to assess signatures. Protocols that verify the origin of the information transmitted across the network randomly encrypts messages sent by the UAV and decrypts those sent back by the GCS, requiring a key pairing [20]. When compared to data stored in the logs, successful verification will enable UAS operation. However, unsuccessful values will prevent any further communication between the UAV and GCS. A small hash can be utilised to facilitate this whilst overcoming cryptographic overheads [20].

2.5.3.4 Network IDS

Real-time network traffic analysis is required to detect, identify and mitigate intrusions against UAVs whilst in operation [20]. ML techniques within IDS increase UAS security significantly and ML algorithms can be implemented to address vulnerabilities with an F1 score across UAV platforms of 99.73% [43]. These approaches can be split into rule-based intrusion, in which specific rules are applied with regards to the expected UAS behaviour, signature-based which depends on attack signatures and anomaly-based that is based on learning or filtering mechanisms [43].

However, limited research from the security community due to sensitivity of information reduces the data. Additionally, large overheads required for data transfer, and time for analytics, will results in systems requiring lightweight versions to reduce computation time and choosing appropriate, highly valuable data sets [20]. The UAS network IDS should be the main mechanism for detection and protection of the seven platform-specific signature-based attacks related to CIA which may have impacted the researcher's freedom of movement within the UAS [20].

2.5.3.5 Software and OS

Software attacks include OS exploits, collecting data on video feeds and system ID spoofing. The vulnerability scan conducted returned UAS OS and associated service, version, potential issues and known vulnerabilities that could be exploited. These attacks, if conducted, are based around injecting malicious software into the UAS such as Maldrone malware. To mitigate these, adopting the authorisation mechanisms for UAV system resources and applying software-based attestation have proved effective [43]. Additionally, robust firewall implementation and use of IDS with anti-virus provides real-time management although at additional computational cost [43].

2.5.3.6 Port Scanning

Without FTP authentication, threat actors are able to connect to the system and extract vast amounts of data. This could pertain to personal information, media, logs and imagery. This initial limited access could be further exploited by editing key configuration files or uploading malicious files to attain further privilege [17]. Maldrone is a virus designed for UAS to infect the flight controller and provide control to the attacker [43]. Additionally, attackers can access the HTTP Webserver without credentials, providing another ability to gather personal, platform and service information and provides the capacity to upload files to the server.

A credential system within the FTP and HTTP services, as well as shutting these services down during flight operations, would reduce the risks presented [2]. If these services are required during flight for anything other than updates, then manufacturers should prevent access that requires no credentials. This could be achieved by stopping the services entirely, data backed up offline or by blocking inbound ports. Iptables could be configured to allow packets when services are in legitimate use and drop all packets for services that are not [10]. MFA could be adopted on servers or services that available to the user. Furthermore, utilising a SSH tunnel for all traffic transmitted to connect the UAV and GCS ports may alleviate potential issues [10].

2.5.3.7 Data and Video

An exploit was engineered to divert the video feed from the UAV to another device in the network by exploiting the open RTSP port. This exploit prevented the main use case for a UAV by blocking the video feed and causing the UAV to potentially be inoperable. Confidentiality is attained through encryption and due to resource limitation, AES is not recommended for video feeds. LWC must be implemented to through AEGIS or Chacha20 types [2]. Researchers have designed a secure video communication system through a chaos-based encryption which has excellent real time performance where the video feed is embedded within a single command link between UAV and GCS [2].

2.6 Conclusions and Future Work

The research was an investigation into the security and potential issues of COTS sUAS. The aim of was to assess the effectiveness of cyber-attacks on COTS sUAS that implement WiFi 802.11 transmission type through a combination of low-cost devices and open-source tools to enhance device cyber security. To enable the research, a test environment was created consisting of virtual machines, a network adaptor and three COTS UAS. An investigation was then conducted on known wireless protocol security issues through various stages of testing and analysis.

On testing conclusion, it was apparent basic cyber security attacks were effective on the UAS due security measures, issues or poor risk analysis and subsequent implementation. The UAS tested were susceptible to DoS attacks through Beacon Floods and Deauthentication packet injection, preventing communication and leading to device inoperability. Deauthentication reason codes could be amended to increase efficiency of attack and when combined with multiple requests transmitted simultaneously, enabled the WPA2 handshake to be captured even when devices implemented PMF due to potential jamming or compatibility issues. The UAS network password was recovered through hash cracking and access authorised for the researcher where issues such as anonymous FTP login, HTTP, RTSP exploit and known CVE issues were identified.

Based on testing results, various measures have been suggested to mitigate the issues identified specifically with regards to UAS and the challenges of limited computational resources, to enhance cyber security. This includes configuring mandatory PMF, frequency hopping, enhancing port security and increasing password complexity.

Many ideas for further research arose throughout the development and conducting of the research. One possible future area could be applying the same or similar attack framework to other COTS sUAS to identify vulnerabilities. Due to the device's resource limitation and minimum security requirements, it is likely this methodology would be successful against a multitude of procured systems. Additionally, this research could be applied to routers that implement similar protocols to ascertain the effectiveness of the attacks with more powerful devices that do not possess the computational constraints of UAS. Furthermore, the attacks conducted could be potentially automated to conduct the attack through a python script.

References

1. Adil M et al (2023) A systematic survey: security threats to UAV-aided IoT applications, taxonomy, current challenges and requirements with future research directions. IEEE Trans Intell Transp Syst 24(2):1437–1455. https://doi.org/10.1109/TITS.2022.3220043
2. Aissaoui R et al (2023) A survey on cryptographic methods to secure communications for UAV traffic management. Veh Commun 44:100661. https://doi.org/10.1016/J.VEHCOM.2023.100661

3. Aljohani M, Mukkamala R, Olariu S (2024) Autonomous strike UAVs in support of homeland security missions: challenges and preliminary solutions. IEEE Access 12:90979–90996. https://doi.org/10.1109/ACCESS.2024.3420235
4. Altawy R, Youssef AM (2016) Security, privacy, and safety aspects of civilian drones: a survey. ACM Trans Cyber-Phys Syst 1(2). https://doi.org/10.1145/3001836
5. Amoordon A et al (2022) A single supervised learning model to detect fake access points, frequency sweeping jamming and deauthentication attacks in IEEE 802.11 networks. Mach Learn Appl 10:100389. https://doi.org/10.1016/J.MLWA.2022.100389
6. Anathi M, Vijayakumar K (2020) An intelligent approach for dynamic network traffic restriction using MAC address verification. Comput Commun 154:559–564. https://doi.org/10.1016/J.COMCOM.2020.02.021
7. Anthi E et al (2019) A supervised intrusion detection system for smart home IoT devices. IEEE Internet Things J 6(5):9042–9053. https://doi.org/10.1109/JIOT.2019.2926365
8. Arikumar KS et al (2022) Enhancing the security of WPA2/PSK authentication protocol in Wi-Fi networks. Procedia Comput Sci 215:413–421. https://doi.org/10.1016/J.PROCS.2022.12.043
9. Arteaga SP et al (2019) Analysis of the GPS spoofing vulnerability in the drone 3DR Solo. IEEE Access 7:51782–51789. https://doi.org/10.1109/ACCESS.2019.2911526
10. Astaburuaga I et al (2019) Vulnerability analysis of AR.Drone 2.0, an embedded Linux system. In: 2019 IEEE 9th annual computing and communication workshop and conference (CCWC), pp 666–672. https://doi.org/10.1109/CCWC.2019.8666464
11. Cacacho HRG et al (2019) Breaking the password security standards using offline attacks and public user attributes. In: 2019 IEEE 11th international conference on humanoid, nanotechnology, information technology, communication and control, environment, and management (HNICEM), pp 1–5. https://doi.org/10.1109/HNICEM48295.2019.9072807
12. Ch R et al (2020) Security and privacy of UAV data using blockchain technology. J Inf Secur Appl 55:102670. https://doi.org/10.1016/j.jisa.2020.102670
13. Chatzoglou E, Kambourakis G, Kolias C (2022) How is your Wi-Fi connection today? DoS attacks on WPA3-SAE. J Inf Secur Appl 64:103058. https://doi.org/10.1016/J.JISA.2021.103058
14. Chen Y et al (2024) Improving physical layer security for multi-UAV systems against hybrid wireless attacks. IEEE Trans Veh Technol 73(5):7034–7048. https://doi.org/10.1109/TVT.2023.3337154
15. Chriki A et al (2019) FANET: communication, mobility models and security issues. Comput Netw 163:106877. https://doi.org/10.1016/j.comnet.2019.106877
16. Colter J et al (2022) Testing the resiliency of consumer off-the-shelf drones to a variety of cyberattack methods. In: 2022 IEEE/AIAA 41st digital avionics systems conference (DASC), pp 1–5. https://doi.org/10.1109/DASC55683.2022.9925879
17. Dey V et al (2018) Security vulnerabilities of unmanned aerial vehicles and countermeasures: an experimental study. In: 2018 31st international conference on VLSI design and 2018 17th international conference on embedded systems (VLSID), pp 398–403. https://doi.org/10.1109/VLSID.2018.97
18. Dwivedi A, Sharma SK (2023) Multilayer security approach for Internet of Things using Omura approach. In: 2023 5th international conference on advances in computing, communication control and networking (ICAC3N), pp 1385–1390. https://doi.org/10.1109/ICAC3N60023.2023.10541565
19. Gabrielsson J, Bugeja J, Vogel B (2021) Hacking a commercial drone with open-source software: exploring data privacy violations. In: 2021 10th Mediterranean conference on embedded computing (MECO), pp 1–5. https://doi.org/10.1109/MECO52532.2021.9460295
20. Hadi HJ et al (2023) A comprehensive survey on security, privacy issues and emerging defence technologies for UAVs. J Netw Comput Appl 213:103607. https://doi.org/10.1016/J.JNCA.2023.103607
21. Halbouni A, Ong L-Y, Leow M-C (2023) Wireless security protocols WPA3: a systematic literature review. IEEE Access 11:112438–112450. https://doi.org/10.1109/ACCESS.2023.3322931

22. He D, Chan S, Guizani M (2017) Communication security of unmanned aerial vehicles. IEEE Wirel Commun 24(4):134–139. https://doi.org/10.1109/MWC.2016.1600073WC
23. Hiergeist S, Schwierz A, Zeitler A (2017) COTS components for a large scale UAS demonstrator datalink system. In: 2017 IEEE/AIAA 36th digital avionics systems conference (DASC), pp 1–8. https://doi.org/10.1109/DASC.2017.8102017
24. Höflinger F et al (2013) A wireless micro Inertial Measurement Unit (IMU). IEEE Trans Instrum Meas 62(9):2583–2595. https://doi.org/10.1109/TIM.2013.2255977
25. Hooper M et al (2016) Securing commercial WiFi-based UAVs from common security attacks. In: MILCOM 2016—2016 IEEE military communications conference, pp 1213–1218. https://doi.org/10.1109/MILCOM.2016.7795496
26. Hoque N, Rahbari H, Rezendes C (2022) Systematically analyzing vulnerabilities in the connection establishment phase of Wi-Fi systems. In: 2022 IEEE conference on communications and network security (CNS), pp 64–72. https://doi.org/10.1109/CNS56114.2022.9947252
27. Jiang P, Wu H, Xin C (2022) A channel state information based virtual MAC spoofing detector. High-Confid Comput 2(3):100067. https://doi.org/10.1016/J.HCC.2022.100067
28. Kampourakis V et al (2022) WPAxFuzz: sniffing out vulnerabilities in Wi-Fi implementations. Cryptography 6(4). https://doi.org/10.3390/cryptography6040053
29. Karmakar G et al (2022) Security of Internet of Things devices: ethical hacking a drone and its mitigation strategies. In: 2022 IEEE Asia-Pacific conference on computer science and data engineering (CSDE), pp 1–5. https://doi.org/10.1109/CSDE56538.2022.10089255
30. Karmakar R, Kaddoum G, Akhrif O (2024) A PUF and fuzzy extractor-based UAV-ground station and UAV-UAV authentication mechanism with intelligent adaptation of secure sessions. IEEE Trans Mobile Comput 23(5):3858–3875. https://doi.org/10.1109/TMC.2023.3284216
31. Kim Y, Cho K, Kim S (2024) Challenges in dynamic analysis of drone firmware and its solutions. IEEE Access 12:106593–106604. https://doi.org/10.1109/ACCESS.2024.3425604
32. Krasnyánszki B, Brassai ST, Németh A (2024) UAV weaknesses against deauthentication based hijacking attacks. In: 2024 IEEE 22nd world symposium on applied machine intelligence and informatics (SAMI), pp 493–498. https://doi.org/10.1109/SAMI60510.2024.10432859
33. Krishna CGL, Murphy RR (2017) A review on cybersecurity vulnerabilities for unmanned aerial vehicles. In: 2017 IEEE international symposium on safety, security and rescue robotics (SSRR), pp 194–199. https://doi.org/10.1109/SSRR.2017.8088163
34. Kumar N, Chaudhary A (2024) Surveying cybersecurity vulnerabilities and countermeasures for enhancing UAV security. Comput Netw 252:110695. https://doi.org/10.1016/J.COMNET.2024.110695
35. Kunertova D (2023) Drones have boots: learning from Russia's war in Ukraine. Contemp Secur Policy 44(4):576–591. https://doi.org/10.1080/13523260.2023.2262792
36. Lamers E et al (2021) Securing home Wi-Fi with WPA3 personal. In: 2021 IEEE 18th annual consumer communications and networking conference (CCNC), pp 1–8. https://doi.org/10.1109/CCNC49032.2021.9369629
37. Lin N et al (2023) Network threat analysis and protection technology of UAV system. In: 2023 IEEE 3rd international conference on electronic technology, communication and information (ICETCI), pp 236–240. https://doi.org/10.1109/ICETCI57876.2023.10176672
38. Lindroos S, Hakkala A, Virtanen S (2021) A systematic methodology for continuous WLAN abundance and security analysis. Comput Netw 197:108359. https://doi.org/10.1016/J.COMNET.2021.108359
39. Liu Z et al (2015) Rise of mini-drones: applications and issues. In: Proceedings of the 2015 workshop on privacy-aware mobile computing. Association for Computing Machinery (PAMCO '15), New York, NY, pp 7–12. https://doi.org/10.1145/2757302.2757303
40. Lv Z (2019) The security of Internet of drones. Comput Commun 148:208–214. https://doi.org/10.1016/j.comcom.2019.09.018
41. Ly B, Ly R (2021) Cybersecurity in unmanned aerial vehicles (UAVs). J Cyber Secur Technol 5(2):120–137. https://doi.org/10.1080/23742917.2020.1846307
42. Marais S, Coetzee M, Blauw F (2021) Simultaneous deauthentication of equals attack. In: Wang G, Chen B, Li W, Di Pietro R, Yan X, Han H (eds) Security, privacy, and anonymity

in computation, communication, and storage. Springer International Publishing, Cham, pp 545–556
43. Mekdad Y et al (2023) A survey on security and privacy issues of UAVs. Comput Netw 224:109626. https://doi.org/10.1016/J.COMNET.2023.109626
44. Nassi B et al (2021) SoK: security and privacy in the age of commercial drones. In: 2021 IEEE symposium on security and privacy (SP), pp 1434–1451. https://doi.org/10.1109/SP40001.2021.00005
45. Oudina Z et al (2024) Empirical analysis of the security threats and risks that drones face, represent, and mitigation. In: 2024 6th international conference on pattern analysis and intelligent systems (PAIS), pp 1–8. https://doi.org/10.1109/PAIS62114.2024.10541193
46. Park KY, Kim YS, Kim J (2012) Security enhanced IEEE 802.1x authentication method for WLAN mobile router. In: 2012 14th international conference on advanced communication technology (ICACT), pp 549–553
47. Pratama D et al (2023) Behind the wings: the case of reverse engineering and drone hijacking in DJI enhanced Wi-Fi protocol. http://arxiv.org/abs/2309.05913
48. Restituyo R, Hayajneh T (2018) Vulnerabilities and attacks analysis for military and commercial IoT drones. In: 2018 9th IEEE annual ubiquitous computing, electronics and mobile communication conference (UEMCON), pp 26–32. https://doi.org/10.1109/UEMCON.2018.8796596
49. Sarabakha A, Suganthan PN (2023) anafi_ros: from off-the-shelf drones to research platforms. In: 2023 international conference on unmanned aircraft systems (ICUAS), pp 1308–1315. https://doi.org/10.1109/ICUAS57906.2023.10155881
50. Shafik W, Mojtaba Matinkhah S, Shokoor F (2023) Cybersecurity in unmanned aerial vehicles: a review. Int J Smart Sens Intell Syst 16(1). https://doi.org/10.2478/ijssis-2023-0012
51. Singhai R, Sushil R (2023) An investigation of various security and privacy issues in Internet of Things. Mater Today: Proc 80:3393–3397. https://doi.org/10.1016/j.matpr.2021.07.259
52. Singhal G, Bansod B, Mathew L (2018) Unmanned aerial vehicle classification, applications and challenges: a review. https://doi.org/10.20944/preprints201811.0601.v1
53. Snehal Samanth PKV, Balachandra M (2022) Security in Internet of Drones: a comprehensive review. Cogent Eng 9(1):2029080. https://doi.org/10.1080/23311916.2022.2029080 (Edited by Chen J)
54. Thankappan M, Rifà-Pous H, Garrigues C (2022) Multi-channel man-in-the-middle attacks against protected Wi-Fi networks: a state of the art review. Expert Syst Appl 210:118401. https://doi.org/10.1016/j.eswa.2022.118401
55. Tian J et al (2022) Adversarial attacks and defenses for deep-learning-based unmanned aerial vehicles. IEEE Internet Things J 9(22):22399–22409. https://doi.org/10.1109/JIOT.2021.3111024
56. Tsao K-Y, Girdler T, Vassilakis VG (2022) A survey of cyber security threats and solutions for UAV communications and flying ad-hoc networks. Ad Hoc Netw 133:102894. https://doi.org/10.1016/j.adhoc.2022.102894
57. Villa TF et al (2016) Development and validation of a UAV based system for air pollution measurements. Sensors 16(12). https://doi.org/10.3390/s16122202
58. Warnakulasooriya K, Segev A (2024) Attacks, detection, and prevention on commercial drones: a review. In: 2024 international conference on image processing and robotics (ICIPRoB), pp 1–6. https://doi.org/10.1109/ICIPRoB62548.2024.10543316
59. Westerlund O, Asif R (2019) Drone hacking with Raspberry-Pi 3 and WiFi pineapple: security and privacy threats for the Internet-of-Things. In: 2019 1st international conference on unmanned vehicle systems-Oman (UVS), pp 1–10. https://doi.org/10.1109/UVS.2019.8658279
60. Wu Q, Zhang R (2019) Intelligent reflecting surface enhanced wireless network via joint active and passive beamforming. IEEE Trans Wirel Commun 18(11):5394–5409. https://doi.org/10.1109/TWC.2019.2936025
61. Yaacoub JP et al (2020) Security analysis of drones systems: attacks, limitations, and recommendations. Internet Things 11:100218. https://doi.org/10.1016/J.IOT.2020.100218

62. Yan S et al (2019) Hiding unmanned aerial vehicles for wireless transmissions by covert communications. In: ICC 2019—2019 IEEE international conference on communications (ICC), pp 1–6. https://doi.org/10.1109/ICC.2019.8761271
63. Yang Z, Xu W (2024) Physical layer security performance optimization of unmanned aerial vehicle-assisted wireless network. In: 2024 7th international conference on advanced algorithms and control engineering (ICAACE), pp 1600–1605. https://doi.org/10.1109/ICAACE61206.2024.10548447
64. Zhi Y et al (2020) Security and privacy issues of UAV: a survey. Mob Netw Appl 25(1):95–101. https://doi.org/10.1007/s11036-018-1193-x
65. Zhong C, Yao J, Xu J (2019) Secure UAV communication with cooperative jamming and trajectory control. IEEE Commun Lett 23(2):286–289. https://doi.org/10.1109/LCOMM.2018.2889062

Chapter 3
Unmanned Autonomous Vehicles for Emergency Response and Disaster Relief

Dhana Lakshmi Mariyala and Umair B. Chaudhry

Abstract Unmanned Autonomous Vehicles (UAVs) are increasingly being recognized as transformative tools in emergency response and disaster relief operations due to their rapid deployment capabilities, versatility, and ability to access hazardous or remote areas. This chapter presents a comprehensive literature review exploring the evolving role of UAVs in enhancing disaster response efficiency, particularly in real-time situational awareness, resource delivery, and search and rescue missions. Special attention is given to the integration of artificial intelligence (AI) for autonomous pathfinding, obstacle detection, and collision avoidance—critical functionalities that improve operational reliability in dynamic and unpredictable environments. Through a synthesis of existing research and case studies, the chapter identifies key technological advancements while also examining current challenges such as regulatory limitations, coordination complexities, energy constraints, and data privacy concerns. The chapter concludes by highlighting the future potential of intelligent UAV systems as integral components of resilient emergency response frameworks and emphasizes the need for further interdisciplinary research to overcome prevailing limitations and drive scalable, ethical adoption in global disaster management efforts.

Keywords Drones · Disaster recovery · Search and rescue · Emergency response

3.1 Introduction

3.1.1 Background of the Study

UAVs have become the paradigm-changing tools in Emergency Response and Disaster Relief. In traditional disaster response, limited accessibility to the affected areas, slow information gathering, and difficulties delivering aid in real-time have

D. Lakshmi Mariyala · U. B. Chaudhry (✉)
Northumbria University London, London, UK
e-mail: umair.chaudhry@northumbria.ac.uk

U. B. Chaudhry and H. Jahankhani, *Autonomous Revolution: Strategies, Threats and Challenges*, Advanced Sciences and Technologies for Security Applications,
https://doi.org/10.1007/978-3-031-96782-5_3

often hampered progress. UAVs have significant advantages in these cases as they can quickly get to potentially dangerous or unreachable places, data on crucial subjects, and perform any kind of task automatically [32]. Through machine learning and artificial intelligence (AI), these vehicles are able to rapidly analyse data, optimise routes and make real-time decisions for maximum success.

Due to the integration of AI, UAVs on air, water, and ground self-determine paths with complex algorithms of path finding and collision avoidance that often led to a path that can negotiate through unpredictable environments in times of disaster. There has been development of UAVs which can dynamically adapt to change in their surroundings, that is, avoiding obstacles and rerouting when necessary.

In addition, UAVs' self-learning capabilities allow the search, rescue, and supply distribution pilots to get better at trajectory optimisation, making for more efficient search, rescue, and supply distribution operations [22]. AI-equipped UAVs potentially cut down on human intervention and risks while maximising response efficiency, and thus become increasingly valuable assets in the management of emergencies and in delivering quick and quick relief to affected populations. UAVs will transform emergency response and disaster relief on air, water, and ground. Rapid deployment, access to hazardous areas, and real-time data make it possible to apply in search and rescue, supply delivery, and infrastructure assessment. UAVs help cut down human risk and improve response efficiency in general, especially in complex, remote environments. Currently, advancements in AI and sensors have led to those vehicles being a vital part of disaster management, making these vehicles quicker, more accurate and safer in lifesaving critical situations.

Natural disasters are becoming more frequent and severe, and more needs to be done to make our response to these disasters more efficient and innovative. Limited access to impacted zones and long delays in gathering real-time data are just some of the challenges traditional disasters have had to face [46]. However, promising solutions are UAVs, enhanced with AI for self-navigation and decision-making. The AI integration of UAVs allows them to optimise relief efforts and autonomously adapt to avoid obstacles rapidly. The study will investigate the application of AI-driven UAVs on air, water and ground in emergency response to address the critical gap in research of their practical application and effectiveness in real-world scenarios. The study analyses their capabilities and limitations to help support the adoption of UAV technology for more effective, timely disaster relief and emergency response strategies.

In conclusion, this study is positioned to explore the critical and evolving role of unmanned autonomous vehicles (UAVs) in emergency response and disaster relief operations. By addressing four key research objectives and corresponding questions, the investigation aims to establish the significance of UAV deployment in high-risk scenarios, while also analysing how artificial intelligence and machine learning can enhance their responsiveness, adaptability, and operational efficiency. The study specifically focuses on understanding how AI-driven self-learning systems can optimise pathfinding and collision avoidance in complex and dynamic environments. Through this exploration, the research aspires to contribute valuable insights into the development of more effective, autonomous UAV solutions capable of improving disaster relief outcomes in real-world applications.

3.2 Methodology

This work adopted a dual-hybrid research design—gathering both qualitative and quantitative data to assess the use of UAVs in emergency relief. This approach enables a simultaneous examination of theory and practice, which could be beneficial.

The qualitative data was obtained from practising professionals in the fields of disaster management and UAV [18]. These interviews were particularly effective in establishing the practical realism when UAVs are used in these operations. Secondary data was obtained from journals, papers, and documented articles on the use of UAVs in disaster management. Data collected from the interviews were analysed qualitatively and thematic analysis was used to analyse the results giving themes such as flexibility, production capacity, and safety [54]. Qualitative and quantitative data from case studies were statistically analysed to assess the UAV performance about different parameters which include response time, coverage area and resource utilisation.

The study also entailed analysing other documented cases, for instance, the use of UAVs in the Nepal earthquake and during Hurricane Harvey. These cases offered research-based proof of the successes of UAVs in various disaster situations [61]. Real-life experiences of UAV operations during simulated disaster scenarios also contributed to the strengthening of the data collected based on field observations highlighting the strengths and weaknesses of UAVs.

3.3 Surveying the Literature

Thus, this section examines the application of unmanned autonomous vehicles (UAVs) in disaster response operations by combining a literature review and the author's findings. There is a brief and critical review summary of the related works, a presentation of the research method, activities implemented, and the possibility of the chosen approach. The various barriers faced during the research are described together with suggestions of how these barriers could be overcome.

Literature on UAVs in disaster management is very rich in concern regarding their effectiveness in improving efficiency, safety and flexibility in disaster situations. The following subsections critically review key themes from the literature.

3.3.1 UAVs in Enhancing Disaster Response Efficiency

The use of UAVs has been developed in the last decade as an effective tool in increasing the efficiency of the disaster response curve. They entail fast deployment, continuous data gathering and self-sufficiency; features that directly counter most of the shortfalls of conventional disaster relief mechanisms. Khan et al. [28, 29] state that UAVs can rapidly reach the scene of a disaster to perform assessment and facilitate situational awareness by making use of images as well as other sensors. This capability helps to minimize delays as well as make better quick decisions whenever there is an emergency [28, 29]. Furthermore, UAVs are indispensable in terms of the efficiency of resource distribution during and after disasters. In this way, it allows responders to strategically allocate resources by identifying genuine needs in the given area based on the summaries that they provide [43]. For example, UAVs used during Hurricane Harvey in 2017 were able to offer aerial information for the better determination of resource deployment to affected areas and thus reduced response time.

Moreover, according to Daud et al. [15], drones or UAVs have become indispensable in improving efficiency in disaster response. The advantage of performing many of their functions independently is the opportunity to give the specifics of the disaster's assessment quickly and accurately, determining the degree and scale of the disaster. As compared to more traditional response schemes, UAVs are not hindered by traffic jamming or damaged infrastructure, which allows them access points other conventional responders cannot. For instance, as found by Kallas and Napolitano [26], after the Beirut explosion in 2020, UAVs offered photogrammetry that enabled the first responders to decide on the degree of the calamity. At the same time, UAVs have significant potential benefits, though in practice they are accompanied by technical and organisational limitations to the efficiency. Chang et al. [12], demonstrated that UAVs depend on a set of coordinates that have been set by GPS to navigate and perform their reconnaissance, the GPS might be affected because of disasters say earthquakes or hurricanes (Fig. 3.1).

Such a limitation can lead to failure in accurately or timely mapping areas affected by disasters hence the importance of designing robust navigation systems that can work in isolation. UAVs are also very vital in directing the resource input and transportation of goods and services [20]. Categorised in a manner that identifies areas needing immediate response based on real-time data they are valuable tools in the responder's inventory. For instance, during the 2021 floods in Germany, UAVs assisted in needy point locations via affected, isolated communities and subsequently provided relevant aid and relief goods delivery points. However, as per Mohsan et al. [41], strategic problems, including lower payload capacity, hinder their effectiveness in moving large amounts of resources. Such a lack makes it clear that, for example, modular designs of UAVs that enable increased capacities for the carriage of payloads and other equipment are necessary. One potential of UAVs is within the real-time surveillance and information gathering which improves knowledge of the environment.

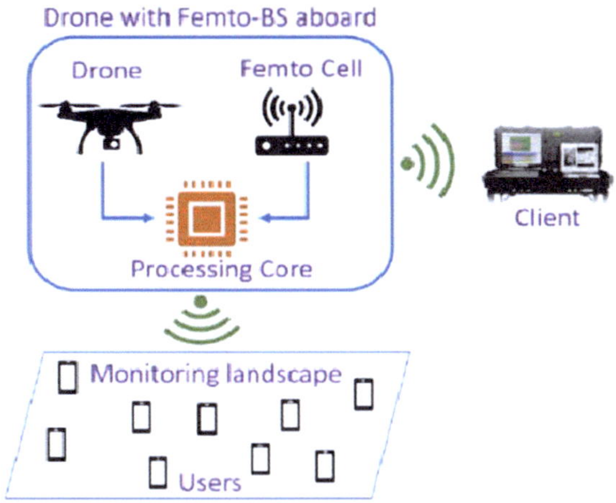

Fig. 3.1 High level schematics of data sensing system (*Source* [10])

Such drones come with HD cameras, thermal imagers, and Lidars; therefore, they offer both visual and thermal surveys [30]. However, they claim that it can result in risky over-dependency with technological functionalities and systems thus marginalising the role of human professionals in decision-making processes. Meanwhile, according to Elmeseiry et al. [17], the huge amount of data that may be received and recognised can also be problematic if not well-screened or pre-processed in a way that will not congest the response teams. One drawback of using UAVs is that their performance is relatively higher in areas where there is infrastructural development than in rural or remote areas. The imagination of limited communication networks, the harsh topography, and the poor availability of skilled operators restrict its application in these regions. These challenges underscore the need for concrete UAV interventions and education with a focus on the increased possibility of geographically applicable solutions.

Technology in the form of unmanned autonomous vehicles has now become a common feature of disaster responses due to robust performance in sensitive and dangerous scenarios. One of the reasons the community depends on them is that they can quickly get to areas that might otherwise be very difficult or dangerous when responding to disasters. Conventional top-down approaches to disaster management involve implementing on-the-ground strategies which can be limited by issues of infrastructure in the disaster area or geographical hindrances. UAVs overcome these challenges by offering a bird's eye view of conditions, to help inform where issues exist, and response assets can be best utilized.

UAVs have the special strength of being able to obtain and process data promptly. Intelligent with better sensors and cams, these cars collect minutiae that improve awareness of the surroundings. This information is most useful for defining the target groups, controlling the environment and planning the actions which would be taken. As argued by Wu et al. [66], UAVs provide an efficient means of collecting and conveying information and facilitate action, which is precise and swift.

Furthermore, in disaster situations, UAVs provide greater efficiency in terms of resource utilization [28, 29]. Helping responders, they depict areas impacted and establish the level of harm for assisting in deciding where to begin help. UAVs minimise redundancies and guarantee that scarce resources, including medical equipment and food, are delivered to areas that require the mentioned commodities. Such efficiency is especially important during big disasters because less money is available and every minute is critical.

However, challenges are inherent in the use of UAVs in disaster response as depicted herein below. The use of such battery life and payload capacity can be hindered by technical challenges such as battery capacity and load-carrying capacity. In addition, the incorporation of UAVs in the currently established emergency structures entails massive capital investment in training and infrastructure. On this basis, the potential benefits are, in this view, many times more valuable than these limitations [41].

Therefore, UAVs improve the capacity to respond to disasters since the traditional models of managing crises result in those inefficiencies while UAVs allow a proactive approach to future disasters. These features are useful in the current world that requires quick reactions and relies on real-time information and accuracy of distribution of resources in disasters.

3.3.2 Integration of AI in UAV Operations

Innovations like smart systems that sit in the UAVs have made them more effective where the environment is unpredictable [31]. Advanced AI capabilities can help UAVs analyse millions of sensor inputs simultaneously and decide, for example, where to fly next to avoid an object or which route to take. In their work, Vanderhorst et al. [65] outline that AI-supported UAVs are capable of making real-time adjustments to the course and operation based on changes to the environment, including weather conditions, or the appearance of risks. Furthermore, AI-supported UAVs perform well in cases of searches and rescues [6]. They are capable of recognizing survivors in disaster-affected areas including cases such as poor lighting or lots of bush. As stated by Lyu et al. [37], increased capabilities are not only beneficial for the speed and efficiency of the rescue operation but also beneficial to the human life threats.

The incorporation of AI into UAV systems has revolutionised concepts of emergency disaster management. AI improves the capabilities of UAVs to make them independent and to utilize specific strategies in dynamic disaster scenes. For instance, as studied by Chiguware [13], AI-based UAVs can examine images captured by satellites and other sensors to forecast the extent of fire danger, or flooding. It all depends on the quality of the data sets that are used to produce any such predictions. A low quality of data entails incorrect predictions, which, in turn, can compromise response efforts. Another innovative use of AI in UAVs is in independence and impediment evasion autonomously [8]. AI techniques make it possible to navigate UAVs to avoid objects on their path that can be fatal to the devices and their control systems. This capability is especially important in disasters where there are unstable structures at the scene.

However, implementations of these algorithms might experience a drop in performance due to poor weather conditions like forming rain, storms or other strong winds that would prevent sensors from being precise, as found by Tlili et al. [64]. AI also provides UAVs with the capability to undertake real-time risk analysis of situations. By considering the environmental conditions, and the UAV's operational characteristics, the AI systems can adapt the UAV flight profiles on the fly, to improve the chances of the completion of the mission. For instance, Leung [35], revealed during the 2021 Haiti earthquake, the AI technology that controls the UAVs redirected them to avoid sites with aftershocks and land instability. However, legal questions come into the challenge of who has the authority to delegate important decisions to AI systems especially when dealing with risking the lives of the people. In addition, it demonstrates how AI improves UAV performances in cases of Search and Rescue operations (Fig. 3.2).

Advanced image recognition algorithms mean UAVs are capable of identifying survivors under unfavourable circumstances such as under thickets, or where visibility is low [62]. Positive and negative cases are still a limitation, which makes the need to enhance the efficiency of AI algorithms a continuous process. Nevertheless, incorporating AI in UAVs is still complicated. Expensive AI structures which are incorporated in UAVs, and the skills required, make AI UAVs impractical in deprived areas. also exposed to risks of security breaches including hacking or data breaches, which are operational risks to the mission. To overcome these challenges there is needed more strict rules in the sphere of cybersecurity and international cooperation to define the rules of using AI with UAVs.

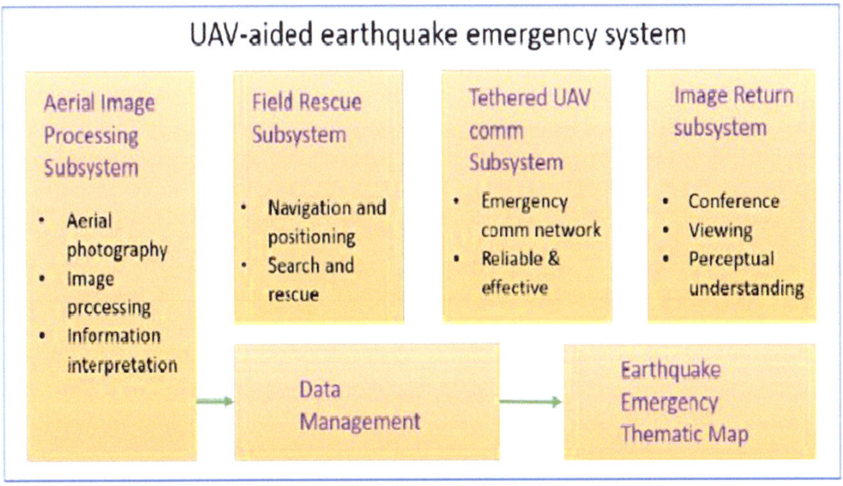

Fig. 3.2 UAV-aided earthquake emergency system (*Source* [10])

The incorporation of AI in UAV operations has changed their functionality and efficiency in disaster management in an unprecedented way. AI is used in UAVs to enhance the capability of processing data in real-time thus improving ethical decision-making as well as operational and independence. The transition from human-operated control to more automatic and intelligent control of UAVs enables the UAVs to change based on changing conditions, something that is important in dynamic disaster response scenarios.

AI-enabled UAVs work best on scenarios for continuous assessment and monitoring of the environment, identification of obstacles and the best pathfinding [45]. Compared with ground coordinate data used in traditional rigid-scheme UAV platforms, an AI-assisted UAV can adjust the course of flight on the go. For instance, such procedures involve the control of UAVs which were stationary in an area due to, for example, blocked roads or shifting debris; in the realization of this, the control of UAVs is automated by AI algorithms in a way that the UAVs are redirected to the correct path without requiring input from an actual human being. The outstanding flexibility offered by such designs minimizes time wastage and hence offers effective and efficient closure of sensitive operations.

Further, AI enhances the efficiency of guiding UAVs in performing difficult tasks like looking for survivors, assessing the level of devastation, and assessing the hazards that prevail around the scene. Using the deeper aspects of image recognition and sensor interfaces AI tries to analyse the optical and thermal data for signs of occupancy or structural damage [9]. These provide useful knowledge that helps the rescue groups to concentrate on areas of greatest necessity leading to the enhancement of the general efficiency of the operation.

However, there are obstacles to AI implementation in the UAV environment: high costs of development, the necessity of sufficiently rich training datasets, and problems associated with stable operation in various harsh conditions. However, there is still a need for a consideration of ethical and legal implementations of AI technologies in disaster circumstances.

AI integration increases the overall improvement of UAV operations in disasters through flexibility, increased accuracy and productivity. Though limitations apply, future developments in AI technologies will only build on the foundations and make UAVs inevitable tools for contingency management.

3.3.3 Machine Learning in Pathfinding and Collision Avoidance

Machine learning (ML) is another factor important in improving UAV performance. By having UAVs on board with predefined ML algorithms, the latter can recognize previous experiences and the efficiency of the UAV in dealing with obstacles increases with time. As suggested by Singh and Hooda [53], it is noted that using ML-based path planning systems, UAVs can find their paths through disaster areas in cities and avoid such obstacles as buildings, debris, and others. As opined by AlHinai [3], this capability is more important in operations such as search and rescue where the time of attending to the survivors could be the difference between life and death. Due to the ability to adjust the paths constantly and the non-interference with other UAVs, the UAVs with ML systems minimize risks and maximize efficiency [51].

ML is a very important component of UAV advancements in terms of its path planning and avoidance mechanism making it very much useful during emergencies. In addition to extending the capability of UAVs to discuss lessons from previous experiences, the ML algorithm enhances the overall manoeuvring capability of UAVs in complex and challenging environments [38]. For instance, in case of collapsed buildings or debris on the site, the ML-powered UAVs can continue functioning without interference. However, as per Linardos et al. [36], the expected ML algorithms could also be hampered by inadequate training data sets, particularly for disaster events which are rare and may lead to wrong navigation when the data is limited. One of the critical opportunities for utilising ML in UAV control is the capability to use adaptive path planning. Moreover, mechanical systems could be preset to recognise certain scenarios and subsequently synchronise the movements of the UAV but with the help of ML algorithms UAVs are capable of adapting their flight patterns concerning the context.

This capability is critical, especially in disaster-hit regions where physical structures impact the accessibility of the environment. Now, Souli et al. [57], argue that such flexibilities may lead to growth in computational needs leading to enhanced hardware and energy which might not always be reachable during disaster operations. ML also improves the ability of UAVs to avoid collisions which is useful in crowded areas with plenty of debris. From various sensory devices, the inputs can be analysed by ML algorithms and used to prevent collisions. For instance, as mentioned by Rezaldi et al. [47], UAVs which were used during tsunami disasters in Indonesia in 2018 proved useful by flying through crowded areas to drop aid. Nonetheless, difficulties in sensing and hence getting false signals or sensing disruptions triggered by external factors such as smoke or dust do affect the results of collision avoidance systems.

Another great use of ML is in the area of swarming, or cooperation of multiple UAVs to accomplish common goals [7]. The use of ML algorithms allows UAV swarms to self-organize their tasks and reduce the chances of inter-swarm interference in large-scale missions. Nevertheless, the task of maintaining coherent communication and coordination within the swarms of UAVs is still a technical problem which sometimes may be aggravated by the weak signals. However, as per Adil et al. [1], the use of ML-enabled UAVs to support disaster management has issues related to accountability and reliability. Some of the misjudgements which come with the ML systems include misidentification of obstacles to name but a few, or wrong selection of routes all in situations that would cause a catastrophe. These risks can however be prevented through the development of sound validation procedures and redundancy systems.

Advanced UAV flight planning and avoidance, critical in disaster relief, have been improved by machine learning (ML). These conventional UAVs are typically operated on a path that is defined before flying, and may not be very responsive to changes on the ground. On the other hand, UAVs with the help of ML algorithms adapt to new data and previous experience and become better at navigating through complicated and unpredictable disaster environments [33].

Mobility in emergencies particularly depends on the ability to pass through the densely built-up urban environment, ruins, or steep terrains. As per the statement, of Alsamhi et al. [5], the use of ML, Algorithms allows UAVs to make path decisions based on such data received from the environment within a limited period. These systems anticipate inflight challenges and respond proactively altering flight paths in real-time and consequently avoiding disruption. For example, if the UAV is flying through an area where there is a collapsed building or an object is moving along a similar path to the UAV, the system can easily reprogram the UAV to avoid danger and still stick to the necessary objectives.

Prevention of impacts with other objects is the other important area of application of ML in dealing with UAVs. By integrating the results of analysis of the data obtained with the help of cameras, lidar and radar, the ML models determine the presence of obstacles to the UAV's flight path [33]. They even define stances on what ought to be done to ensure that one is forced to avoid the other subject area encountering the other while at the same time maintaining the mission objectives set. Unlike rule-based systems, ML models can learn from a given context and are capable of developing improvements in identifying and evaluating new types of risks such as differential gradation of weather and other inconveniences such as the presence of debris on the roads.

Another obvious advantage of ML about the use of UAVs is the issue of identifying the changes in the overall efficiency, and, in particular, the measure of the mission completion. As stated by Chandran and Vipin [11], ML also reduces the risks or slights at operations such as collisions that may reduce access to the intended aid or data collection by UAVs. However, challenges such as restrictions on computation and the desire for large training sets stay with the algorithm. However, when implementing a COTS-based system deployment, particularly one that is integrated with ML technologies, legislation provisions and restrictions of UAV operation should be considered.

ML has emerged as a means of improving path-finding and avoidance bearings on drones. This flexibility in deterioration and enhancement of different conditions underlying entails safer and more efficient operations during disasters.

3.3.4 AI-Integrated Self-learning Trajectory Systems

Self-learning and AI-integrated trajectory systems in self-reliant UAVs present a considerable level of reality which can transform disaster relief horizon. According to Naing [44], these systems allow UAVs to learn from the environment, find solutions as they fly, and change accordingly based on the surrounding environment to overcome the limitations of preprogrammed systems. Of particular importance is the ability to

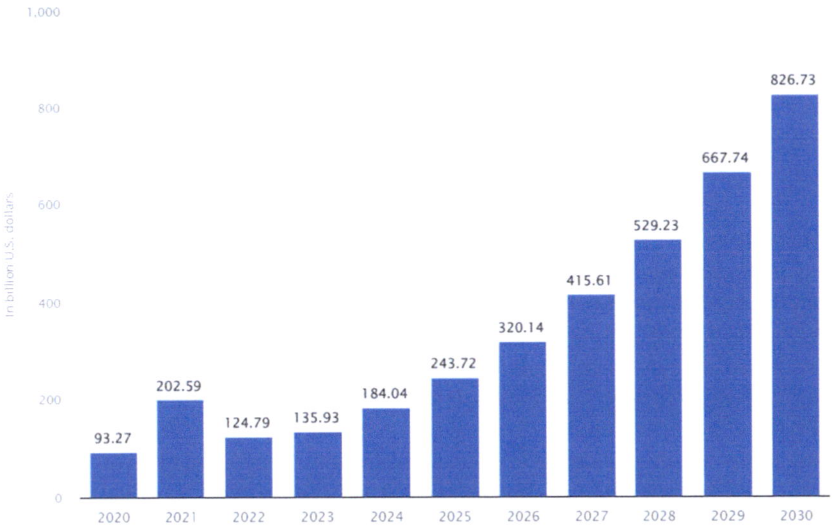

Fig. 3.3 Artificial intelligence (AI) market size worldwide (*Source* [58])

manoeuvre in disaster areas where regular, pre-programmed drones would not be able to cope with physical environments such as debris, fallen buildings or even abrupt weather changes. The benefits of these systems are not simply deep but numerous. These UAVs develop and correct their paths independently, which means that human engagement is limited, and operations are made more effective. In this regard, as reported in Statista [58], the market for AI grew above 184 billion U.S. dollars in 2024, which is expected to reach at around 826 billion U.S. dollars by 2030 (Fig. 3.3).

For instance, as per AlMahamid and Grolinger [4], they can speed up in assessing whether there are survivors in the disaster scene, areas that would necessitate speedy delivery of resources and which airspace is free from collision. Using this capability, it becomes possible to ensure the delivery of the necessary resources to the affected regions for golden-hour rescues that may need the use of the life-saving resources. Furthermore, as mentioned by Sacco et al. [49], self-learning systems can be integrated to become more efficient in the following missions of UAVs. However, there are certain limitations and constraints concerning such an approach. Self-learning trajectory systems require sophisticated AI implementation and solutions, and computational resources which complicate and contribute to the high cost of UAVs, as found by Huang et al. [23]. This can make them restricted in areas that are prone to disasters but economically not well off. Also, these systems work presupposing effective data input to teach or train and can be effective.

Consequently, their effectiveness might drop when there is scarce or unreliable data, and outcomes may vary between inefficient or simply failing. Autonomous decision-making is the fourth issue of major concern in military robotics, and it has many flaws: there is no accountability for the decisions made [59]. When one disaster zone is preferred over another by an AI decision, one has questions to ask about equity, morality and accountability. Besides, these systems could be vulnerable to cybersecurity threats since the adversary would target the AI reliance of the systems. These effective ideas of AI-integrated self-learning trajectory systems make a huge contribution to disaster relief, and trying to analyse and resolve issues such as cost, ethical issues and technical challenges can contribute to the development of these plans.

Robotic drones with the capability of Self Learning Trajectory System AI are expected to have a distinct impact on disaster relief operations. Compared to traditional methods, integrating these UAVs into applications benefits from sophisticated methods to control adaptable, precise and efficient functions. As argued by Al Shafian and Hu [2], the application of their learning intelligence and their capacity to optimize the paths instantaneously makes them relevant in instances where the incident is intricate and dynamic such as disasters.

Perhaps the most notable advantage of AI-integrated trajectory systems is the ability to modify flight paths in real-time. The existing generation of UAVs is usually equipped with predetermined flight paths that may be unusable in quickly evolving surroundings. On the other hand, self-learning systems, based on processed sensor data, determine obstacles, assess risks and determine new courses of action immediately. This adaptability is beneficial to guaranteeing that completely self-sufficient UAVs can enter and operate proficiently in ruined constructions, intense woods, or some other testing territory, all of which enormously grow their operational range.

The fourth benefit of these systems is with the aspects of resource management where they are effective in providing resources to the right areas. Due to their ability to undertake several missions and learn from them, UAVs can make better decisions and direct their attention to areas of priority [14]. For example, they can find out where to deliver the aid or the places where people are alive more easily and accurately. This has the effect of targeting more directly and reduces leakage and hence increases the effectiveness of the interventions made.

Further, self-learning trajectory systems are useful in minimizing human intervention where UAVs are operated in dangerous zones [37]. This feature not only improves safety but makes it easier for rescue teams to plan and concentrate on more vital issues while tackling the problem. It is also self-governing, and it means that UAVs can be expanded to a large scale of aircraft and used during large-scale disasters without employing a large number of people.

However, there are disadvantages associated with it, the reliability of the system, high development cost and ethical issues. However, the existing challenges stress that these barriers will one day be overshadowed by the technological progression in AI and machine learning, leading to constant implementation.

Thus, surviving UAVs with self-learning AI trajectory control is an improvement the disaster relief industry has not seen before. It improves the speed, accuracy and efficiency of response efforts, through flexibility to learn and work autonomously making them a viable solution for future disasters.

3.3.5 Feasibility of the Research Method

The above-identified research approach can be used to support the study of the use of UAVs in disasters, though not without some difficulty. This section assesses the practicability of the approach and the suggestions for further study.

The research identified that the use of mixed methods provided the opportunity to investigate the chosen topic in depth. The qualitative insights from the interviews with industry professionals were complemented by case studies, which served to support the theoretical recommendations [19]. There were challenges in scrutinizing the important data and engaging key professionals in the industry. Some occupational groups were not easily accessible for interviews because of issues of confidentiality, which also constrained the acquisition of more first-hand information.

Observations in the field also had some drawbacks, and scenes that were reconstructed for study purposes cannot mimic real-life situational conditions [16]. At the same time, these challenges were offset by the strengths of the used methodology which offered a valuable insight into both the theoretical and practical aspects of using UAVs in disaster management.

3.4 Challenges and Limitations

Although the use of UAVs in disaster management is highly effective, some challenges still emerge as follows. Barriers in terms of initial high costs, extensive technical requirements and the requirements for skilled operators in many cases impede the diffusion process [50]. Besides, legal limitations and corporate 2: Legal Framework: The legalities of operation and deployment of UAV technology should also be considered due to issues of security of data and privacy.

A common gap that is identified from the literature is the scarce research focused on the qualitative analysis of the consequences of the use of UAVs in real-world disasters in the long run [56]. Real-life experience with these theoretical concepts is insufficiently investigated and represents a major gap in the existing body of knowledge. This research aims to fill this research gap by assessing the strengths and weaknesses of UAVs in emergency response and relief missions.

Challenges and limitations of UAVs restrict their capability to optimise emergency response and disaster relief. One limitation is the prohibitive cost of sophisticated UAV systems especially when they include artificial intelligence and machine learning [48]. These costs typically encompass not only the actual UAVs used but also those of the related equipment, for example, ground control stations, and communications systems. It is a reasonable assumption, therefore, that this financial requirement may hamper usage in poor areas, thereby widening the gap in disaster mitigation equipment availability. Certainly, as per Heidari et al. [21], technical issues of UAV operation are one of the main concerns. The use of UAVs in disasters thus demands trained personnel, proper procedures of UAV management, and sound software regimes. Where there is low technical capability, these requirements may hamper the use of UAVs and slow down much-needed relief operations.

According to Soori et al. [55], the legal and regulatory frameworks that tend to govern the use of UAVs also emerge as key challenges affecting the use of UAVs. Challenges that arise from areas of operation, information security, and operational issues make the application of UAVs in disaster management complicated. For instance, the absence of homogeneous regulation always slows down the authorisation of UAV flights in times of emergency, therefore limiting their application. Also, the operation of UAVs can be limited by non-technical factors some of which include extreme weather conditions [63]. Disabilities caused by high wind gusts, intense precipitation, and account for the low use of UAVs in specific disaster circumstances. This is a limitation of the existing UAV technology, and thus an improvement on weather-proof UAVs and their general operational durability must be made.

Furthermore, the current literature does not offer any data regarding the long-term effects of UAV use in disaster operations [34]. Despite these improvements, it is crucial to underline the fact that UAVs' benefits over the short run, long-term consequences like environmental effects and societal acceptance have not been thoroughly examined. Filling this gap through adequate research and evaluation is critical for enhancing the desirable, long-term incorporation of UAVs into disaster response activities.

There is a lot that can be achieved using UAVs and yet the use of UAVs in disaster response scenarios has a lot of challenges and limitations. They originate from technical problems, operations restrictions and legal requirements that shape their effectiveness. Understanding these matters is crucial to the advancement of UAV use in disasters.

The first of these technical concerns is the dormant aspect of batteries in UAVs. An extension of time in the disaster area may involve extended flight duration but most of the UAVs are constrained by battery. As stated by Mohsan et al. [42], charging the battery or getting a replacement disrupts the flow of important missions hence low efficiency. Payload concerns also ensure that UAVs cannot carry too much assistance, goods, or sensors, thereby limiting the number of supplies they transport or the information they collect.

Concerning the practical concerns, there are also obstacles and difficulties for UAV utilization. As per the statement of Sziroczak et al. [60], severe weather conditions, including wind gusts, rains or any other dangerous weather states will affect UAV stability or performance. Moreover, due to the constantly changing and unforeseeable environment and the potential obstacles of the disaster areas, UAVs have to be performed accurately and flexibly. Moreover, despite the recommendation capabilities being adaptable through the use of AI and machine learning, recommendations are sometimes conditioned by technical glitches within an IT system.

Additional barriers to UAVs are regulatory and logistical, hailing from the industry and technology. As opined by Hussain et al. [24], the arrangements of airspace are usually different from one region to another and this makes the coordination of disaster response over national borders to be challenging. Compliance with the legal and ethical requirements such as data protection and security also comes into question. Furthermore, the incorporation of UAVs into the current disaster management frameworks demands a huge investment in training of personnel and equipment, which may be very costly, especially in the developing world.

Finally, most people have not embraced UAVs, and there is a lack of confidence in them. This is unusual because technology items can create concerns about observation, mistreatment or reliance that would delay technology adoption. People should be informed about the use of big data and analytics and their advantages and disadvantages.

3.5 Conclusion

The discovery of this study reveals that UAVs have the capability of enhancing disaster response and relief practices. Integration of further AI in UAVs highlights great performances in dynamic environments, data gathering, and resource management [27]. The case study shows the huge untapped potential of AI including UAVs in emergency management and response to make its operation faster, more effective and safer. These technologies do not only increase organizational productivity but also mitigate risks that may otherwise be fatal to human beings. The research also shows that there are factors that may hinder UAV use, including high costs, technical challenges and regulatory issues. There must be policy formulation and implementation of policies that steer the deployment of UAVs in cases of disaster [39]. Additional resources to enhance training and other capacity development will enhance the implementation of UAV technology to complement the existing emergency response.

UAVs enable systems that dynamically coordinate in disaster response and emergency management due to their ability to execute a set of tasks that traditional fixed-form response systems cannot execute. These systems can quickly evaluate disaster effects and reach places inaccessible by conventional methods, such as the 2020 Beirut blast, where UAVs delivered essential visualised data to emergency responders. The combination of AI and ML has further improved UAV features such as; self-navigation, risk assessment, and improved search and rescue [48]. For instance, during the 2021 Haiti earthquake, the self-controlled UAVs adjusted their flight trajectories based on the detection of the components of earthquake aftershocks and unstable ground.

However, several challenges hinder UAV operation and efficiency in disaster response. The technical constraints include over-dependency on GPS, limited payload control and uncertain circumstances such as extreme weather conditions. As studied by Mohan et al. [40], high novelties costs may limit the application of advanced UAV systems to regions with weak economic potential, especially when AI and ML-integrated systems are utilised. However, regulatory measures and demand for qualified staff are unique operation issues. These groups are not without their weaknesses. In actualising a more effective and efficient means of disaster relief, the focus should hence be placed on the continued development of autonomous UAVs which incorporate self-learning trajectory systems for trajectory determination and modification concerning natural disasters [52]. Nevertheless, precautions must be taken in the organisation to take heed of ethical questions and questions of legal responsibility regarding fully automated decision-making systems.

Future work can address these blurry areas of research with increased help from UAVs because the technology is not yet completely solved by these strategic interventions for helping in disaster management. This research plays an important role in integrating the theoretical developments and identifying the practical applications of UAV technology to enhance disaster response strategies to promote timely response in the future. As UAV technologies are further developed, the application of the equipment for the purpose of disaster management will greatly transform the field and offer new approaches to the problems of contemporary rescue operations. Future research should focus on collecting data from industry partners to have access to the data and research findings used in practice. Therefore, broadening the focus of the study to cover different geographic and socio-demographic settings may help increase external validity [25]. Furthermore, it may be possible to enhance the topography of a field through advancement in avails able to offer better simulations to make observations that would otherwise be difficult, thus making field observations more accurate.

References

1. Adil M, Song H, Mastorakis S, Abulkasim H, Farouk A, Jin Z (2023) UAV-assisted IoT applications, cybersecurity threats, AI-enabled solutions, open challenges with future research directions. IEEE Trans Intell Veh 9(4):4583–4605
2. Al Shafian S, Hu D (2024) Integrating machine learning and remote sensing in disaster management: a decadal review of post-disaster building damage assessment. Buildings 14(8):2344
3. AlHinai YS (2020) Disaster management digitally transformed: exploring the impact and key determinants from the UK national disaster management experience. Int J Disaster Risk Reduct 51:101851
4. AlMahamid F, Grolinger K (2022) Autonomous unmanned aerial vehicle navigation using reinforcement learning: a systematic review. Eng Appl Artif Intell 115:105321
5. Alsamhi SH, Shvetsov AV, Kumar S, Shvetsova SV, Alhartomi MA, Hawbani A, Rajput NS, Srivastava S, Saif A, Nyangaresi VO (2022) UAV computing-assisted search and rescue mission framework for disaster and harsh environment mitigation. Drones 6(7):154
6. Anandaraj APS, Dhivya R, Ateeq K, Subramaniam S (2024) A survey of AI integration in unmanned aerial vehicles (UAVs) using digital twin technology: advancements and applications. In: Digital twin technology and AI implementations in future-focused businesses. IGI Global, pp 14–26
7. Azoulay R, Haddad Y, Reches S (2021) Machine learning methods for management UAV flocks—a survey. arXiv:2108.13448
8. Banafaa MK, Pepeoğlu Ö, Shayea I, Alhammadi A, Shamsan ZA, Razaz MA, Alsagabi M, Al-Sowayan S (2024) A comprehensive survey on 5G-and-beyond networks with UAVs: applications, emerging technologies, regulatory aspects, research trends and challenges. IEEE Access 12:7786–7826
9. Bayomi N, Fernandez JE (2023) Eyes in the sky: drones applications in the built environment under climate change challenges. Drones 7(10):637

10. Bushnaq OM, Mishra D, Natalizio E, Akyildiz IF (2022) Unmanned aerial vehicles (UAVs) for disaster management. In: Nanotechnology-based smart remote sensing networks for disaster prevention. Elsevier, pp 159–188
11. Chandran I, Vipin K (2024) Multi-UAV networks for disaster monitoring: challenges and opportunities from a network perspective. Drone Syst Appl 12:1–28
12. Chang Y, Cheng Y, Manzoor U, Murray J (2023) A review of UAV autonomous navigation in GPS-denied environments. Robot Auton Syst 104533
13. Chiguware T (2024) Assessing the relevance of emerging technologies in disaster mitigation and relief efforts. In: The Palgrave handbook of language and crisis communication in Sub-Saharan Africa, pp 539–554
14. Damaševičius R, Bacanin N, Misra S (2023) From sensors to safety: Internet of Emergency Services (IoES) for emergency response and disaster management. J Sens Actuator Netw 12(3):41
15. Daud SMSM, Yusof MYPM, Heo CC, Khoo LS, Singh MKC, Mahmood MS, Nawawi H (2022) Applications of drone in disaster management: a scoping review. Sci Justice 62(1):30–42
16. Eigege CY, Daundasekara SS, Gomez ML, Walton QL, Hernandez DC (2022) Conducting research six feet apart: the feasibility of transitioning qualitative research to meet the emerging research needs during a pandemic. Int J Qual Methods 21:16094069211069442
17. Elmeseiry N, Alshaer N, Ismail T (2021) A detailed survey and future directions of unmanned aerial vehicles (UAVs) with potential applications. Aerospace 8(12):363
18. Ganesha HR, Aithal PS (2022) How to choose an appropriate research data collection method and method choice among various research data collection methods and method choices during Ph.D. program in India? Int J Manag Technol Soc Sci (IJMTS) 7(2):455–489
19. Gatto NM, Campbell UB, Rubinstein E, Jaksa A, Mattox P, Mo J, Reynolds RF (2022) The structured process to identify fit-for-purpose data: a data feasibility assessment framework. Clin Pharmacol Ther 111(1):122–134
20. Gupta A, Afrin T, Scully E, Yodo N (2021) Advances of UAVs toward future transportation: the state-of-the-art, challenges, and opportunities. Future Transp 1(2):326–350
21. Heidari A, Jafari Navimipour N, Unal M, Zhang G (2023) Machine learning applications in internet-of-drones: systematic review, recent deployments, and open issues. ACM Comput Surv 55(12):1–45
22. Hildmann H, Kovacs E (2019) Using unmanned aerial vehicles (UAVs) as mobile sensing platforms (MSPs) for disaster response, civil security and public safety. Drones 3(3):59
23. Huang X, Liu J, Jia C, Wang Z, Li W (2023) Online self-learning attitude tracking control of morphing unmanned aerial vehicle based on dual heuristic dynamic programming. Aerosp Sci Technol 143:108727
24. Hussain A, Li S, Hussain T, Lin X, Ali F, AlZubi AA (2024) Computing challenges of UAV networks: a comprehensive survey. Comput Mater Continua 81(2)
25. Johnson E, Parrish R, Nelson G, Elias K, Kramer B, Gaviola M (2020) Expanding pharmacotherapy data collection, analysis, and implementation in ERAS® programs—the methodology of an exploratory feasibility study. In: Healthcare, vol 8, no 3. MDPI, p 252
26. Kallas J, Napolitano R (2023) Image-based 3D modeling as a damage prioritization tool for historic buildings in post-disaster areas: the case of the 2020 Beirut blast. J Cult Herit 62:314–321
27. Karunarathna I, Gunasena P, Hapuarachchi T, Gunathilake S (2024) The crucial role of data collection in research: techniques, challenges, and best practices. UVA Clin Res 1–24
28. Khan A, Gupta S, Gupta SK (2022) Emerging UAV technology for disaster detection, mitigation, response, and preparedness. J Field Robot 39(6):905–955
29. Khan AJ, Ul Hameed W, Iqbal J, Shah AA, Tariq MAUR, Ahmed S (2022) Adoption of sustainability innovations and environmental opinion leadership: a way to foster environmental sustainability through diffusion of innovation theory. Sustainability 14(21):14547
30. Kovanič Ľ, Topitzer B, Peťovský P, Blišťan P, Gergeľová MB, Blišťanová M (2023) Review of photogrammetric and LiDAR applications of UAV. Appl Sci 13(11):6732

31. Kuru K (2021) Planning the future of smart cities with swarms of fully autonomous unmanned aerial vehicles using a novel framework. IEEE Access 9:6571–6595
32. Kyrkou C, Theocharides T (2019) Deep-learning-based aerial image classification for emergency response applications using unmanned aerial vehicles. In: CVPR workshops, pp 517–525
33. Kyrkou C, Kolios P, Theocharides T, Polycarpou M (2022) Machine learning for emergency management: a survey and future outlook. Proc IEEE 111(1):19–41
34. Lahmeri MA, Kishk MA, Alouini MS (2021) Artificial intelligence for UAV-enabled wireless networks: a survey. IEEE Open J Commun Soc 2:1015–1040
35. Leung J (2023) Technology in disaster management. In: Disaster management and emergency medicine in the Asia-Pacific Region: current practices and future directions, p 249
36. Linardos V, Drakaki M, Tzionas P, Karnavas YL (2022) Machine learning in disaster management: recent developments in methods and applications. Mach Learn Knowl Extract 4(2)
37. Lyu M, Zhao Y, Huang C, Huang H (2023) Unmanned aerial vehicles for search and rescue: a survey. Remote Sens 15(13):3266
38. MahmoudZadeh S, Yazdani A, Kalantari Y, Ciftler B, Aidarus F, Al Kadri MO (2024) Holistic review of UAV-centric situational awareness: applications, limitations, and algorithmic challenges. Robotics 13(8):117
39. Matthew UO, Kazaure JS, Onyebuchi A, Daniel OO, Muhammed IH, Okafor NU (2021) Artificial intelligence autonomous unmanned aerial vehicle (UAV) system for remote sensing in security surveillance. In: 2020 IEEE 2nd international conference on cyberspace (CYBER NIGERIA). IEEE, pp 1–10
40. Mohan M, Richardson G, Gopan G, Aghai MM, Bajaj S, Galgamuwa GP, Vastaranta M, Arachchige PSP, Amorós L, Corte APD, De-Miguel S (2021) UAV-supported forest regeneration: current trends, challenges and implications. Remote Sens 13(13):2596
41. Mohsan SAH, Khan MA, Noor F, Ullah I, Alsharif MH (2022) Towards the unmanned aerial vehicles (UAVs): a comprehensive review. Drones 6(6):147
42. Mohsan SAH, Othman NQH, Li Y, Alsharif MH, Khan MA (2023) Unmanned aerial vehicles (UAVs): practical aspects, applications, open challenges, security issues, and future trends. Intel Serv Robot 16(1):109–137
43. Munawar HS, Hammad AW, Waller ST, Thaheem MJ, Shrestha A (2021) An integrated approach for post-disaster flood management via the use of cutting-edge technologies and UAVs: a review. Sustainability 13(14):7925
44. Naing KM (2022) Design of autonomous sustainable unmanned aerial vehicle—a novel approach to its dynamic wireless power transfer
45. Politi E, Purucker P, Larsen M, Reis RJD, Rajan RT, Penna SD, Boer JF, Rodosthenous P, Dimitrakopoulos G, Varlamis I, Höß A (2024) Enabling technologies for the navigation and communication of UAS operating in the context of BVLOS. Electronics 13(2):340
46. Rajan J, Shriwastav S, Kashyap A, Ratnoo A, Ghose D (2021) Disaster management using unmanned aerial vehicles. In: Unmanned aerial systems. Academic Press, pp 129–155
47. Rezaldi MY, Yoganingrum A, Hanifa NR, Kaneda Y, Kushadiani SK, Prasetyadi A, Nugroho B, Riyanto AM (2021) Unmanned aerial vehicle (UAV) and photogrammetric technic for 3D tsunamis safety modeling in Cilacap, Indonesia. Appl Sci 11(23):11310
48. Rezwan S, Choi W (2022) Artificial intelligence approaches for UAV navigation: recent advances and future challenges. IEEE Access 10:26320–26339
49. Sacco A, Esposito F, Marchetto G, Montuschi P (2022) A self-learning strategy for task offloading in UAV networks. IEEE Trans Veh Technol 71(4):4301–4311
50. Samarakkody A, Amaratunga D, Haigh R (2023) Technological innovations for enhancing disaster resilience in smart cities: a comprehensive urban Scholar's analysis. Sustainability 15(15):12036
51. Schlenoff CI, Kootbally Z, Rachakonda P, Lightman S, Vassilev A, Wollman DA, Griffor E (2024) Standards and performance metrics for on-road automated vehicles

52. Singh R, Bhushan B (2021) Evolving intelligent system for trajectory tracking of unmanned aerial vehicles. IEEE Trans Autom Sci Eng 19(3):1971–1984
53. Singh HD, Hooda Y (2023) Resilience of digital society to natural disasters. In: Technological prospects and social applications of society 5.0. Chapman and Hall/CRC, pp 185–199
54. Singh N, Sahu O (2022) Feasibility assessment for E-commerce: a data collection from developing country (Ethiopia). MethodsX 9:101639
55. Soori M, Arezoo B, Dastres R (2023) Artificial intelligence, machine learning and deep learning in advanced robotics, a review. Cogn Robot 3:54–70
56. Soto-Vergel AJ, Velez JC, Amaya-Mier R, Pardo M (2023) Transforming ground disaster response: recent technological advances, challenges, and future trends for rapid and accurate real-world applications of survivor detection. Int J Disaster Risk Reduct 98:104094
57. Souli N, Karatzia M, Georgiades C, Kolios P, Ellinas G (2024) Mission-critical UAV swarm coordination and cooperative positioning using an integrated ROS-LoRa-based communications architecture. Comput Commun 225:205–216
58. Statista (2024) AI market size worldwide from 2020–2030 [Online]. https://www.statista.com/forecasts/1474143/global-ai-market-size. Accessed 8 Jan 2025
59. Stumborg MF, Roh B, Rosen M, Bendett S, Pollpeter K, Center for Naval Analyses (2021) Dimensions of autonomous decision-making: a first step in transforming the policies and ethics principles regarding autonomous systems into practical systems engineering principles. DRM-2021-U-030642-1 Rev Center for Naval Analysis
60. Sziroczak D, Rohacs D, Rohacs J (2022) Review of using small UAV based meteorological measurements for road weather management. Prog Aerosp Sci 134:100859
61. Taherdoost H (2021) Data collection methods and tools for research; a step-by-step guide to choose data collection technique for academic and business research projects. Int J Acad Res Manag (IJARM) 10(1):10–38
62. Tarr AA, Tarr JA, Thompson M, Ellis J (eds) (2021) Drone law and policy: global development, risks, regulation and insurance. Routledge
63. Telikani A, Sarkar A, Du B, Shen J (2024) Machine learning for UAV-aided ITS: a review with comparative study. IEEE Trans Intell Transp Syst
64. Tlili F, Ayed S, Fourati LC (2024) Advancing UAV security with artificial intelligence: a comprehensive survey of techniques and future directions. Internet Things 101281
65. Vanderhorst HR, Suresh S, Renukappa S, Heesom D (2021) Strategic framework of unmanned aerial systems integration in the disaster management public organisations of the Dominican Republic. Int J Disaster Risk Reduct 56:102088
66. Wu Q, Xu J, Zeng Y, Ng DWK, Al-Dhahir N, Schober R, Swindlehurst AL (2021) A comprehensive overview on 5G-and-beyond networks with UAVs: from communications to sensing and intelligence. IEEE J Sel Areas Commun 39(10):2912–2945

Chapter 4
Self-piloted Unmanned Vehicles in the Quantum Realm

Sanjay Kumar Kodur and Umair B. Chaudhry

Abstract This research investigates the potential of quantum computing to elevate self-piloted unmanned vehicles (SPUVs) by directly addressing some of the most significant outstanding challenges. These revolve around the crucial need for optimization, fast and efficient data processing, and solid security in these systems. When it comes to making them fully autonomous and capable of operating in any environment, these remain largely unsolved, classical problems. SPUVs operate in the realm of complexity. The paths they must follow in route planning can involve an exponential number of moves and branched decisions, whereas the fusion of the vehicle's many sensors must take into account a virtually limitless set of potential measurements and signals from the environment. As these systems make ever more autonomous decisions and as they expand over vast regions of Earth, sea, and sky they are bound, under the laws of physics, to "hit the wall" of computability.

Keywords Drones · Quantum computing · Quantum algorithms · Unmanned aerial vehicles

4.1 Introduction

Reducing human involvement and improving efficiency in our industries have seen great gains from the emergence of self-piloted unmanned vehicles a new breed of machines found across the spectrum from autonomous drones to ground-based systems [21]. Still, even these advanced automated systems face significant challenges in sheer computational power when it comes to performing real-time navigation and making decisions that also need to be secure [8]. After all, our lives are at risk if an unmanned vehicle misjudges a situation and runs into a person or makes a turn off a path where it should have stayed on the ground or in the air. Until recently, the most advanced command-and-control systems for any kind of vehicles were digital

S. K. Kodur · U. B. Chaudhry (✉)
Northumbria University London, London, UK
e-mail: umair.chaudhry@northumbria.ac.uk

U. B. Chaudhry and H. Jahankhani, *Autonomous Revolution: Strategies, Threats and Challenges*, Advanced Sciences and Technologies for Security Applications,
https://doi.org/10.1007/978-3-031-96782-5_4

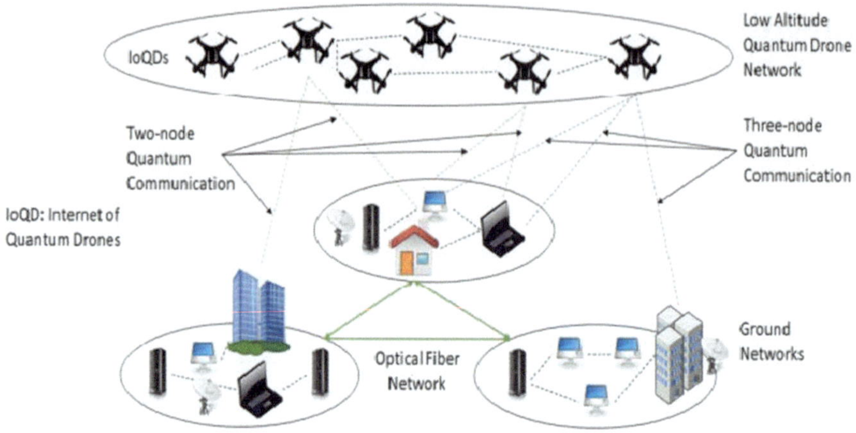

Fig. 4.1 Quantum computing

computers. Today, however, we're holding out hope for quantum computers to come to the rescue [42] (Fig. 4.1).

Applications of quantum computing to self-driving cars have yielded some initial results and highlighted several potential areas of improvement [30, 31]. For instance, researchers have pointed out that the basic arithmetic operations required for route optimization an essential part of the navigation problem tend to scale poorly with classical algorithms [10]. They have suggested that quantum algorithms, which are designed to carry out these kinds of operations in a more parallel fashion, could enable a self-driving car to navigate a complex and dynamic environment much more efficiently than could a V2I or V2V scenario based upon classical route optimization methods. As quantum machinery keeps moving forward, the integration of quantum computing with self-driving car technology is poised to make these vehicles far more efficient and secure. Research investigates these possibilities and presents several quite plausible scenarios that could play out when real-world autonomous systems begin to rely on quantum power plants.

Unmanned vehicles that navigate by themselves need advanced computational methods to accomplish three critical tasks navigation, obstacle detection, and data processing [23]. These tasks occur in real time, and when you consider doing them not just for a single vehicle but for many in a complex environment like air traffic or urban ground operations, the demand on the computing system becomes immense [22]. Even today's most powerful conventional computers would face serious challenges in meeting this demand and would likely become the controlling bottleneck in the next-generation architectures needed for fully autonomous vehicles. Quantum computing, with its very different approach and potentially vast increase in capacity, has raised hopes that it could revolutionize the unmanned vehicle systems [36].

This research aims to explore the potential of quantum computing to boost the capabilities of self-piloted unmanned vehicles in terms of navigation, data processing, and security. More concretely, we are looking for quantum algorithms and frameworks that have the potential to enable these sorts of improvements. We are especially interested in their potential to enhance real-time decision-making, sensor fusion (making sense of what the vehicle's various sensors are "telling" it), and cybersecurity. If we can sort through these issues and find some "answers," our next step will be to suggest how the development of these algorithms and frameworks could make a real difference in the functioning of autonomous vehicles. This study investigates how the nascent field of quantum computing could integrate with the navigation systems of self-piloted, unmanned vehicles. Specifically, it looks at the families of quantum algorithms that might be most useful if not already useful in this context. The algorithms include the Quantum Approximate Optimization Algorithm (QAOA) and other forms of quantum machine learning, which could, in theory, enable much smarter real-time decision-making and route planning, as well as more robust data fusion from the many sensors these vehicles are likely to use. Because QAOA and its ilk are not really useful if the vehicle can't communicate securely with other vehicles or with the infrastructure around it, we also consider the application of quantum cryptography in this context.

4.2 Literature Review

Combining quantum computing with self-piloted unmanned vehicles is a state-of-the-art leap forward. The computational functions that autonomous vehicles require are ripe for a transformation that could be instigated by the novel capabilities of quantum computing. Those functions include real-time data processing route optimization, which often hinges on a type of decision-making that can benefit from quantum parallelism and an area that is quite essential to the next generation of autonomous vehicles, namely, machine learning. This literature review looks at recent research to see where promising synergies might be forming and to visualize better the performance and safety of unmanned vehicles in the near future.

4.2.1 Quantum Computing and Autonomous Navigation

Obtaining and maintaining autonomy in navigation for self-piloted vehicles is essential, as the task of navigating a vehicle demands the simultaneous execution of several commands [13]. These commands fall into two broad categories: decision-making and path planning. Decision-making involves determining the appropriateness of vehicle control commands based on sensor input, while path planning concerns the

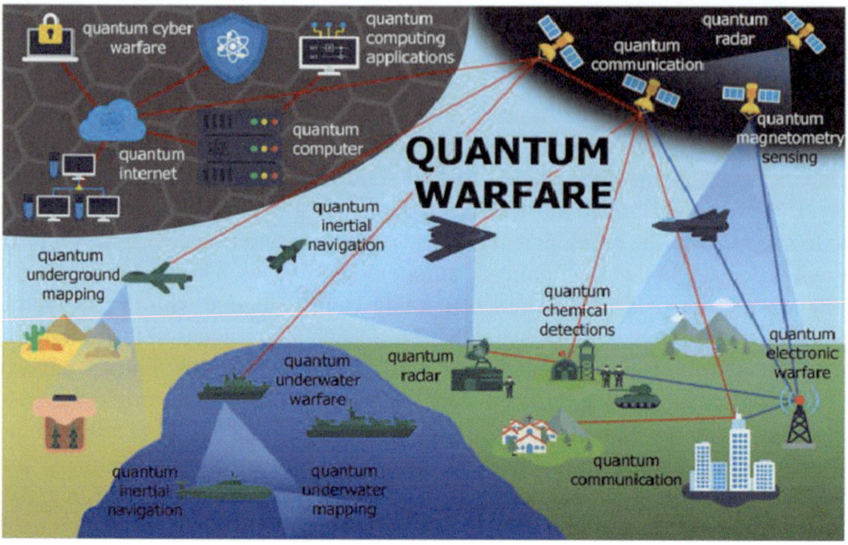

Fig. 4.2 Quantum warfare20

generation of commands to move the vehicle to a given destination [34]. Although path planning and decision-making can, in principle, be done on a computer, it has been found that when using quantum computing in conjunction with the QAOA, one can find near-optimal solutions to the aforementioned two problems in a more efficient manner than in using traditional computing or even high-performance computing (Fig. 4.2).

Furthermore, the coordination of multiple vehicles can be vastly enhanced through quantum computing, enabling fleets of cars, trucks, and aircraft to navigate safely and without colliding with one another a capability essential for logistics and military applications. Path optimization, as it's often referred to, can be done on a massive scale [17]. Indeed, using a fleet of drones to deliver packages is in many ways just an up-scaled problem that the delivery trucks we use now solve but with far more routing opportunities and thus far more safely.

4.2.2 *Autonomous Navigation Challenges and Quantum Computing Applications*

Aspect	Category	Challenges	Quantum computing applications	Potential improvements
Decision-making	Real-time processing	Processing large sensor data sets Handling uncertainties in data	Quantum Machine Learning (QML) Quantum Neural Networks (QNN)	Faster data processing More accurate decision-making under uncertainty
Path planning	Optimization	Complex route optimization Dynamic obstacle avoidance	Quantum Approximate Optimization Algorithm (QAOA) VQE	Near-optimal route selection Reduced energy and time for pathfinding
Fleet coordination	Multi-agent systems	Collision-free navigation Real-time synchronization	Quantum algorithms for coordination Quantum communication	Efficient vehicle communication Scalable fleet management
Sensor fusion	Data integration	Integrating heterogeneous data Noise and error handling	Quantum state estimation Noise reduction algorithms	Improved accuracy of sensor data Robust performance in noisy environments
Logistics and delivery	Large-scale systems	Complex delivery route mapping Handling vast regions	Quantum routing optimization Hybrid quantum–classical algorithms	Scalable optimization Enhanced efficiency in drone and vehicle delivery networks
Military applications	Defence operations	Navigating hostile environments Secure communications	Quantum cryptography Quantum-aided decision support systems	Enhanced operational safety Secure and reliable data transmission
Environmental navigation	Unstructured terrain	Difficulties in unknown or unstructured terrains	Quantum simulation for terrain analysis	Improved adaptability to unknown terrains

4.2.3 Sensor Fusion and Data Processing

Vehicles that drive themselves depend on sensor fusion for the accurate and real-time situational awareness needed to navigate safely. They do this by taking the data from a cornucopia of sensors like LiDAR, radar, and GPS and calculating with it to create the environmental model the vehicle requires to understand exactly where it is and what's going on around it [33]. However, traditional methods of data processing can't keep up with the demands of processing the huge amounts of data that the very smart and very powerful sensors put out, especially when the vehicle finds itself in complex and unpredictable environments [33]. Quantum machine learning, or QML, is another approach to the problem. And by a large number of demonstrations, it's more effective (Fig. 4.3).

The fast, fine-tuned operation of self-driving cars could be taken to another level with the use of quantum computers. In dense urban areas or industrial zones, those vehicles need to make a lot of decisions safely and quickly [11]. If an autonomous car can operate on the same time scale as a human brain, and if it truly has infinite horsepower, then how much safer might those cars be at not just driving but also detecting objects and mapping where they are and the environment around them.

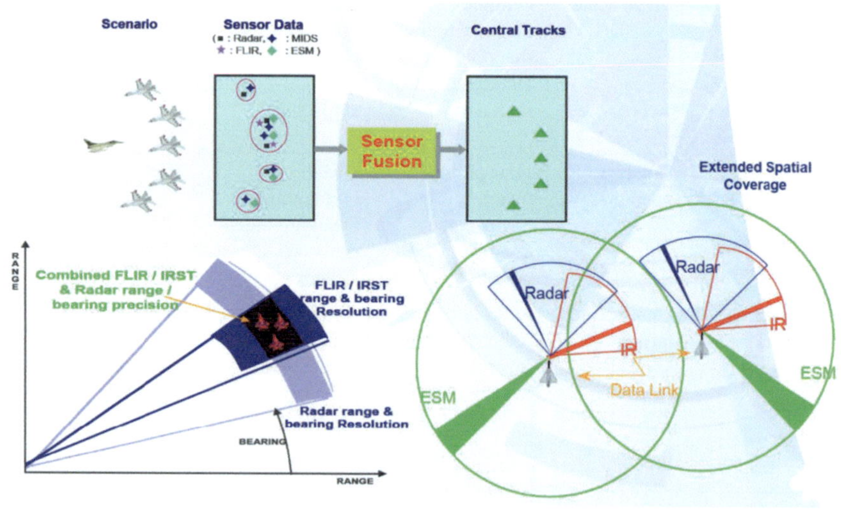

Fig. 4.3 Sensor fusion15

4.2.4 Machine Learning (QML) and Quantum Computing Address Current Challenges

Aspect	Traditional approaches	Challenges	Quantum computing advantage	Impact on autonomous vehicles
Sensor fusion	Combines data from LiDAR, radar, GPS, and cameras to create an environmental model	High computational demand Struggles in unpredictable environments	QML enhances the accuracy and speed of sensor data integration through quantum algorithms	Real-time, high-accuracy situational awareness in complex environments
Data processing	Processes large volumes of sensor data in real-time to ensure safe navigation	Limited by classical computation Latency in decision-making Poor scalability for dense regions	Quantum computing processes vast datasets faster, reducing latency and increasing precision in decision-making	Enables fast, efficient operations in urban and industrial areas
Decision-making	Relies on algorithms to determine appropriate control actions	Slow reaction times in high-stakes scenarios Suboptimal control in real-time	Quantum optimization (e.g., QAOA) finds near-optimal solutions quickly, outperforming traditional	Safer, quicker decision-making in dynamic scenarios
Path planning	Determines the best route for reaching the destination while avoiding obstacles	Struggles with complex multi-vehicle coordination and large-scale optimization	Quantum computing enables large-scale path optimization with better fleet coordination	Efficient route planning for single vehicles and fleets, enhancing logistics and safety
Multi-vehicle coordination	Ensures safe operation of multiple vehicles (e.g., drones or cars) without collisions	High computational overhead for real-time fleet management Collision risks in dense operations	Quantum computing provides scalable solutions for simultaneous multi-vehicle coordination	Improved logistics and military applications with enhanced safety and efficiency

(continued)

(continued)

Aspect	Traditional approaches	Challenges	Quantum computing advantage	Impact on autonomous vehicles
Object detection and mapping	Identifies objects and creates a map of the environment to avoid collisions	Limited detection accuracy in complex or dense environments	Quantum algorithms process environmental data more accurately, enabling better mapping and object detection	Enhanced ability to navigate safely in dense and unpredictable environments

4.2.5 Advancements in Quantum Hardware and Algorithms

The current state of quantum hardware and algorithms has advanced greatly and has begun to shape the potential of quantum computing for previously intractable problems in various domains autonomous systems included. Here, the authors look into the latest happenings in quantum processors and algorithms and consider what it all might mean for real-world problem-solving in places like autonomous navigation, sensor fusion, and even quantum-safe cybersecurity.

4.2.6 Quantum Processors: Pioneering Hardware Developments

The bedrock of quantum computing is quantum processors, which allow for the manipulation of qubits using superposition, entanglement, and other quantum principles. In the last five years, significant progress has been made toward increasing the fidelity and coherence times of quantum hardware and toward making it scalable.

Google's Sycamore processor achieved "quantum supremacy" in 2019 when it solved a problem far faster than any classical supercomputer could handle. Since then, newer versions of Sycamore have emphasized increasing the number of qubits and improving error correction. According to Arute et al. [5], these enhancements are crucial for addressing practical optimization challenges, such as the ones met by the autonomous navigation systems.

IBM has also made remarkable advancements with its IBM Quantum roadmap. This is a program that lies under IBM Research. Since 2019, it has been headed by physicist, engineer, and inventor Jerry Chow. The roadmap lays out a strategy for reaching major milestones over the next several years. It aims to deliver processors with more than 1,000 qubits by 2025. As reported by Chow et al. [12], these developments contribute directly to the possibility of putting into practice sophisticated quantum algorithms for the coordination and routing of groups of autonomous vehicles.

Another important development is the appearance of trapped-ion quantum processors, which have much longer coherence times than superconducting qubits. For example, IonQ's Harmony processor has been applied with great success in quantum machine learning, a field that offers ample opportunity for the amount of "data" to be processed by the high-dimensional "vectors" that define the kind of elementary operations performed by these technologies [25].

4.2.7 *Innovations in Quantum Algorithms*

Equally revolutionary is the creation of quantum algorithms. They let quantum systems tackle problems that are beyond the reach of classical computers. For self-driving cars, some of these algorithms are proving to be especially powerful.

4.2.8 *Types of Quantum Algorithms, Their Functionalities, Applications in Autonomous Systems, and Their Potential Benefits*

Algorithm	Functionality	Application in autonomous systems	Quantum advantage
Quantum Approximate Optimization Algorithm (QAOA)	Solves combinatorial optimization problems by approximating near-optimal solutions	Path planning, decision-making, and vehicle routing	Efficiently solves large-scale optimization problems faster than classical methods
Variational Quantum Eigen solver (VQE)	Solves optimization problems using a hybrid quantum-classical approach to find minimal energy states	Energy-efficient routing for fleets of drones and electric vehicles	Optimizes large and complex systems more efficiently than classical optimization techniques
Quantum Neural Networks (QNNs)	Enhances machine learning tasks by utilizing quantum systems for processing high-dimensional data	Sensor fusion, data integration from multiple sources, and environmental modelling	Processes and integrates data from multiple sensors more effectively than classical neural networks

(continued)

(continued)

Algorithm	Functionality	Application in autonomous systems	Quantum advantage
Grover's algorithm	Provides a quadratic speedup for unstructured database search and pattern recognition tasks	Object detection, obstacle recognition, and real-time environment analysis	Significantly improves search efficiency, enabling real-time decision-making and object detection
Shor's algorithm	Factors large integers efficiently, solving problems related to encryption and decryption	Quantum-safe cybersecurity for protecting autonomous systems	Enables the development of quantum-resistant cryptographic protocols to secure vehicle communication systems
Amplitude amplification	Increases the probability of finding a correct solution in probabilistic algorithms	Enhancing sensor accuracy and noise reduction in data from autonomous systems	Boosts the reliability of sensor data, improving decision-making accuracy
Quantum walk algorithms	Models traversal problems on graphs, offering efficient solutions for network-based problems	Multi-vehicle coordination and traffic flow optimization	Reduces computation time for large-scale navigation and coordination tasks
Quantum Fourier Transform (QFT)	Transforms data for solving problems like period finding and frequency analysis	Signal processing in autonomous vehicles, e.g., LiDAR and radar data analysis	Provides faster and more precise frequency analysis compared to classical Fourier transforms
Hybrid quantum-classical models	Combines classical computation power with quantum algorithms for enhanced scalability	Route optimization, traffic flow management, and decision-making in complex environments	Overcomes limitations of current quantum processors by leveraging classical systems for preliminary data processing

4.2.9 Quantum Approximate Optimization Algorithm (QAOA)

The Quantum Approximate Optimization Algorithm (QAOA) has been a fast-developing topic in recent years, and it is now gaining quite a bit of prominence [16]. Therefore, in light of this recent uptick in interest, we felt it necessary to provide a basic overview of the QAOA algorithm and what it is fundamentally all about. Indeed,

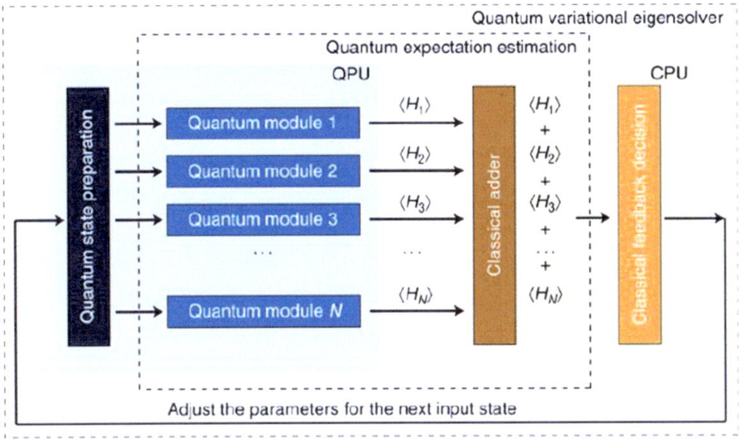

Fig. 4.4 Quantum variation eigen solver [29]

the QAOA has emerged as a potential (if still somewhat nuanced) front-runner for fulfilling the promise of quantum computers, particularly for combinatorial optimization problems. These are mathematical problems where one must find the best object from a finite set of objects.

4.2.10 *Variational Quantum Eigensolver (VQE)*

The energy consumption of autonomous fleets has been optimized using the Variational Quantum Eigensolver (VQE). According to Kandala et al. [18, 19], this quantum algorithm takes a hybrid quantum–classical approach that allows it to solve large, complex systems much more efficiently than classical methods. That makes it an excellent candidate for applying to large, complex optimization problems like drone swarms and electric vehicle routing (Fig. 4.4).

4.2.11 *Quantum Neural Networks (QNNs)*

Quantum neural networks have transformed machine learning into something that can handle the much larger datasets of today's world. In many ways, they are still in their infancy and far from being widely available tools [37, 38]. And yet, their use in sensory fusion tasks, where they combine data from multiple sources, has already shown them to be superior to classical neural networks in this and in other aspects of integrating the kind of multi-source data so important for the next generation of AI.

4.2.12 Grover's Algorithm

Database search applications, including object recognition and obstacle detection, have benefitted from Grover's algorithm. By providing a quadratic speedup relative to classical methods, Grover's algorithm has increased the efficiency of these applications to the point that we now might consider self-piloted vehicles to have real-time decision-making capabilities [7].

4.3 Implications for Autonomous Systems

Far-reaching effects can be expected from the progress of both quantum hardware and algorithms, and this is especially true concerning autonomous systems. Quantum power means better-than-ever computing, for which we're going to need new, hybrid quantum–classical frameworks to keep our code flowing across the computing continuum from today's supercomputers to tomorrow's quantum Erector Set. These computational enhancements will be applied primarily to three areas of autonomous systems research and development: navigation, sensor fusion (the way we perceive our environment), and cybersecurity (the way we ensure our systems can't be hacked).

4.3.1 Comparative Analysis Quantum Versus Classical Computing

The fundamental differences between quantum computing and classical computing, in terms of operational principles, computational efficiency, and area of applicability, make both types of computing unique in their own right. Still, an analysis of their comparative strengths and weaknesses along with the application of each type of computing in specific scenarios gives a comprehensive view of the two types of technologies. The advancements and research of the past five years, which has taken both types of technology to the next level, is emphasized in this section to shed light on the era of evolving computing technologies that we find ourselves in (Fig. 4.5).

4.3.2 Foundations of Quantum and Classical Computing

The traditional computer operates using binary logic, processing all its information in terms of bits whose values are either 0 or 1. In contrast, a quantum computer makes use of the strange and powerful phenomena of quantum mechanics most

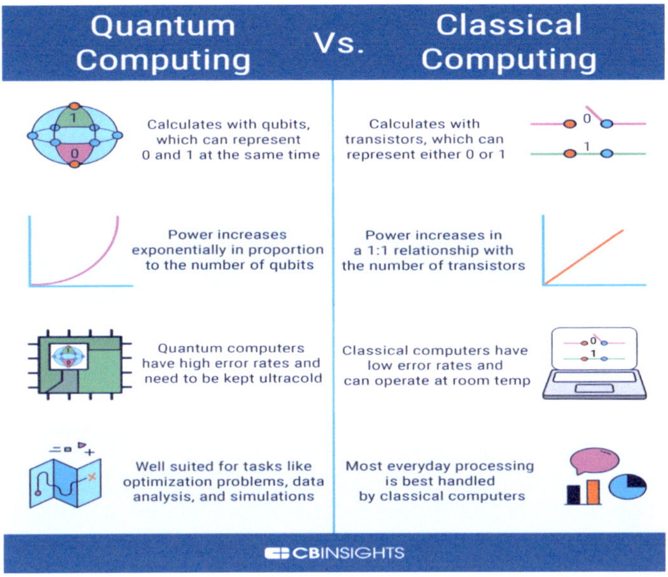

Fig. 4.5 Quantum computing versus classical computing [9]

notably, superposition and entanglement. In a quantum computer, a basic unit of information, called a qubit, can represent a 0, a 1, or both 0 and 1 at the same time. This ability to represent multiple values simultaneously underlies the power of a quantum computer to solve certain kinds of complicated problems much faster than a classical computer can. Research by Nielsen and Chuang (2021), shows that when we consider the number of qubits we work with, the scaling up of quantum states with qubits happens in an exponential fashion. This gives us a computational power that is way beyond what classical systems can offer, even when we use the same dimensionality of structures as we find in the brain. And, of course, we know that classical systems are good at structures that work deterministically and at structures that we can count on to do what they're supposed to do again and again, because we can either keep them running in a steady state or because we can turn them off and on in a reliable way (Fig. 4.6).

4.3.3 Computational Efficiency and Performance

Different domains have proved the computational efficiency of quantum computing. A good example is Shor's algorithm. This quantum algorithm for integer factorization demonstrates an exponential speedup over classical methods, such as the RSA algorithm (Research by Patel et al. [27]), that we used to rely on for Autonomous

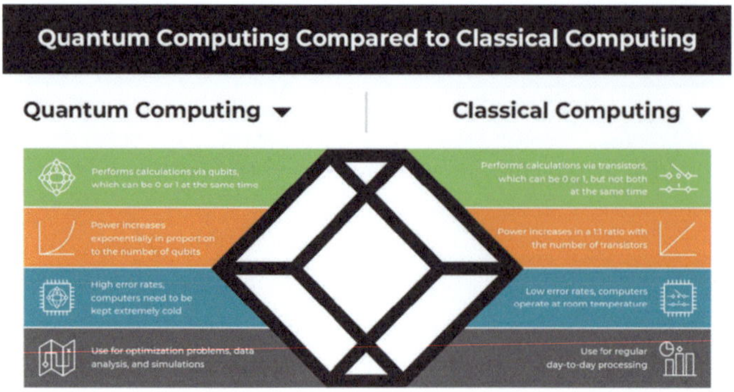

Fig. 4.6 Foundations of quantum computing [6]

navigation is another domain where we can see some promise from quantum algorithms. Two in particular and Grover's search algorithm have potential for solving path optimization and real-time decision-making problems. But classical algorithms that we do have, even the decent ones, don't serve us well in this domain because they struggle and fail to handle the combinatorial complexity of the problems we're throwing at them [16]. Nonetheless, quantum computing encounters problems associated with noise, decoherence, and error correction that constrain not only their current scalability but also their performance. In contrast, classical computing enjoys the benefits of decades of optimization. It has robust error correction mechanisms, and a mature software ecosystem that gives it reliable and efficient performance [1].

4.3.4 Applications in Data Processing and Machine Learning

Data processing and machine learning can benefit from the unique advantages that quantum systems present. Among them are the kinds of algorithms we can run on them like the Quantum Support Vector Machines (QSVM) or Quantum Neural Networks (QNN) which are fundamentally different from the workhorse algorithms of classical computation [37, 38] (Fig. 4.7).

One reason for this difference is that QNNs are able to use quantum entanglement, and the kinds of superposition states that Qubits can exist in, to perform pattern recognition with a level of parallelism that our best classical systems can't match [18, 19].

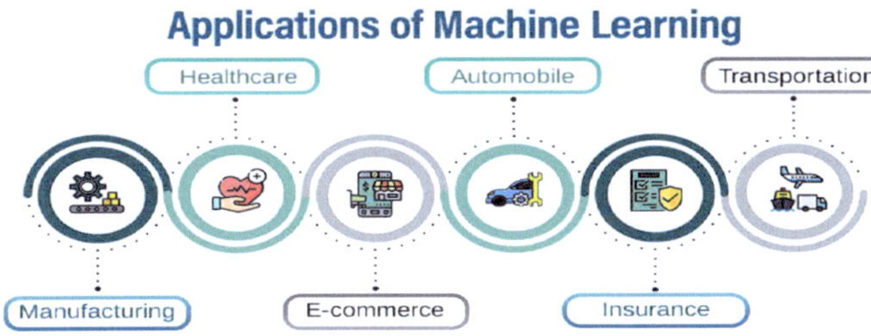

Fig. 4.7 Applications of ML [2]

4.3.5 Security and Cryptography

Cybersecurity has both opportunities and threats posed by quantum computing. Shor's algorithm, for instance, threatens classical encryption methods, while certain techniques of quantum cryptography like Quantum Key Distribution promise unbreakable security based on the principles of quantum mechanics [30, 31].

Most of our current systems like those at the NSA and in banking or corporate settings rely on either RSA or AES. Both of these are classical encryption protocols that are robust against classical attacks but vulnerable to quantum based decryption. We are now in the research and development stage of creating post-quantum cryptographic protocols to replace them (Chen et al. 2021).

4.3.6 Scalability and Hardware Considerations

The challenge of making quantum computers larger and more powerful hinges on the delicate nature of their basic units, or qubits, and on the extreme conditions required to keep them working reliably. These include temperatures very close to absolute zero [26]. Advances in quantum computing using trapped ions or designs based on superconducting qubits seem to be overcoming these problems and producing devices that are much less fragile and noisy than past efforts.

Moore's Law has guided improvements in classical computing for nearly half a century. If a similar trend were to occur in the new field of quantum computing, we would be in for some very exciting and powerful devices in the coming decades [32].

4.3.7 Cost and Accessibility

At present, constructing and maintaining quantum systems is prohibitively expensive, which restricts their use to large corporations and research entities. Classical systems, on the other hand, are available at a low cost and in great variety, making them the fit-for-priority system choice for most computing applications [1].

Remote-access cloud infrastructures, like those developed by IBM and Google, are paying accessibility dividends by letting a far greater number of researchers and developers work with actual quantum processors. In this manner, platform-provided access to quantum resources really is a democratic enabler of the nascent Q computing economy [12].

4.3.8 Future Prospects and Hybrid Models

Computing's future is in the integration of quantum and classical systems. Hybrid models take the best from both worlds, using prequantum computing for preprocessing and exerting the full power of postquantum algorithms on the hardest part of the problem. Research by Perdomo-Ortiz et al. [28]. This potential using quantum power in decision-making and prequantum power for the sensor suite in autonomous vehicles optimizes performance.

The applications of quantum computing and algorithms still cover only a few problem types, and their use is limited to areas where their unique capabilities confer a decisive edge. Even so, this nascent technology's synergy with classical computing promises to drive innovation in many fields: logistics, healthcare, the defence industry, and others [16].

4.3.9 Quantum Cryptography for Secure Communications

The growing reliance of self-driving cars on communications with other vehicles and with roadside infrastructure makes cybersecurity an ever more pressing concern. The classical methods of encryption that we know are not perfect. One reason they're not perfect is that if you have a very powerful computer, such as a quantum computer, you can use it to solve the kinds of mathematical problems that serve as the basis for the algorithms we use to encrypt our data [39]. The promises of quantum cryptography, which is more secure because it uses the laws of quantum mechanics instead of the laws of classical physics, mean that (or at least very much hope that) our future cars and road systems will be communicating with one another in a half-secure, half-public way that eavesdroppers can't really take advantage [4].

Moreover, highlight how absolutely imperative it is for QKD to secure command and control systems for autonomous vehicles [35]. This is where next-level data protection and communication integrity are of the utmost importance. Their work demonstrates how QKD-based protocols can keep the hack-prone communications used by vehicles safe from interference an application all the more vital because the communications could be controlling a vehicle in defence situations or for public safety.

4.3.10 *Quantum Simulations for Testing and Development*

Autonomous vehicles must be tested and validated, but doing so in the traditional, tried-and-true manner is both very expensive and very time-consuming. The truly game-changing aspect of quantum computing, though, is its ability to work with simulations of the virtual worlds in which the vehicles must operate [3]. These virtual worlds can be made to include all sorts of scenarios imagine the range of meteorological events, let alone the kinds of car-to-car and car-to-pedestrian interactions that one ought to be able to work up to in a well-functioning society [41].

High-dimensional environments can benefit from quantum computing. Exhaustive testing is necessary prototypes are too impractical. For test environments, an exhaustive virtual test is becoming standard fare [3]. This is how you get to safety–critical applications. At the same time, virtual tests tend to flag potential issues. If you're using an environment for development, identification of issues is a good sign. If you're using it for pre-deployment, it's also a good sign.

4.4 Conclusion, Challenges and Future Work

4.4.1 *Challenges in Implementing Quantum Computing*

The challenges of implementing quantum computing span problems that afflict all kinds of computers, albeit in more severe forms. They include the limitations of current conceptual hardware designs, issues with integrating the vastly different qubits of a quantum computer into a working whole, and scalability: if we can build a few dozen qubits into a functioning device, how do we construct one with thousands or millions of qubits. Although the developments for quantum computing seem promising, there remain significant hurdles to overcome before we can realize the many possible applications in self-piloted unmanned vehicles. Today's quantum hardware is constrained by the number of qubits it has and the ability to use those qubits to represent and process information [39]. Any number of storage elements or processors, if poorly designed and integrated into a system, will yield a platform that is almost impossible to work with. A platform that works efficiently when all its parts work together is a good first step toward a future where self-piloted vehicles could utilize quantum computing to navigate and perform other tasks [11] (Fig. 4.8).

Research keeps pushing forward with the idea of using quantum computing in the realm of autonomous vehicles. For one thing, the researchers are still working on hardware and algorithms [41]. They're trying to scale up the quantum processors make them bigger and better come up with even more improved error correction, and hybrid computing systems that are potentially makeable with our current knowledge of physical systems and materials since there are still enough unknowns in these areas to potentially cover the computational needs of the quantum parts of the system.

Fig. 4.8 Challenges of quantum computing [14]

4.4.2 Hardware Limitations

The practical implementation of quantum computing is chiefly hindered by the hardware. Today's quantum processors are limited by their coherence times, conditioning, and noise susceptibility according to Kjaergaard et al. (2020).

For example, superconducting qubits work only at or near absolute zero, making for a very resource-intensive system. By contrast, we know that trapped-ion qubits have much higher coherence times [26]. Yet, when we consider the thousandfold increase we would need to see in trapped-ion systems to get up to, say, 1,000 qubits, we realize that we are going to require some major breakthroughs in trapped-ion technology just to scale up to that level.

4.4.3 Error Correction and Noise Mitigation

The practical implementation of quantum computing is chiefly hindered by the hardware. Today's quantum processors are limited by their coherence times, conditioning, and noise susceptibility. For example, superconducting qubits work only at or near absolute zero, making for a very resource-intensive system as per Fowler et al. (2021).

By contrast, we know that trapped-ion qubits have much higher coherence times. Yet, when we consider the thousandfold increase we would need to see in trapped-ion systems to get up to, say, 1,000 qubits, we realize that we are going to require some major breakthroughs in trapped-ion technology just to scale up to that level [18, 19].

4.4.4 Integration with Classical Systems

Both technical and operational hurdles must be overcome to merge quantum systems with contemporary classical computing infrastructures. One of the biggest problems concerns the very low level of accessibility for potential users of quantum computers. As reported by Videla et al. (2020) these users must be able to write code that the quantum system can understand. Yet, the very nature of a quantum computer means that code-writing must happen at a much lower level of electronical fidelity than is the case with classical computers.

Another big integration problem is achieving cloud interoperability, which is crucial to using cloud-based quantum computers. Cloud interoperability between quantum systems and classical systems is necessary for running meaningful computations on the cloud. Both accessibility and interoperability are vital to enabling cloud-based quantum computing (Cross et al. 2022).

4.4.5 Challenges in Quantum Hardware and Algorithm Integration

Implementation of practical applications using quantum hardware and algorithms has made great strides but still faces some obstacles. The most significant issue is the limited number of qubits available in contemporary processors. This constraint directly affects the range and scale of problems that can be tackled and limits the nature of algorithms that can run on the nascent quantum mode of computation (Fig. 4.9).

Going beyond the immediate qubit problem, researchers in quantum error correction are searching for ways to make a fragile quantum computation robust against the kinds of perturbations that a real-world apparatus is bound to experience. One proposed solution to this situation the problem of not having enough qubits to work with and the problem of having too many qubits to work on without first solving

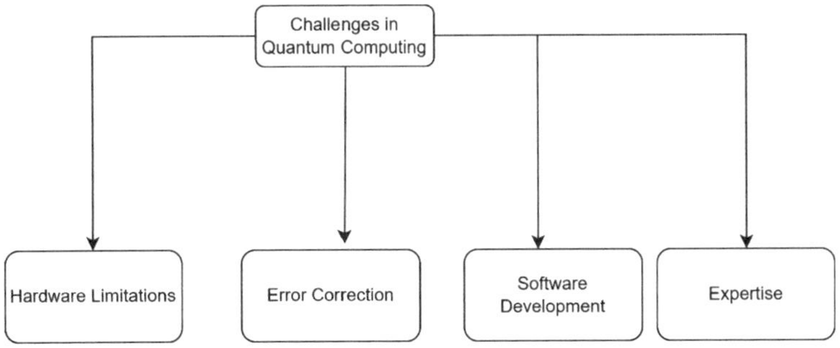

Fig. 4.9 Challenges in quantum computing

the error problem has been the hybrid quantum–classical framework. For example, [28], proved the successfulness of hybrid models in maximizing route planning for self-driving cars.

Another obstacle is the need for a common quantum programming language and tools. Although platforms like Qiskit and Cirq have made life a little easier for the algorithm developer, the absence of a common standard for languages and tools to use stalls the interoperability of quantum systems. Solving this problem is a key to getting widespread acceptance of quantum systems in the autonomous vehicle world.

4.4.6 Future Directions in Quantum Hardware and Algorithms

The future of quantum hardware and algorithms is promising, with ongoing research focused on overcoming current limitations and unlocking new capabilities. Some of the key areas of develoment are presented in Fig. 4.10.

4.4.7 Hardware Efficiency, Scalability, Error Correction and Algorithm Optimization

Build processors that possess thousands of qubits. That's the target in initiating the next phase of quantum computing, which is pushing toward executing more complex algorithms. Not too many companies are really working on this, but IBM and Righetti are the two at the forefront of this big effort.

To enhance the reliability of quantum computing, we are developing sophisticated error-correcting codes, like surface codes.

New quantum algorithms designed for particular applications are being developed and investigated by researchers. These include using quantum devices to handle traffic flow optimization and using them for real-time decision-making in autonomous vehicles.

The latest designs for qubits, like topological qubits, seek to boost the stability and coherence of quantum systems, making them much better suspects for looking into potential real-world applications.

4.4.8 Ethical and Societal Challenges

The ethics and societal impacts of quantum computing are a huge topic of concern to many people right now. Most significant, in my opinion, is the issue of who gets to use and control such a powerful system. If a nation-state or corporation gets a

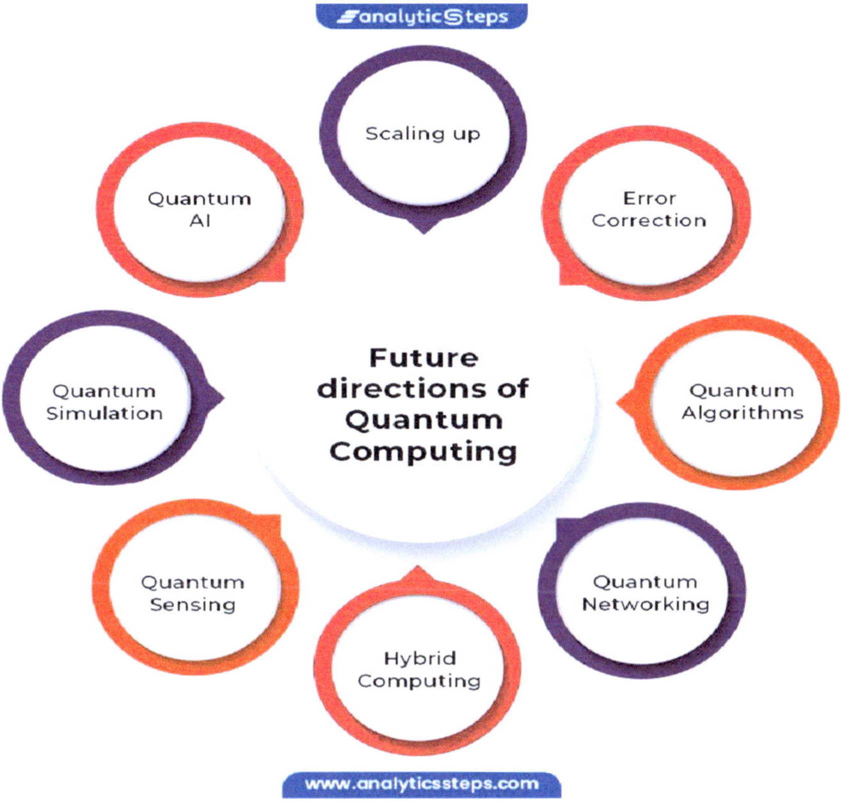

Fig. 4.10 Future directions of quantum computing [24]

quantum computer and uses it for nefarious means, all bets are off for our society and its cybersecurity. As noted by Mosca (2021), quantum systems could, in theory, break classical encryption protocols, which is why we need to start transitioning to "quantum-safe" cryptographic methods as soon as possible.

4.4.9 Research Gap and Future Work Areas

The potential of quantum computing to improve autonomous navigation is known, but major holes in the research remain. Most of the work centres on singular quantum algorithms, and these studies do not tend to consider the next level up the integration of several such algorithms into a real-world, autonomous navigation system at something resembling the human scale. Rarely, too, is this hypothetical system pushed to the limit to see if it could work in the kinds of environments that strain current navigation systems: places where many objects are moving in many directions and at many different speeds. With this far from being a comprehensive look at the potential of quantum computing to improve navigation, our study emerges across several dimensions to better understand this potential and how to harness it.

As a groundbreaking technological advance, quantum computing is ready to reshape the future of computation. It is based on the fundamental principles of quantum mechanics superposition, entanglement, and tunnelling and offers transformations that resolve problems too complex for classical computing systems. These promises range from the future of cryptography to optimization and artificial intelligence. Yet, fulfilling these promises demands the overcoming of numerous technical and practical challenges, and at least a couple of them are of great ethical interest.

In the past few years, the field has seen acceleration and attention like never before, with major milestones being reached by the likes of IBM, Google, and Righetti in quantum hardware and algorithms. Perhaps the most significant single event in all of this recent Vigor was Google's announcement in 2019 that it had achieved quantum supremacy meaning that its quantum system was able to perform a calculation that no classical supercomputer could pull off, despite many hours of effort. This computational test was designed to show off the power of a particular quantum system, called Sycamore, and indeed it did.

The power of quantum systems to solve intricate problems was brought to light by quantum algorithms like Shor's and Grover's. Shor's algorithm, for instance, threatens to do away with computationally intensive encryption systems and instead may necessitate a transition to more "quantum-resistant" forms of cryptography. Grover's algorithm, while not as well known, is similarly impressive and has some equally ominous implications if one is to believe that quantum computing will, indeed, take over from classical computing.

References

1. Aaronson S, Chen L (2021) The computational complexity of quantum algorithms. Nat Phys 17(4):313–320
2. Alagar [Alagar]. (2025, January 27) Machine learning applications and examples. Retrieved June 27, 2025, from https://iabac.org/blog/machine-learning-applications-and-examples
3. Amin M, Smith J, Ramakrishna P (2020) Quantum simulations for autonomous vehicle testing in virtual environments. J Quantum Comput Appl 6(3):215–225
4. Andersen J, Li Q, Zhang X (2021) Quantum cryptography for secure V2V and V2I communications in autonomous systems. IEEE Commun Mag 59(6):64–70
5. Arute F, Arya K, Babbush R et al (2021) Quantum supremacy using a programmable superconducting processor. Nature 574(7779):505–510
6. BBTPS (2025) Quantum Computing vs. Classical Computing. [online] Available at: https://www.blackbox.com/insights/blogs/detail/tps/2024/03/12/more-than-a-bit-about-quantum-qubit-computing
7. Biamonte J, Wittek P, Pancotti N et al (2021) Quantum machine learning. Nature 549(7671):195–202
8. Blanco-Justicia A, Domingo-Ferrer J, Martínez S (2021) Security and privacy in deep learning models for autonomous vehicles. IEEE Trans Int Veh 6(2):249–259

9. CBInsights (2021) What Is Quantum Computing? [online] CB Insights Research. Available at: https://www.cbinsights.com/research/report/quantum-computing/
10. Chen X, Wang H (2021) Quantum-enhanced machine learning in multi-sensor fusion for autonomous systems. J Quantum Inf Sci 13(2):125–137
11. Chen L, He D, Zhou M (2022) Quantum-enhanced data processing for autonomous vehicles. IEEE Trans Quantum Comput 3(1):78–89
12. Chow JM, Gambetta JM, Steffen M (2022) The IBM quantum roadmap: pioneering quantum advantage. Quantum Sci Technol 7(1):014002
13. Das P, Ramachandran S, Patel R (2021) Quantum optimization algorithms for navigation in autonomous vehicles. J Quantum Inf Sci 9(4):310–322
14. Engineering NKV (2024, May 6) What is Quantum Computing? A complete Guide. VLink. https://vlinkinfo.com/blog/what-is-quantum-computing
15. Eurofighter Typhoon (n.d.) Academic Dictionaries and Encyclopedias. https://de-academic.com/dic.nsf/dewiki/414137
16. Farhi E, Goldstone J, Gutmann S (2020) A quantum approximate optimization algorithm. Quantum Inf Process 19(8):275–290
17. Hernandez R, Li Y (2020) Quantum algorithms for multi-vehicle coordination in autonomous fleets. IEEE Access 8:131223–131234
18. Kandala A, Mezzacapo A, Temme K et al (2021) Hardware-efficient variational quantum eigensolver for small molecules and quantum magnets. Nature 549(7671):242–246
19. Kandala A, Mezzacapo A, Temme K et al (2021) Hybrid quantum-classical frameworks for energy optimization. Phys Rev X 11(2):041015
20. Krelina M (2021) Quantum technology for military applications. EPJ Quantum Technology 8. https://doi.org/10.1140/epjqt/s40507-021-00113-y
21. Kumar A, Sharma DK, Wang H (2022) Quantum computing applications in unmanned aerial vehicles and cyber-physical systems. J Quantum Inf Sci 13(3):47–65
22. Li B, Chen D (2022) Quantum data processing for sensor fusion in unmanned vehicles. IEEE Trans Quantum Eng 15(3):180–195
23. Martinez A, Zhao Q, Patel T (2021) Enhancing real-time navigation in autonomous vehicles with quantum computing. J Auton Syst Res 9(2):120–135
24. Mishra U (n.d.) What is quantum computing? Working and Requirement | Analytics Steps. https://www.analyticssteps.com/blogs/what-quantum-computing-working-requirement
25. Monroe C, Kim J, Rice J (2020) Scaling the IonQ quantum computer: challenges and future directions. npj Quantum Inf 6:79
26. Monroe C, Kim J, Rice J (2021) Trapped-ion quantum computing: challenges and opportunities. npj Quantum Inf 7(1):91
27. Patel D, Singh R, Kumar A (2020) Quantum algorithms for autonomous navigation: a review. J Quantum Comput Appl 12(3):215–230
28. Perdomo-Ortiz A, Benedetti M, Realpe-Gómez J et al (2020) Opportunities and challenges for hybrid quantum-classical machine learning. Quantum Sci Technol 3(3):030502
29. Peruzzo A, McClean J, Shadbolt P, Yung M, Zhou X, Love PJ, Aspuru-Guzik A, O'Brien JL (2014) A variational eigenvalue solver on a photonic quantum processor. Nat Comm 5(1). https://doi.org/10.1038/ncomms5213
30. Pirandola S, Andersen UL, Banchi L et al (2020) Advances in quantum key distribution. Nat Rev Phys 2(12):680–694
31. Pirandola S, Andersen UL, Banchi L, Berta M, Bunandar D, Colbeck R, Wehner S (2020) Advances in quantum cryptography. Adv Opt Photonics 12(4):1012–1236
32. Preskill J (2022) Quantum computing in the NISQ era and beyond. Quantum Sci Technol 7(1):014002
33. Rahmani F, Zhang L, Chen J (2021) Sensor data fusion in autonomous systems with quantum machine learning. J Auton Veh Technol 12(5):345–356
34. Rajasekhar M, Bhandari N, Yoo H (2022) Advancements in quantum algorithms for autonomous navigation. J Comput Quantum Inf 14(2):177–189

35. Rajput H, Wang T (2021) Quantum key distribution protocols for cybersecurity in autonomous vehicles. Quantum Secur Auton Syst 4(3):93–104
36. Ramirez C, Lee J (2023) The role of quantum machine learning in autonomous navigation systems. Quantum Comput Appl J 7(4):245–260
37. Schuld M, Sinayskiy I, Petruccione F (2022) Quantum neural networks: a review of methods and applications. Quantum Mach Learn J 8(1):55–78
38. Schuld M, Sinayskiy I, Petruccione F (2022) Quantum neural networks for supervised learning. Phys Rev A 101(3):032308
39. Singh A, Bhushan S (2021) Quantum cryptographic methods for securing autonomous vehicle communications. J Quantum Inf Technol 8(3):245–255
40. Sutikno T (2023) Quantum drones and the future of military warfare. Int J Informatics Comm Technol (IJ-ICT) 12(3):293. https://doi.org/10.11591/ijict.v12i3.pp293-299
41. Tan W, Tran K, Seo Y (2022) Quantum simulation techniques for autonomous vehicle testing environments. Int J Quantum Comput Appl 5(4):299–309
42. Zhou Y, Wang Y, Deng F (2021) Quantum algorithms for graph optimization and their applications in smart transportation. IEEE Access 9:28701–28716

Chapter 5
The Future Cyber Threat from Drones Against the United Kingdom's Key Assets

Wayne Gill and Hamid Jahankhani

Abstract During the ongoing Russia-Ukraine war, there has been an increasing reliance on the use of drones in all aspects of the conflict. As the war continues, the prospect of NATO entering into conflict with Russia has never been higher. At the same time, the cyber threat landscape continues to change, and the conflict has brought together these two elements; drones and cyber-attacks. This report investigates the probability and impact of a drone-initiated cyber-attack against the United Kingdom's (UK) Ministry of Defence (MoD) and national assets and highlights which assets are the most likely to be targeted. Clear motivating factors driving this paper are the increasing threat of international terrorism to the United Kingdom and the increasing threat from state-level actors which has been exacerbated by the ongoing Russia-Ukraine conflict that continues to present a geopolitical risk that threatens the wider NATO-Russia relationship. Several analysis methods were used to understand the impact and likelihood of this type of attack against UK assets, along with a small lab to highlight the vulnerabilities of Wi-Fi networks. The results of the analysis show different variables, but the key indication is that it is likely that there will be increased hostilities and proxy conflicts between the two nations (Russia and Ukraine), resulting in an increased level of cyber-attacks and an increased likelihood of a drone-initiated cyber-attack against key UK MoD and national assets.

Keywords Done · Cyberwar · Counter-drone technology · WiFi · Cyber attack · Mitre ATT&CK · NATO

5.1 Literature Review

A cyber-attack can be difficult to quantify but it is quite clearly a malicious and deliberate attempt to breach or circumvent the security controls of a computer system, network, or device. Those controls could be a combination of physical, technical,

---

W. Gill · H. Jahankhani (✉)
Northumbria University, London, UK
e-mail: Hamid.jahankhani@northmbria.ac.uk

U. B. Chaudhry and H. Jahankhani, *Autonomous Revolution: Strategies, Threats and Challenges*, Advanced Sciences and Technologies for Security Applications,
https://doi.org/10.1007/978-3-031-96782-5_5

procedural or policy controls. A cyber-attack's aims will vary and be dependent on a combination of the threat actors' motivations and abilities. However, the primary purpose of an attack will initially be to (but not limited to) compromise the availability, integrity or availability of a system, network, or device to cause damage, disrupt operations (denial of service), or gain unauthorised access to sensitive data. The motivations and rationale behind such an attack are wide and varied, which can range from script kiddies trying to showcase their ability to a purely financially driven cybercriminal (for example a ransomware attack), all the way to a state-sponsored attack designed to destabilise a nation-state (for example, through an attack on its critical national infrastructure) which could be of a standalone nature or as prelude to, or part of an armed (in the conventional sense) conflict.

The scope, scale and technical sophistication of cyber-attacks have increased year after year. This has seen the subsequent increase in the impact of such an attack, an example being WannaCry ransomware that hit the NHS in 2017 [37]. Cyber-attacks continue to be an ongoing and significant risk, not only to individuals, but to organisations big and small, and even to nation-states. As we have seen, cyber-attacks can take many forms, with currently the most prevalent being through malware ingestion, cross-site scripting, phishing/vishing campaigns and denial-of-service attacks. These types of attacks can target a wide range of systems and devices, all the way from personal/individual computers and IoT devices to large-scale networks and critical national infrastructure. The subsequent impacts of a successful attack can have severe consequences, including financial losses, loss of operation, reputational damage, the compromise/loss of sensitive information and ultimately the loss of life.

One of the key aspects of a cyber-attack is the potential to impact connected systems, where an attack will move to all interconnected systems to achieve maximum impact or to use the initial system as a starting point to attack a specific connected system.

The interconnected nature and attack surface of modern systems has made them more susceptible to attack, as demonstrated by the increasing frequency of cyber incidents in critical national infrastructure sectors [2]. The attacks on cyber-physical systems are a growing issue. With an attack on this type of system, an attacker can exploit vulnerabilities in both sides; the cyber and the physical elements, an attack of this nature can be used to target critical national infrastructure and lead to widespread disruption, loss of service and physical damage [12].

The biggest concern has to be the threat that cyber-attacks have against national security and international relations. Cyber-attacks are used to carry out espionage, sabotage, or even (conventional) warfare without the need to commit troops and military assets to the target area or even to the nation being attacked. This is an ever-growing concern for governments around the world [64]. An attack at this level can be used to target, as previously discussed, critical national infrastructure such as the energy grids, rail and road networks or financial institutions. This is assessed as a credible threat to the stability, safety and security of a nation-state and can be seen in the ongoing conflicts being played out on the world stage. As a consequence, there is an increasing requirement to address the role of cybersecurity to address the increased threat from cyber-attacks at all levels and in particular; at the state-sponsored level in both national defence strategies and international agreements with our allies [40].

The constantly expanding nature of cyber threats and the increasing sophistication of attacks have made their detection and mitigation increasingly challenging. However, the use of Artificial Intelligence along with machine learning combined with data-driven processes is proving to be successful in providing real-time identification and subsequent response to cyber-attacks [62]. The leveraging of these technologies can analyse large data sets of cyber threat intelligence at incredible speed and identify anomalies and indicators that may indicate an attack is already in progress [69]. This type of technology can almost deliver a self-healing and self-securing network, it offers a higher degree of resilience, in particular in complex interconnected networks [38].

In order to effectively detect, defend, mitigate and recover from cyber-attacks it takes a strategy that includes multiple elements including technical and physical controls, policies and procedural controls and ongoing user awareness and training. The appropriate selection and combination of these measures will be used to protect data, systems, and networks from unauthorised access, control or destruction [19]. It is also crucial to support continual assurance from the implementation of regular security assessments, penetration testing, and ongoing security awareness training [9]. From a strategic perspective; governments and regulatory bodies play an increasingly vital role in the setting of cybersecurity standards and governance, and with that comes the ability to effectively and securely share information, along with effective cybersecurity compliance and cyber resilience [68].

Cyber-attacks are a significant and ever-evolving threat in today's digital world. As technology continues to evolve, there is a corresponding ever-increasing reliance on it. This, along with the internet and the expanding IoT means that the likelihood of cyber-attacks will only continue to grow and also, the impact of an attack will continue to grow and be more widely felt. With the widespread adoption of technology and the IoT, the attacks will impact a greater portion of society and will affect everyone, from individuals, businesses, governments and nations alike. It is vital to not underestimate the impacts and to understand the full scope of cyber-attacks, their potential likelihood and their impact, in order to build the appropriate strategies and frameworks to prevent and respond to them. The key principles in the defence of cyber-attacks are the investment in cyber security technologies and practices, the development of cyber security frameworks and collaboration at the industry, national and international levels.

5.1.1 Adversaries

At a state level, the main adversaries of the United Kingdom (in a National Security) context appear to be crystal clear, namely, China and Russia being the main two and followed closely by Iran. In 2010, Chatham House produced a report that discusses the National Security of the United Kingdom in Cyberspace, the threats and its responses [15]. The report looks at the threat from four distinct levels: the lower-level individual criminal domain, the serious and organised crime domain, the ideological and political extremism domain to the higher level of the state-sponsored cyber-attacks domain. The report clearly articulates that the principal state-sponsored attacks would be from Russia and China. With regards to Iran, the UK Government has been explicitly clear with its viewpoint, stating that they will hold them to account for their nuclear activity and that they are instrumental in the worsening of the global security environment, posing a threat to not only the UK but the 'West' in general [26].

Being an integral part of the North Atlantic Treaty Organisation (NATO) adds to the complexity of defining exactly who are the adversaries of the UK who would pose this type of threat. It is argued that NATO is a relic of the Cold War [21], which may be true but, it is interesting to note that the events of September 11 in New York prompted the first discussion by NATO under Article 5 [48], which essentially states that an attack on one member of NATO constitutes an attack on them all. Therefore, it can be argued that the adversaries of any member state of NATO is consequently a potential adversary of the UK. Furthermore, in the Balkans conflict, it is widely regarded that the NATO alliance's decisive aerial intervention led to stability in the region [77]. We find ourselves in a new era within NATO and the Russia-Ukraine war with the ever-growing pressure for NATO to join the fight. NATO is deemed to be more important on the World stage than ever before and is being increasingly viewed as the eventual deciding factor in the outcome of the ongoing war.

However, Article 5 is suitable for assessing both state and non-state acts of aggression aligned with traditional warfare as the Article was written in light of the tactics and technology of World War II. Although NATO has invoked the application of the Article in a post-Cold War era, it cannot suitably apply the principle of collective self-defence against the ever-increasing application of cyber-attacks on the battlefield [31]. Amendments are required to the Article to define what constitutes a cyber event and clear guidelines for determining the origins of a cyber-attack.

Currently, NATO will find it challenging to clearly determine where an attack originated from and, if the attack was authorised by the host nation. Also, without clear and definite classifications for the differing types of cyber-attacks, there will be challenging and unhelpful discourse between its various members. Furthermore, with any uncertainty in identifying cyber-attacks that equate to an armed attack in the traditional sense, NATO will be challenged in deciding on an appropriate and proportionate armed (or otherwise) response following the principles of jus in bello.

By developing and introducing a set of standards and clear unambiguous rules before a NATO member state is the recipient of a major cyber-attack, NATO may be able to stop or minimise the level of such an attack. As well as serving as a deterrent to a possible attack of this nature. It will also reduce the time in which NATO determines an effective and proportionate response.

Clearly, with the UK being a member state of NATO, the unambiguous clear identification of the UK's adversaries is not without challenge. However, given the current geopolitical climate, it would be prudent to focus on the ever-increasing threat from the Russian State and to be fully cognisant of both its intent and its capability.

5.1.2 Cyber-Attacks on Allies

The current and potential increase in cyber-attacks against NATO member states has become an increasingly significant issue and a concern within NATO which has been exacerbated by the ongoing Russia-Ukraine war. The war has seen a dramatic surge in both the use of drones on the battlefield and the use of cyber-attacks. In the context of war; these types of attacks pose a severe threat to life, as well as physical infrastructure, data management, various communication networks and cyber-physical systems linked to critical national infrastructure [55]. The effects of cyber-attacks by state actors and state-sponsored threat actors, particularly those coordinated by Russia, have been identified as ongoing threats to NATO member states and importantly, not just to Ukraine [46].

The highest profile and most commonly known state-sponsored cyber-attack to date is the attack carried out by Russia against Estonia in 2007. The cyber-attack targeted Estonian public and commercial institutions [30] in Tallinn, Estonia, and marked a distinct turning point in cybersecurity and international relations. This was the first attack of its nature that could be directly attributed to the Russian state. This attack has been referred to as the first real war in cyberspace. The attack was a politically motivated attack that aimed to force Estonia to change its policy over the location of a war memorial and the attack was an indicator of how cyber-attacks could be used as a weapon in diplomatic disputes [28]. This attack in 2007 was the start of a new era of cyber warfare, with incidents like the Stuxnet attack on Iranian nuclear centrifuge systems highlighting the growing sophistication and impact of cyber weapons and most importantly, the willingness to deploy an attack of this nature [52]. This attack was the catalyst that led to the formation of the NATO cyber security centre in Tallinn. It highlighted the need for international (within the confines of NATO) collaboration in defending against cyber warfare and future cyber threats [33]. The attack also raised several questions regarding how traditional laws and conventions on war and sovereignty would apply in the context of cyber-attacks and cyberspace, given that an attack of this nature can be initiated from almost anywhere and is certainly not restricted by geographical borders and boundaries. The attack challenged the interpretation of international law, particularly with respect to any potential response by NATO and the use of force, it blurred the lines between conventional and the cyber warfare [27]. The attack clearly highlighted the vulnerability of national infrastructure and research has shown how hostile threat actors can attack the vulnerability of a connected system and move latterly to target and manipulate national infrastructure [16]. The attack is seen as a wake-up call for governments and organisations around the world to assess their cybersecurity and implement controls to strengthen their defences against an attack of this nature [43]. It marked a significant turning point highlighting both the susceptibility of national infrastructure to attack, and more concerning; the ability and intent of nation-states to carry out offensive cyber operations of this nature, it indicated the changing dynamic of future conflict. The attack emphasised the requirement of a robust cybersecurity strategy that requires global collaboration, not only between governments but also with industry. It highlighted the need to review conventional frameworks and international laws regarding the use of force and what would be a justified proportionate response to future state-sponsored cyber-attacks. The attack was seen as ground-breaking and highlighted the complexities and changing face of warfare, diplomacy, restraint and the ever-changing threat landscape on the global stage.

In the transition from conventional warfare to a more hybrid approach that utilises the cyber domain there is a shift that sees cyber-attacks, espionage and sabotage (non-traditional threats) becoming more prevalent and being used alongside conventional warfare at all stages of conflict. Historically, it is recognised that those non-traditional threats were often seen as the precursor to conventional warfare with the aim to provoke unrest and to undermine the governance and leadership of a nation. NATO's ongoing participation in cyber defence has been acknowledged as critical due to the significant potential impact and probability of cyber-attacks on member nations [20].

The legal foundation for employing defensive cyber measures by NATO against non-state actors has been thoroughly examined. NATO can employ cyber countermeasures in specific situations, such as when a nation-state is unable to control non-state actors within its boundaries or when it anticipates a significant number of cyber-attacks that require proactive self-defence [8]. After the 2014 Wales Summit, it was recognised that cyber threats play a crucial role in the concept of 'collective defence' as outlined in Article 5 of the NATO Treaty [81]. The cyber domain has been a focal point for NATO due to its strategic significance and the aggressive efforts of countries like Russia and China to utilise cyberspace for military gain [56]. The widespread recognition of state-level cyber-attacks has significantly influenced public support for UK and NATO countermeasures. These now involve actively monitoring networks [3] and an increase in offensive cyber operations, which are now publicly acknowledged.

NATO is highlighting the importance of the requirement to deal with cyber threats in the current and future security landscapes. The prevalence of hybrid warfare and misinformation methods, as observed in recent conflicts like the Russia-Ukraine war, underscores the need to effectively address this danger posed by hostile nation-states [17]. Cyber resilience, as defined by NATO, refers to the capacity to effectively identify, detect, respond and recover from intentional attacks or issues that cause a denial or interruption in service or capability [47]. It is evident that skilled threat actors, whether they are nation-states or not, may quickly initiate conflicts by employing specialised operations, cyber-attacks, and disinformation campaigns. The goal of these actions is to minimise both the unintended physical harm caused and the financial and physical losses suffered by the aggressor.

The importance of coordination and collaboration in cyberspace cannot be underemphasised in order to enhance cyber defences, this is a fundamental aspect of NATO's mission of collective defence [60]. The Cyber Defence Policy and its strategy for addressing hybrid threats are crucial elements in building resilience against emerging and changing threats [63]. NATO's adoption of a comprehensive strategy to tackle 21st-century challenges demonstrates its recognition of the dynamic geopolitical landscape and its influence on global risk and security management.

The escalating cyber-attacks targeting NATO pose a growing challenge that necessitates a synchronised and all-encompassing response. By fully comprehending the nature of cyber threats, NATO (and the UK) can effectively strengthen cyber resilience and safeguard member states against the increasingly pervasive cyber threat present in both modern and conventional security contexts. This can be achieved through the implementation of proactive defensive and offensive measures.

5.1.3 Drone Technology

The advances in drone technology have progressed considerably in recent years, transitioning from its original military purposes to a diverse array of commercial and civilian applications. However, it must be noted that the ongoing Russia-Ukraine war prompted significant advances in drone technology and has seen extensive use of both commercial and military drones on the battlefield. Outside of the battlefield and due to the improvements in drone technology, availability and affordability, drones have been widely embraced in areas like industry, agriculture, environmental auditing, and healthcare [41]. The development of drone technology has transitioned to greater autonomy and drone clustering/swarms [79]. The continuous research and development within the field of copter drones are currently influencing the development and evolution of commercial flight control systems, which shows how the knowledge and progress in one area can influence that of another [57].

Drone technology is regarded by the construction industry as a 'game-changer' as it has the potential to improve performance and delivery at every stage of construction projects [75]. Furthermore, drones have become indispensable instruments for farmers and auditors in the field of agriculture. Drone technological advances are used to improve efficiency and productivity in environmental auditing and crop management [39]. The inclusion of AI into drone technology is revolutionising unmanned aerial vehicles, with developments such as edge computing and swarm intelligence ever expanding the capabilities and potential uses of drones [50]. In modern agriculture, the introduction of nano drones has developed new solutions for many agricultural processes [22].

The extensive use of drones in several sectors highlights the influence of drone technology on conventional methods and ways of working. An example of a key area where drones are providing life-saving assistance is the medical sector. In countries such as Malawi, Madagascar and Senegal drones are being used to enhance medical and healthcare services by transporting life-saving equipment and supplies in areas that lack efficient, low cost and high-speed transport networks [34]. There is further work in a number of countries that is exploring the viability of the use of drones for wider medical logistic and emergency response scenarios [80], and by integrating IoT technology with drones, the potential to optimise healthcare supply chains and improve life-saving capability cannot be underestimated [73]. A further area that is harnessing the advances in drone technology is their use in areas that present a clear risk to life, an example being their use in wind turbine inspections where the requirement for an individual to scale a wind turbine in various conditions has been almost fully removed.

The proliferation of drones across multiple sectors and not just the military has prompted wider discussion regarding the introduction of a global framework to govern the safe and responsible use of drones. Drone use regulations continue to evolve across the globe incoherently, an example of this is the European Union's involvement in the regulation of drone technology that seeks to develop a framework to provide the secure use of and integration of drones into industry and commercial fields [70]. The global use of drones and the disjointed global regulations governing their use may lead to an increase in the use of this technology by threat actors.

The current drone technology exemplifies a dynamic and swiftly progressing domain with immense potential for innovation and influence across several sectors. Drones are revolutionising various industries, including manufacturing, agriculture, and healthcare, by introducing innovative methods that enhance efficiency, safety, and sustainability. The future presents intriguing prospects for additional breakthroughs and uses in both commercial, humanitarian and of course, military environments.

5.1.4 Counter-Drone Technology

Admiral Sir Tony Radkin, the professional head of the United Kingdom's Armed Forces, delivered a speech at Chatham House on 24th February 2024. He unambiguously identified Russia as currently, the most significant threat to the United Kingdom and emphasised that the UK faces daily cyber-attacks. In addition, Radkin stated that the Russia-Ukraine conflict has demonstrated the efficacy of drones and he highlighted the possible vulnerability of the UK to a drone attack, emphasising the requirement to engage in discussions on homeland security [58].

As drones grow more common in public and business contexts, counter-drone technology is essential to protecting airspace [53]. These systems identify, track, and destroy unauthorised or malicious drones that threaten security [65]. The widespread proliferation of drone use in the recreational and commercial sectors is widely accepted and embraced. Their use on the military stage has grown exponentially in recent years and with that also comes their use by non-state threat actors. Drones controlled by hostile (state or non-state sponsored) threat actors can be used to target national infrastructure, public places, military targets etc. without the requirement for the controller to be in the vicinity of the target, which increases the requirement for an effective counter-drone response. To counter the hostile and dangerous uses of drones, anti/counter-drone technology needs to advance in a similar vein [32]. An effective counter-drone capability needs to be reliable, rapidly deployable, and scalable and must be able to be developed/updated in response to advances in drone technology [66] to stay ahead of developing threats and mitigate hostile drone activity [67]. Airports require counter-drone technologies to prevent intentional aerial attacks that could disrupt operations, endanger public safety and impact economic stability [5]. Steps have been taken to integrate counter-drone technology into frameworks like Unmanned Traffic Management (UTM) systems where they can be used to aid decision-making and effectively counter threats [66]. These systems can promptly detect and neutralise unauthorised drones to minimise any interruption/denial of service [10].

Counter-drone technology can include drone identification, drone destruction and taking control of the drone. Artificial intelligence is being incorporated into this technology to aid real-time detection and an automated response [57], furthermore, as we move forward, counter-drone technology aims to counter not only individual drones but also drone swarms [23]. Counter-drone technology is harnessing the power of artificial intelligence coupled with deep reinforcement learning that assists the system to analyse, and counter threats [54]. Drones come in a multitude of different shapes and sizes, with fixed wings or rotors. Some even resemble living creatures like birds or insects. Control frequencies also vary—the majority of drones are controlled via common radio protocols, others via Wi-Fi. As drone technology progresses more drones will fly autonomously and won't be remotely piloted by an individual. All must be detected, and the detection accuracy relies on known drone signatures. To enhance counter-drone technology and to provide a broader range of coverage and hostile drone detection, the integration of various sensors is required; including RF, video, acoustic, and optical [11]. The information from the sensors is used to identify and classify the drone, which is then used against a database of drone signatures to determine the specific characteristics attributed to that drone.

It must be noted that within the UK there are several permanent airspace restrictions that apply to drones (and manned aircraft) and rules put in place by the Civil Aviation Authority that govern their use. Industry and governments must collaborate to implement global frameworks and standards governing the production of, and the use of drones. An effective counter-drone response requires this collaboration in order to stay ahead of the drone threat on and off the battlefield.

5.1.5 The United Kingdom's Key Assets

Articulating the UK MoD and national assets in priority order is subjective and can feasibly shuffle in a transition from a peace to a war footing. There are a number of assets that stand out as being crucial to the nation's national security [51]. The integrated Review is keen to stress the importance of cyberspace and the digital domain in conjunction with ongoing consultations with our Five Eyes Partners and NATO. The UK will continue to act globally as part of NATO and as such the key UK MoD and national assets may be fixed within the UK or be deployed operationally to all four corners of the globe. Some of the clear key defence assets are:

- Defence Logistics and Medical facilities (in a war footing)
- Faslane Naval base
- UK strategic nuclear deterrent
- Ministry of Defence Main building

- Permanent Joint Headquarters (PJHQ)
- GCHQ
- HMS Prince of Wales, and HMS Queen Elizabeth
- UK Special Forces fixed and deployed locations
- Defence communications networks
- Frameworks for Cyber Attacks

The adoption and adaption of a cyber security framework is fundamental in building the cyber resilience of an organisation, and the use of a cyber-attack framework is key to understanding, identifying and addressing a range of diverse cyber threats. A widely used cyber-attack framework is the Mitre ATT&CK, which is a globally accessible knowledge base of adversary tactics and techniques based on real-world observations. It details the tactics, techniques and procedures that are employed by threat actors/adversaries in their cyber-attacks. This is key in developing a coherent cyber defence strategy. There are more nuanced frameworks such as the "Digital-twin Based Cyber-Attack Detection Framework" which supports cyber-physical manufacturing systems [4]. In addition, the adoption of trust-based frameworks that have been developed emphasises the importance of strong defensive systems and controls [6] and are key in protecting critical infrastructure like the electricity grid network.

Frameworks have also been developed using a data-driven approach for networked control systems and frameworks using precise dynamic models which are designed to address advanced threats such as fake data injection, zero dynamics attacks, covert attacks and replay attacks [78]. With the proliferation of IoT devices, frameworks have also been developed to counter the threat in this field [61]. These frameworks can harness deep learning technology in an effort to detect device vulnerabilities and cyber-attacks.

On a bigger scale, the NIST Cyber Security Framework (NIST CSF) is an open-source set of guidelines that can be used that follow the five NIST functions; Identify, Protect, Detect, Respond and Recover [49]. Through the use of frameworks such as the NIST CSF an organisation will understand how robust their cybersecurity defences are and importantly, what is required to strengthen them. Their effective use will defend the organisation against a compromise of confidentiality, integrity or availability and in turn reputational loss and market share. NIST recommend that the framework is not only adopted but it is also adapted to suit the organisation, and by doing so all organisations can evaluate the current cyber security position, highlight areas of weakness and prioritise their cybersecurity remediation plan appropriately.

The area of drones is in part linked to the aviation industry, which has distinct cybersecurity issues. This has led to research into the use of artificial intelligence-empowered defence frameworks to protect modern aircraft from potential cyber-attacks [76]. Using artificial intelligence in this area allows the automation of defensive controls and improves the ability to counter advanced cyber threats. Similarly, artificial intelligence and machine learning have been leveraged to develop frameworks to help defend autonomous vehicles from cyber-attacks.

Adopting and implementing cyber security frameworks is a key step for organisations to enhance their protection against cyber-attacks. Through a comparison of frameworks such as NIST and ISO27001 [1], organisations can determine the most appropriate framework that matches their particular needs, which will allow them the ability to implement strong cybersecurity solutions customised to their specific requirements. Cyber-attack frameworks are crucial in strengthening defences against various cyber threats, including both conventional network breaches and advanced cyber-physical attacks. Organisations can strengthen their resilience and ability to withstand cyber threats and protect vital systems from threat actors (including state-sponsored) in cyberspace by leveraging data-driven strategies, machine learning algorithms, and dynamic game models.

5.1.6 Drones and the Cyberthreat

As the capability, affordability and availability of both commercial and hobbyist drones are increasing, there is an increasing likelihood of drones being used for malicious purposes, one use may be to combine the use of a drone with a cyber-attack. The reporting of such an attack is very scarce. However, there is evidence of successful cyber-attacks through the use of drones, one of which was the attack against a US financial institution [14]. This paper seeks to understand the likelihood of such an attack taking place against key UK defence assets and the potential ramifications of such an attack along with the development of a framework to defend against this threat.

5.2 Research Method

Initial methodology applied in the delivery of this paper was the analysis of secondary data. The analysis was carried out using various methods, which are discussed below. It must be noted that the author and any members of the military who assisted with the analysis, annually complete Unconscious Bias training. The analysts were also briefed on several biases that could influence their analysis. The analysis was supplemented by a lab exercise exploring the security of Wi-Fi networks. The analysis and exercise were used to develop a framework to defend against a drone-initiated cyber-attack on key UK MoD and national assets.

5.2.1 SWOT Analysis

SWOT analysis is a widely adopted tool that can be used across various domains and not just for an individual sector or industry. It is used to help inform an organisation of both internal capabilities and limitations as well as external factors that may affect the organisation [42]. SWOT is broken down into four areas.

- Strengths (Internal): Elements of the organisation/business that are under its control, that give it an advantage over others (competitors/enemies etc.)
- Weaknesses (Internal): Internal factors organisation/business under its control that adversely affect the organisation's ability to succeed or meet its goal.
- Opportunities (External): Factors that are outside the control of the organisation/business that could be used to assist it in meeting its aims and goals.
- Threats (External): Factors that are outside the control of the organisation/business that may prevent it from meeting its goals.

Using a SWOT analysis can provide an informed evaluation of both internal and external factors and allow an organisation to make informed decisions that appropriately align resources with their aims and objectives, and to identify and subsequently mitigate associated risks. It should not be completed in isolation but should be complemented by the application of further analysis tools and should be revisited and updated to take account of the shifting environment [72]. The effective use of SWOT analysis will provide an auditable view of the organisation's strategic position in whatever environment it resides.

A SWOT analysis was completed against adversary (Russian) drone capabilities.

5.2.2 Red Teaming

Red Teaming is an analysis tool that again, can be used in various fields, it is particularly used in both the military and the cybersecurity fields. The process involves two teams: one red and one blue. The blue team focus on defence and works to protect [35] the network or system (in a cyber security context) or would be our forces in a military context. The Red team, however, would be the enemy/aggressor in a military scenario and a cyber threat actor/team in the cybersecurity context. The red team challenges/attacks the blue team by simulating attacks, exposing weaknesses and identifying vulnerabilities.

Within the cybersecurity context, red teaming is often associated with ethical hacking, where a team of cybersecurity professionals have been authorised to penetrate a network for the network owner to identify and address vulnerabilities before the network, or indeed a patch goes live. The use of red teaming in this way is also used to test an organisation's incident response procedures. Its effectiveness is further highlighted through Capture-the-Flag (CTF) exercises, which enhance the understanding of potential risks and also provide recommendations to strengthen an organisation's cybersecurity posture [18].

For the purposes of this paper, the Red Teaming exercise was carried out by serving members of the British Army, and the MOD Red Teaming Handbook was used as a handrail [44]. Red Teaming activity is an important factor in understanding a network/system or a military force weakness and thereby allows for mitigations to be developed, tested and implemented. It allows for the proactive identification and mitigation of vulnerabilities in a test environment. For the paper, the Red Team were the notional aggressor (in this case Russia) whose aim was to carry out a drone-initiated cyber-attack on key UK MoD and national assets.

5.2.3 Cone of Plausibility

The Cone of Plausibility is another analysis tool that lends itself to scenario generation and analysis where there is a high degree of uncertainty. It is a very useful tool in determining what is plausible [7] and what could happen in the future and differentiating from the unplausible in complex environments. The process considers key drivers (for example PESTLE-M) and allows the application of various assumptions. It identifies a range of outcomes from unplausible to probable scenarios, and if used effectively, can determine the impacts of the plausible scenarios. Its use can allow for strategic planning and risk management by allowing the appropriate use of assets, investments and resources in the appropriate area and aligned to what is plausible and that is more likely to happen. The Cone of Plausibility has been extensively used by military intelligence analysts to inform military planning by understanding the impact that drivers can have on a given scenario and how that can shape the potential outcome.

The Cone of Plausibility was used to develop scenarios to assess the credibility of the threat of a drone-initiated cyber-attack on key UK MoD and national assets by an adversary.

5.2.4 Uncertainty Yardstick

The Defence Intelligence (DI) Uncertainty Yardstick is a very useful tool used to communicate levels of uncertainty. The Uncertainty Yardstick is primarily used in the defence and security area of the UK Government [45]. However, its use has spread to other areas within Government. The tool was developed to assist in the assessment and communication of those levels of uncertainty found in intelligence assessments by using an agreed vernacular aligned to a numerical scoring. Understanding this 'scoring' is key to the interpretation of assessments and aims to remove any ambiguity in a report and removes a degree of subjectivity. The use of the tool Fig. 5.1, allows for the appropriate response to an intelligence assessment and allows for a risk-managed approach to decision-making in an area where there will inevitably be degrees of uncertainty.

The DI Uncertainty Yardstick has been applied throughout this paper.

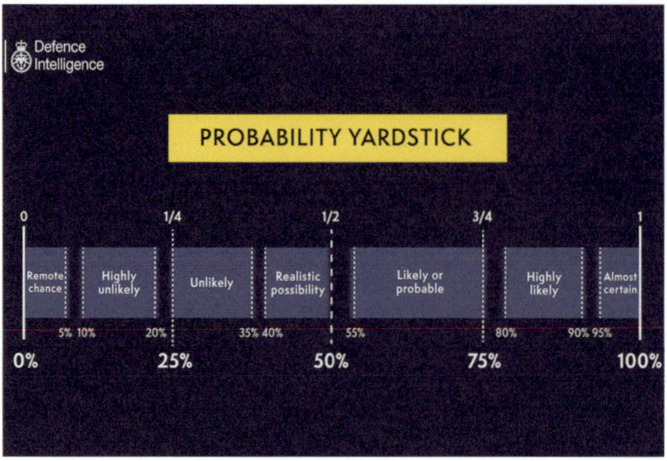

Fig. 5.1 DI uncertainty yardstick [45]

5.2.5 Intelligence Analyst Bias

Understanding that all humans are susceptible to various biases is essential in carrying out analysis work. These biases can impact on decision-making and the output of the analysis. A common bias is confirmation bias, where analysts look for information that fits with their own hypotheses [59]. Another is group think where analysts can be swayed by what their wider cohort believes, which in turn leads to a very subjective scenario generation, interpretation and judgement.

There are numerous biases, and it is vital to understand and try and mitigate them. This is a difficult task as studies have shown that robust attempts to de-bias analysts can lead to the oversimplification of complex intelligence issues which can reduce the effectiveness of the analyst and the subsequent analysis.

It is essential to use a variety of analysis tools and a diverse cohort of analysts who, with an awareness of bias, can effectively moderate and mitigate any influencing bias that can adversely affect the output of the intelligence analysis [29]. It is clear that an awareness of, and subsequent addressing of biases in intelligence analysis is crucial for improving the veracity of assessments and in maintaining the integrity of the analyst.

5.2.6 Cybersecurity Frameworks

There are several different Information Security (IS) Standards, Risk Management Frameworks and Cybersecurity Frameworks available through various sectors, across Government and globally. The MoD and the US Department of Defence specify the use of some specific Standards that are not proprietary and have been independently assured. Based on the business requirement [71], a review and detailed evaluation of different Frameworks and Standards was carried out, which established that the Standards below, are prevalent within the defence arena.

ISO 27001

The British Standards Institute (BSI) ISO/IEC set of Standards have been developed to be independently implemented as part of a larger work stream and have been developed to be mutually supportive of each other. The ISO 27001 standard is the most widely implemented and accepted of all the information security standards. It delivers a framework that initially enables an organisation to develop and implement an Information Security Management System (ISMS) from scratch. The framework then promotes continual improvement through maintenance, monitoring, and review [25].

5.2.7 NIST

The use of various NIST standards and its Risk Management Framework (RMF) is gathering momentum within the UK as it moves to a Secure by Design methodology across the Government [24]. The NIST RMF encompasses the risk management process across the full lifecycle. A key benefit is that the NIST RMF is not prescriptive, it is designed to be firstly adopted and subsequently adapted to the specific needs of any organisation. The NIST 800-53 (Security Controls) are mandatory for the United States Government, they are also being implemented within areas of the UK MoD.

5.2.8 Cyber Essentials Plus

Within the MoD and wider Government, it has been recognised that securing the supply chain is fundamental to National Security. It is a requirement that, as part of the Defence Cyber Protection Partnership (DCPP) initiative, all MoD suppliers must complete a Supplier Assurance Questionnaire (SAQ) along with a supplier Risk Assessment [13, 24], in an effort to mitigate known supply chain risks.

5.2.9 MITRE ATT&CK®

The MITRE ATT&CK framework is an open-source widely adopted resource that offers insight into Adversarial Tactics, Techniques, and Common Knowledge (ATT&CK). It assists in the understanding and categorisation of cyber threats and, provides a clear and structured approach to assessing cyber risk. Its use can lead to an improvement in an organisation's cybersecurity culture.

It is essential that organisations implement and adopt a carefully selected series of standards and frameworks. This will provide the appropriate guidance, best practices and standards to strengthen their security posture in the complex and ever-changing cyber threat landscape. The frameworks above were reviewed and compared to assist in the development of a framework to defend against a drone-initiated cyber-attack on key UK MoD and national assets.

5.2.10 Securing Wi-Fi Networks (Lab)

The global use of Wi-Fi technology ensures that wireless networks are found almost everywhere and in almost every aspect of our lives, WiFi is the mechanism that allows everyday devices to talk to each other over a short distance. A small lab was conducted to investigate the 802.11 (universally 802.11 is referred to as Wi-Fi) wireless network standards, the lab explored some known security weaknesses and was used to develop a framework that could be implemented to secure this type of wireless network.

A lab environment utilising a Raspberry Pi Linux machine and an Alfa AWUS036NHA (Wi-Fi) adaptor was used to exploit a number of known vulnerabilities that prevail in the 802.11 (Wi-Fi) standard. The two devices were also tested using a 50,000 mAh power bank USB fast charger as a power source to simulate being operated on a drone in a cyber-attack scenario. It should be noted that the devices were not tested using a drone.

5.3 Data Analysis and Critical Discussion

5.3.1 Red Teaming

The Red Team were the aggressor (in this case Russia), whose aim was to carry out a drone-initiated cyber-attack on key UK MoD and national assets. The team attempted to adopt the aims and perspective of the Russian state with a measured understanding of their tactics and procedures, to gain a clearer understanding of their strategies, tactics and objectives for this type of attack.

- Objectives
- Critical National Infrastructure—The targeting of essential services such as power grids, transportation networks, financial institutions and water supplies in order to impact the day-to-day routines of the population and undermine the UK Government.
- Military Capability—Destroy, disable or disrupt the UK's military command and control systems, communication networks and logistic chains to weaken military readiness, reaction and response.
- Intelligence Gathering—Ongoing activity that would be increased. The gathering of sensitive/classified information from Government, military, science and industry networks in order to gain a technical and strategic advantage.
- Influence—A misinformation and disinformation campaign aimed to undermine the Government and increase social, economic and political tensions.

5.3.2 Attack Vectors

- Cyber Espionage—The widespread use of drones to intercept communications, gather intelligence on network vulnerabilities and to deploy malware.
- Cyber Attacks—The use of drones to covertly access secure locations and gain access to Wi-Fi and Bluetooth networks.
- Electronic Warfare—The use of electronic warfare to jam or disrupt communication networks and radar systems in order to enable further undetected cyber-attacks.
- Payload Effect—The use of drones to inject a malicious payload into a network or the use of drones to drop infected USB devices into secure areas with the anticipation that they would be connected to a network by a bystander (either state actor influenced or not).
- Attack Phases (Lockheed Martin Cyber Kill Chain)
- Reconnaissance—The use of drones to conduct reconnaissance of target locations using a combination of high-resolution cameras and sensors to gather information on physical security measures, routines and vulnerabilities. Also, to gather network and system information.
- Weaponisation—The use of bespoke or readily available malware to exploit identified vulnerabilities. Configure/load drones with the payload (USB device to be dropped or rogue Wi-Fi access points, mini computing modules running software (Raspberry Pi), signal jammers etc.).
- Delivery—The use of drones to deliver the payload (as above). The use of drones saves committing personnel into the forward area of the attack, and they can avoid detection by using low flight paths, low light and using information gathered in the reconnaissance phase.
- Exploitation—The use of drones will allow close access to the target and where possible the drone can remain in the target area to facilitate data exfiltration and further reconnaissance.

- Installation—The installation of the payload/malware to allow access/control of the target network/system. The establishment of command and control to facilitate further aspects of the attack.
- Command and Control—maintain control of the attack via the deployed malware or the network to facilitate further actions.
- Actions on Objective—The ongoing command and control of the attack to facilitate the exfiltration of data, lateral movement or further activities (disrupt, deny, destroy).

5.3.3 *Targets of Interest*

1. Command and Control—Critical for the coordination and control of military operations and strategic decision-making.
   a. Permanent Joint Headquarters (PJHQ), Northwood
   b. Ministry of Defence Main Building, Whitehall.
2. Nuclear Deterrent—The UK's nuclear deterrent is pivotal to its National Security Strategy.
   a. His Majesty's Naval Base (HMNB) Clyde (Faslane)—Home to the UK's Vanguard-class submarines that carry the UK's nuclear missiles (Trident).
   b. The Atomic Weapons Establishment (AWE) at Aldermaston and Burghfield; is responsible for the design, manufacture and maintenance of the UK's nuclear warheads.
3. Military Airbases—Crucial for any military air campaign.
   a. Royal Air Force (RAF) Brize Norton—Primary UK Armed Forces air transport base.
   b. RAF Lossiemouth—UK base for the (quick reaction) Typhoon fighters.
   c. RAF Coningsby—UK base for Typhoon fighters.
4. Naval Bases—Crucial for the Royal Navy's Surface and submarine fleets.
   a. HMNB Portsmouth—Base for the Queen Elisabeth-class aircraft carriers along with other fighting vessels.
   b. HMNB Devonport—Base supporting both nuclear-powered submarines and surface vessels.
5. Intelligence and Cyber facilities—Intelligence collection, cyber operations and electronic warfare.
   a. Government Communications Headquarters, Cheltenham—UK's signals intelligence and cyber capabilities.
   b. MOD Corsham—A key hub for defence communication and information networks.

6. Space and Satellite assets—critical for global communication, reconnaissance and navigation.
   a. Skynet—UK's military satellite communication system.
   b. Ground Control Stations (GCS)—The command and control of Skynet.
7. Logistics—Critical for sustaining military operations.
   a. Bicester Logistics Depot—Central distribution hub of military supplies and equipment.
   b. Defence Equipment and Support (DE&S) sites—procurement, development and support to military capability.
8. Training and Research Facilities—Advanced training and research locations.
   a. Defence Science and Technology Laboratory (DSTL)—The centre for UK defence technology research and development.
   b. Salisbury Plain Training Area—The largest UK military training area.
9. United Kingdom Special Forces (UKSF)—Elite globally deployable military capability.
   a. Special Air Service—Hereford.
   b. Special Boat Service—Poole.

5.3.4 SWOT Analysis

The SWOT analysis was conducted on the cyber-attack capabilities of the offensive Russian military drone capabilities.

- Strengths

Strategic Advantage—Effective use of drones to conduct offensive cyber operations/attacks can disrupt adversary command and control networks/systems. The use of drone warfare that leverages cyber capability enhances the adversary's ability to conduct asymmetric warfare and counteracts any conventional warfare disadvantages.

Operational Experience—Extensive recent and ongoing experience (Ukraine) in hostile environments in the use of drone technology in cyber operations. A history of successful integration of cyber capabilities using a multitude of differing platforms.

Technical Integration—Russia has invested heavily in integrating cyber-attack capabilities into its drone programme, and has used this approach in electronic warfare, cyber espionage and the interception of communications.

Efficiency—The use of drones is both cost and manpower effective in comparison to conventional warfare.

- Weaknesses

Operational complexity—Integrating cyber-attacks into drone technology adds complexity to offensive operations mission planning and execution. Skilled operators and advanced infrastructure are a requirement (and also a target).

International Sanctions—There will be a reliance on the supply chain and imports to facilitate this type of asymmetric warfare. There will be components subject to international sanctions that will hinder their procurement, and the subsequent development and deployment of drone-based cyber-attacks.

Technological Limitations—Despite being advanced and experienced in these types of attacks, Russia may still lag behind their adversaries in the incorporation of AI and the use of advanced encryption algorithms in their drone technology, and the anti-drone technology employed by them. They will also be limited by the range and endurance of drones which will adversely impact any prolonged cyber-focused operations.

Reliability and Security—Challenging environments can affect the operational effectiveness of drone technology, and vulnerabilities in drones could be exploited by an adversary.

- Opportunities

Geopolitical Influence—Effective use of drone-based cyber-attacks within an asymmetric conflict can strengthen Russia's geopolitical influence and deter a response from its adversaries.

Collaboration—Technologically advanced countries such as China may collaborate with Russia, leading to innovative solutions and a technological advantage, further strengthening its cyber capabilities.

Increased Funding—The current conflict has seen the benefits of the use of drones on the battlefield, an increase in spending can develop the use cases further and expand the development of drone-based cyber-attacks, along with the opportunities to leverage the use of AI and autonomous decision-making in drone warfare.

Operational Use Cases—Expansion of use cases within the field, leading to more sophisticated drone-based cyber-attacks and the development of countermeasures against such an attack.

- Threats

Defensive Cyber Measures—The adversaries' advanced cyber defences and counter-drone technology can mitigate or even neutralise the ability of Russia to carry out a successful attack of this nature.

International Sanctions—The ongoing and potential increase in sanctions levied against Russia may restrict access to key technologies and slow the development of an effective drone-based cyberattack capability that stays ahead of its adversaries' countermeasures.

Competition—The advances made by competing countries in this field, such as the US, Israel and China may outpace Russia and consequently limit any perceived strategic advantage.

Legal and Ethical Challenges—Increased international scrutiny of the collateral damage and consequences of such attacks could have political and diplomatic ramifications and invoke NATO-led repercussions.

5.3.5 Cone of Plausibility

The following analysis using the cone of plausibility explores a range of future outcomes ranging from least, to most likely, between Russia and the UK, whilst considering geopolitical, military, economic and social factors.

- Baseline Scenario—Continuing Tensions Without Direct Conflict

The geopolitical tensions and rivalry between Russia and the UK continue on their current trajectory. Diplomatic tensions persist along with increasing sanctions and proxy conflicts (Russia-Ukraine war) but direct military conflict between the two nations continues to be avoided.

- Key Drivers
- Ongoing Russia-Ukraine war
- Economic and political sanctions
- Disinformation campaigns
- Cyber and traditional espionage
- Continuing Russia-NATO tension

Highly likely. As this scenario aligns with the strategic interests of both nations; the appetite for a direct military conflict between the two nations is not a favourable scenario for either and it aligns with the current state of international and political relations between them.

- Optimistic Scenario—Diplomatic Engagement and De-escalation of Tensions.

Diplomatic relationships between the two nations improve leading to renewed dialogue and a relaxation in sanctions and cooperation on global issues such as climate change.

- Key Drivers
  - Mutual economic and global interests
  - A change in leadership that favours renewed diplomacy
  - Successful diplomatic negotiations and initiatives

Realistic possibility. Although this is assessed as a Realistic Possibility, it will be driven by leadership change and a renewed sense of purpose for the Russian state, a shift in attitude that can align with the global sentiment of cooperation and not conflict.

- Pessimistic Scenario—Increased Hostilities and Proxy Conflicts

Relations between Russia and the UK deteriorate which leads to an increase in military posturing from both sides. An escalation in proxy conflicts in areas within the Middle East and Eastern Europe and a dramatic increase in cyber-attacks from both sides.

- Key Drivers
  - An aggressive and robust change in foreign policy
  - Military incidents and incursions (land/sea/air)
  - An escalation of cyber warfare

Likely/Probable. As the tensions between the two nations continue to escalate combined with the strategic interests of both nations, the likelihood of this scenario increases. This would be exacerbated by an increase in military incidents between them.

- Worst Case Scenario—Direct Military Conflict

A series of provocations, increasingly severe military incidents or a significant geopolitical crisis leads to a direct military conflict between Russia and the UK, with the potential of triggering NATO Article 5.

- Key Drivers
  - Increasing military incidents resulting in misinterpretations and escalations
  - Major geopolitical crises involving NATO that leads to retaliatory action directed at the UK

Highly Unlikely. A direct military conflict between two nuclear powers is considered highly unlikely due to the known projected outcome of such a conflict and the likelihood of total destruction of both nations.

- Wildcard Scenario—High-Scale Cyber and Hybrid Warfare

Russia launches unprecedented levels of cyber-attacks and a hybrid (non-conventional) warfare campaign solely against the UK. It causes widespread disruption to Critical National infrastructure, the economy and key public services.

- Key drivers
  - Non-justified pre-emptive strikes by Russia against the UK
  - The strategic objective of Russia is to weaken the UK economy and undermine the UK Government
  - Offensive cyber capabilities allow high impact that can be difficult to attribute

Likely/Probable. The current geopolitical climate combined with the plausible deniability nature of offensive cyber operations makes this a likely scenario and provides a lower risk to a conventional armed conflict.

It must be noted that these scenarios were generated focusing solely on Russia and the UK. The outcome would be markedly different if focused on Russia and NATO (including the UK).

The analysis indicates both the likelihood of an attack and prioritises which assets would be attacked, highlighting the potential impact. The attacks would target UK MoD Command and Control centres (PJHQ and MoD Main Building), communications networks and Critical National Infrastructure. The attacks would also target GCHQ and MoD Corsham to gather intelligence and cause disruptions in the availability of key networks and systems. The attacks would utilise drones to jam/spoof communication networks and to cause disruptions in the availability of radar and air defence systems. Drones would be used to gather intelligence and disrupt operations at key sensitive sites including DSTL, HMNB at Clyde and Portsmouth, RAF Brize Norton and the key nuclear facilities at Faslane and Aldermaston.

5.3.6 WI-FI Lab

This lab highlights the ease of circumventing some of the universally adopted security controls that are applied to Wi-Fi networks. The lab was completed using a Raspberry Pi Linux machine and an Alfa AWUS036NHA (Wi-Fi) adaptor. The lab was used to exploit a number of known vulnerabilities that prevail in the 802.11 (Wi-Fi) standard. The two devices were also tested using a 50,000 mAh power bank USB fast charger as a power source to simulate being operated on a drone in a cyber-attack scenario. It should be noted that the devices were not tested using a drone (Fig. 5.2).

- Hidden SSID

The use of a Hidden Service Set Identifier (SSID) is a widely used setting that ensures the name of the Wi-Fi network is not publicly sent. It is regarded as a security configuration, but it offers minimal protection. Relying only on a hidden SSID is not a robust security tactic, and it is recommended that it is supplemented with stronger authentication and encryption protocols. Hidden SSIDs are susceptible to de-authentication attacks that are used to force a legitimate user to disconnect from a wireless network/access-point. Once disconnected, a device will attempt to initiate reconnection which will expose the hidden SSID.

To obtain the hidden SSID, the following approach was used:

The wlan0 interface of the Alfa adaptor is put into Monitor mode using airodump-ng wlan0mon, which then prompts it to listen for 802.11 beacon frames broadcast by nearby access points. Figure 5.3 shows the target SSID ('BT-QNAKJC') is hidden.

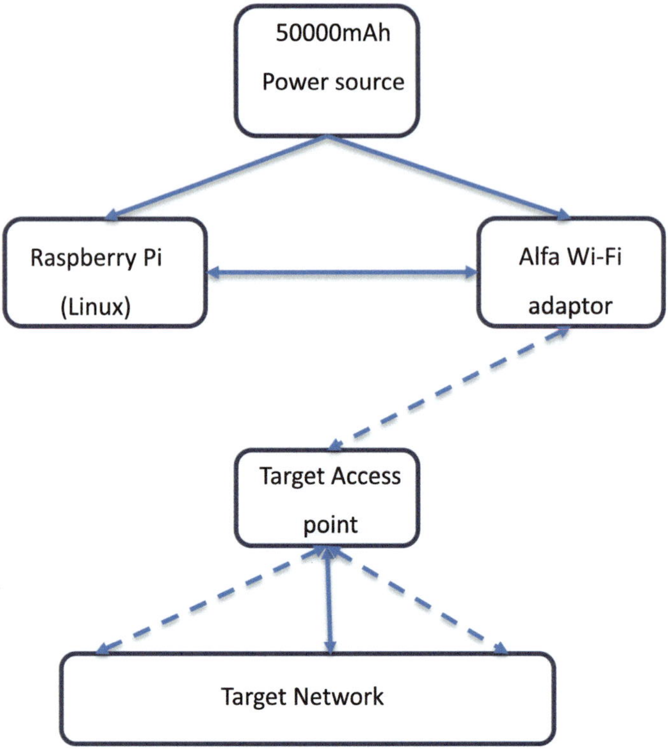

Fig. 5.2 Lab set up

Fig. 5.3 Target WI-FI network SSID hidden

airodump-ng is used with the previously retrieved MAC address, to indicate clients which have legitimately connected using the hidden SSID of the access point (Fig. 5.4).

The client is then manipulated into de-authenticating from the access point through the use of the aireplay-ng command which utilises the MAC addresses from both the access point and client (Figs. 5.5 and 5.6). The client will attempt to re-authenticate, and the SSID is then sent in clear (un-hidden).

Fig. 5.4 airodump-ng command

Fig. 5.5 De-authentication command

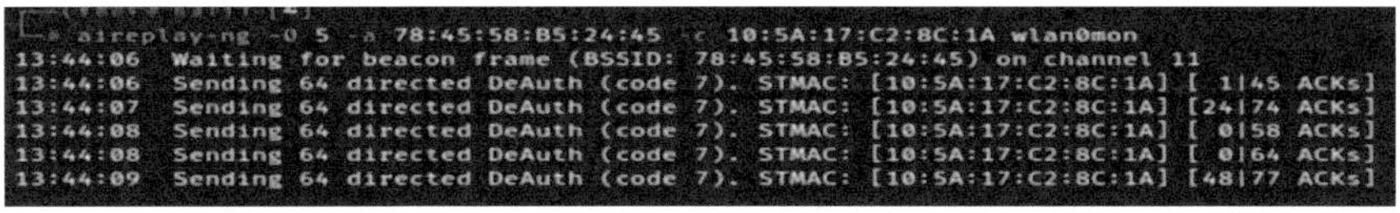

Fig. 5.6 De-authentication request packets

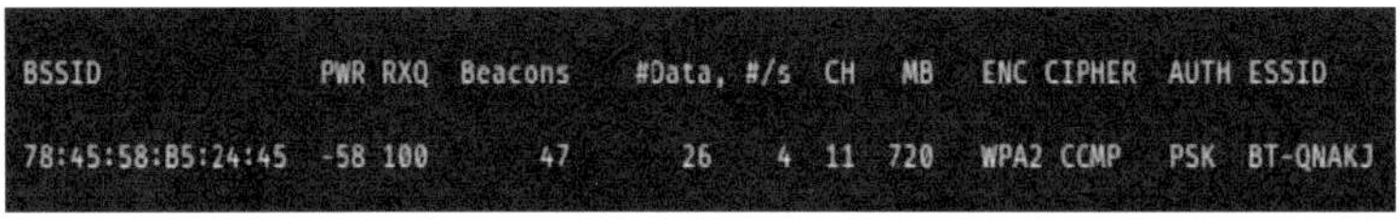

Fig. 5.7 'Hidden' SSID now shown

By leaving airodump-ng running on wlan0mon, the previously hidden SSID is revealed (Fig. 5.7).

- MAC filtering

MAC filtering uses the Media Access Control protocol to manage device access in a wireless network [74]. This generates a list of authorised MAC addresses of devices which are permitted to connect to the network. An attacker can spoof a MAC address and can impersonate a legitimate client/device on the network, or they can intercept traffic from legitimately authorised devices to facilitate network access.

MAC filtering does not encrypt a connection between an access point and a connecting client/device, this renders a network vulnerable to eavesdropping and attacks that can be used to compromise the integrity and confidentiality of the data on it.

MAC address Randomization is disabled to bypass MAC filtering. This is achieved by the editing of the Network Manager configuration file (Fig. 5.8).

Any processes which can prevent the aircrack-ng command from running, are stopped by the use of the sudo airmon-ng check kill command. Following this; the wlan0 interface monitor mode is again enabled by the sudo airmon-ng start wlan0 command (Fig. 5.9).

sudo airodump-ng wlan0mon command captures the 802.11 frames and indicates the MAC address, channel and ESSID of the access point (Fig. 5.10).

sudo airodump-ng --bssid (MAC Address) -c 11 wlan0mon indicates the devices connected to the access point (Fig. 5.11).

The WLAN interface (wlan0) is disabled (down), the interface monitor mode is also disabled, and the Network Manager is restarted. sudo macchanger -m (MAC

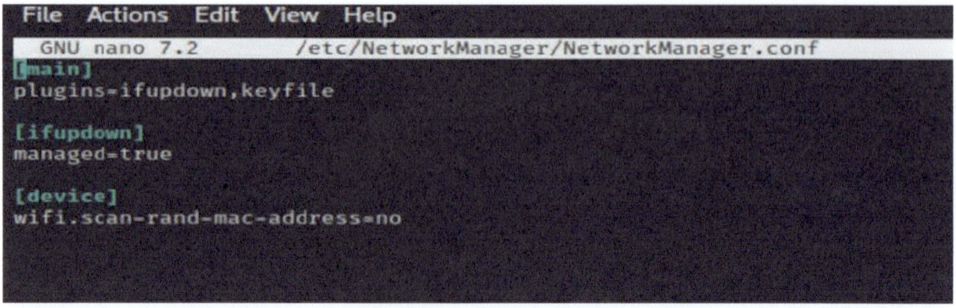

Fig. 5.8 MAC address randomisation disabled

Fig. 5.9 Monitor mode started

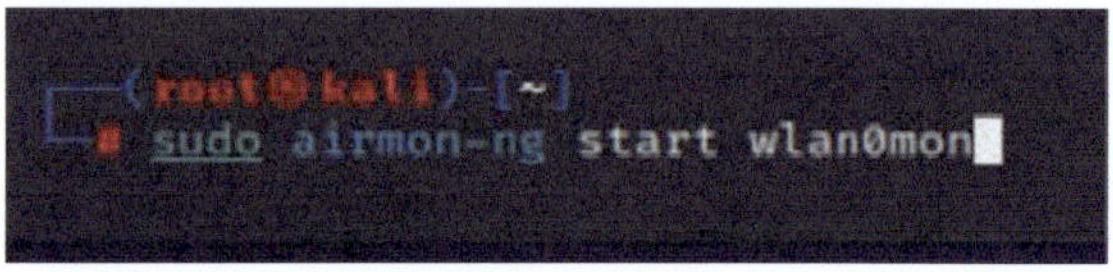

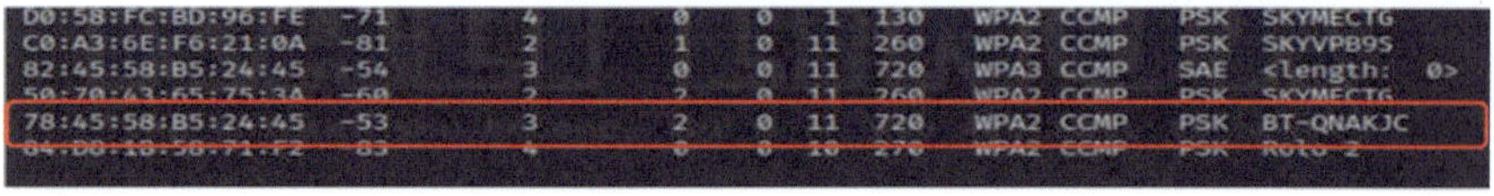

Fig. 5.10 Wi-Fi access point details

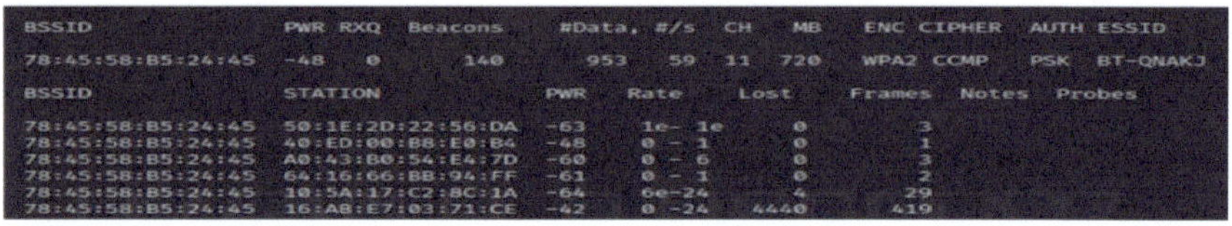

Fig. 5.11 Connected devices

address) wlan0 is used to amend the WLAN MAC address and prompt connection to the access point (Fig. 5.12) with the new MAC address now indicated (Fig. 5.13).

The (wlan0) interface is re-enabled (up) and a brute-force dictionary attack is used to crack the access-point password, as detailed in the following section.

- WEP/WPA2 Wi-Fi encryption

WLAN encryption is used to protect data communicated over Wi-Fi networks. WEP (Wired Equivalent Privacy), a common encryption protocol, is now regarded as being vulnerable, stronger protocols such as WPA2 or WPA3 should be used. However, even a WPA3 vulnerability has been exploited (Dragonblood).

The following process can be used to bypass both WEP and WPA2.

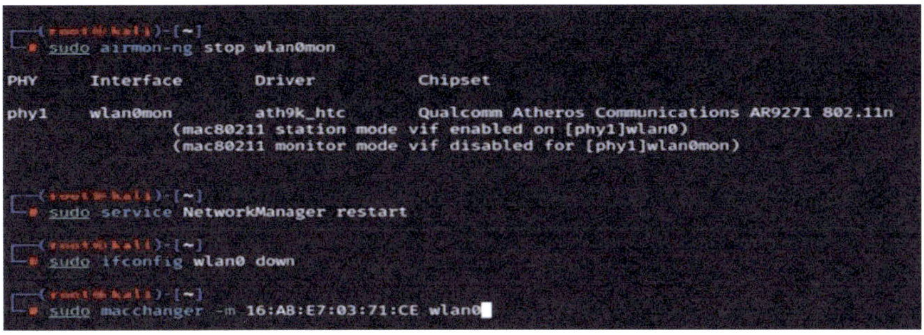

Fig. 5.12 Macchanger command

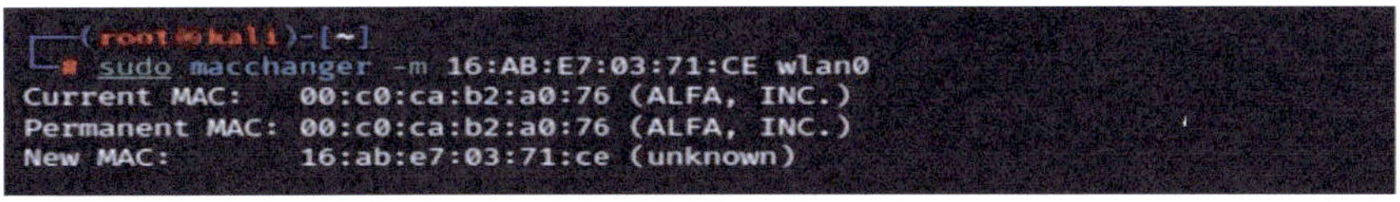

Fig. 5.13 Newly assigned MAC address

Monitor mode again is used on the wlan0 interface with the sudo airmon-ng start wlan0 and the sudo airodump-ng wlan0mon commands, capturing 802.11 frames, and indicating the MAC address of the access point, along with its channel and ESSID (Fig. 5.14). The output indicates that WPA2 is configured.

The handshake packets are used to crack WPA/WPA2. The airodump command is used (Fig. 5.15), with the output and is saved to (cap) file.

De-authentication of the connected (legitimate) client is achieved by the use of the sudo airplay-ng –deauth 50 -a (MAC address) wlan0 command and Wireshark is used on re-authentication to capture the Extensible Authentication Protocol (EAPOL) handshake made between the access point and client (Fig. 5.16).

Cracking the password requires the use of a saved password list, in conjunction with the aircrack-ng command and the previously saved cap file (Fig. 5.17), the cracked password is shown at Fig. 5.18.

- 802.11 (Wi-Fi) client identity vulnerabilities

Fig. 5.14 Access point—WPA/WPA2

Fig. 5.15 Airodump

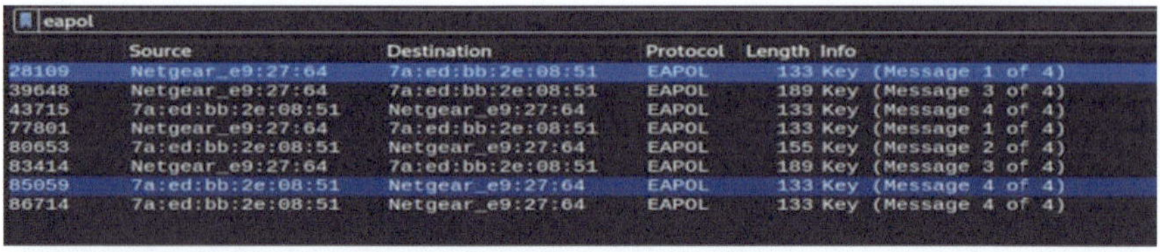

Fig. 5.16 Access point and client handshake

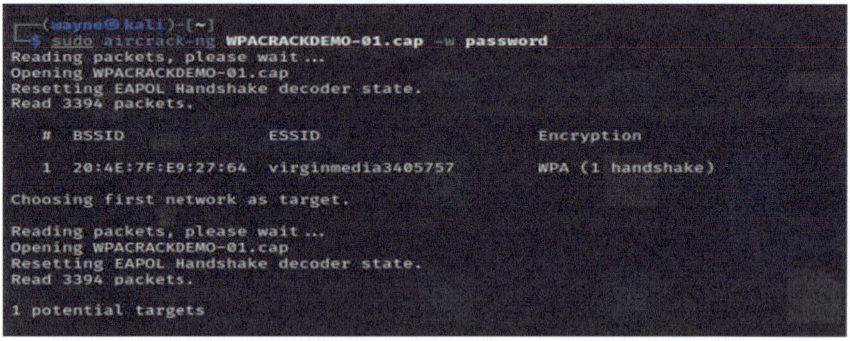

Fig. 5.17 Password cracking commands

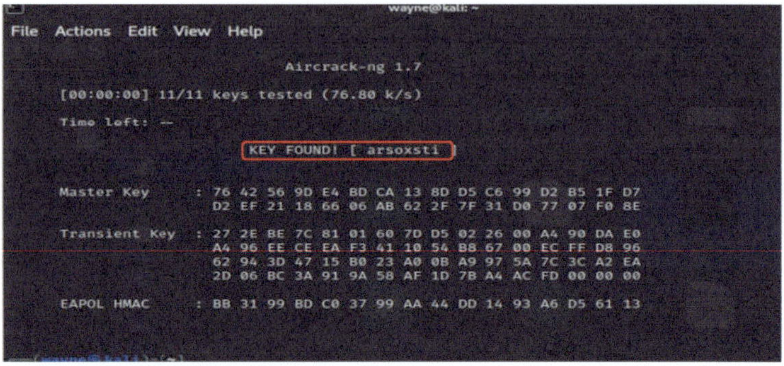

Fig. 5.18 Cracked password identified

An 802.11 Wi-Fi client authentication using Pre-Shared Keys (PSK), that uses a shared passphrase, is extremely vulnerable to straightforward brute-force attacks, that are further simplified if using non-complex, easily accessible or predictable passphrases.

If 802.11 clients insecurely store credential details, they will present themselves as an easy target for attackers, who will seek to obtain the credentials through a variety of means such as; compromised devices, social engineering or phishing attacks [36], leading to unauthorised access to a network.

Moreover, an attacker can create unauthorised (rogue) networks to facilitate Wi-Fi (802.11) clients connecting to a malicious access point. After a connection has been established, it is a straightforward process to capture data in transit from the client. Another attack that can be used to exploit an 802.11 network is a man-the-middle (MitM) attack. In this attack, an attacker would attempt to compromise the confidentiality and/or integrity of the information in transit between the client and the access point over the network by intercepting and/or modifying the data.

- Further Drone (agnostic) enabled attack steps

To enable the attack, a script is used to run the Linux commands at a pre-determined location using the gpspipe tool:

Step 1: Install gpsd and the gpsd-clients package by running: sudo apt-get install gpsd gpsd-clients

Step 2: Create the geofence script to run the attack at the pre-determined location (Fig. 5.19)

Step 3: Include the commands detailed above in the above script

Step 4: Make the script executable: (in bash) by the following command: chmod + x /path/to/geofence.sh

Step 5: Run the geofence.sh script using the following command when it reaches the pre-determined location: nohup /path/to/geofence.sh &

```
#!/bin/bash

# Define target latitude and longitude
TARGET_LAT=51.5074   # Example: London latitude
TARGET_LON=-0.1278   # Example: London longitude
THRESHOLD=0.01       # Threshold for proximity in degrees

# Function to calculate distance between two points
function calc distance() {
    local lat1=$1
    local lon1=$2
    local lat2=$3
    local lon2=$4
    local dlat=$(echo "$lat1 - $lat2" | bc -l)
    local dlon=$(echo "$lon1 - $lon2" | bc -l)
    echo "sqrt(($dlat * $dlat) + ($dlon * $dlon))" | bc -l
}

# Monitor GPS coordinates
gpspipe -w -n 10 | while read -r line; do
    LAT=$(echo $line | jq '.lat')
    LON=$(echo $line | jq '.lon')

    if [[ -n $LAT && -n $LON ]]; then
        distance=$(calc_distance $LAT $LON $TARGET_LAT $TARGET_LON)
        if (( $(echo "$distance < $THRESHOLD" | bc -l) )); then
            echo "Target location reached at $(date)" >> /home/your_username/location_log.txt
            # Execute your desired command here
            /path/to/your_task.sh
            break
        fi
    fi
done
```

Fig. 5.19 Geofence script

This lab shows how a Wi-Fi network can be compromised using readily available tools; it shows that this process can be completed through the use of a connected power source, and it is anticipated that through the secure attachment to a suitable drone, it could be used to target MOD 802.11 Wi-Fi networks.

5.3.7 Cybersecurity Frameworks

The choice of cybersecurity frameworks has grown over the years, some of which are more appropriate to niche areas of technology and development. Some are far more widespread than others; an example being ISO27001, whose adoption has increased by almost 500% in the last decade. Some, however, are mandated to be used within the MoD, who are (along with wider Government) moving to a Secure by Design (SbD) methodology that is centred on NIST. The following framework has been developed taking guidance from both ISO27001 and NIST and is influenced by the MITRE ATT&CK framework and the Lockheed Martin Cyber Kill Chain.

5.3.8 Drone-Initiated Cyberattack Defensive Framework

The following framework was constructed whilst considering the output of the analysis, the Wi-Fi lab and the current anti-drone technology. The framework is designed to counter drone-initiated cyber-attacks on key UK MoD and national assets.

- Cybersecurity strategy

Strengthen cyber defences across all MoD networks and Critical National Infrastructure and increase international collaboration with NATO and partner cyber defence agencies. At a high level, it is recommended that the NIST Cyber Security Framework is utilised as this can be tailored to the MoD requirement and an understanding of the risk appetite and tolerance.

- Network Controls

The ISO27002:2022 control set should be applied to all MoD systems and networks. The Control Set template containing the control types and descriptions is at Annex D and it has been supplemented with mapping to the NIST CSF five core functions shown at Fig. 5.20.

- Wi-Fi Security Measures

Rogue Access Point Detection

The implementation of a Wireless Intrusion Detection System (WIDS) would identify unauthorised access points and a key attack vector.

Function	Category	Category Identifier
Govern (GV)	Organizational Context	GV.OC
	Risk Management Strategy	GV.RM
	Cybersecurity Supply Chain Risk Management	GV.SC
	Roles, Responsibilities, and Authorities	GV.RR
	Policies, Processes, and Procedures	GV.PO
	Oversight	GV.OV
Identify (ID)	Asset Management	ID.AM
	Risk Assessment	ID.RA
	Improvement	ID.IM
Protect (PR)	Identity Management, Authentication, and Access Control	PR.AA
	Awareness and Training	PR.AT
	Data Security	PR.DS
	Platform Security	PR.PS
	Technology Infrastructure Resilience	PR.IR
Detect (DE)	Continuous Monitoring	DE.CM
	Adverse Event Analysis	DE.AE
Respond (RS)	Incident Management	RS.MA
	Incident Analysis	RS.AN
	Incident Response Reporting and Communication	RS.CO
	Incident Mitigation	RS.MI
Recover (RC)	Incident Recovery Plan Execution	RC.RP
	Incident Recovery Communication	RC.CO

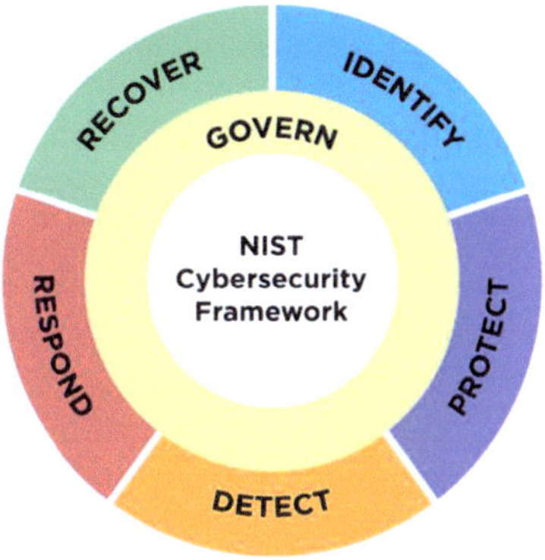

Fig. 5.20 NIST CSF [49]

Network Hardening

The networks should be hardened, taking guidance from NCSC appropriate to the classification of the information within that network, including Access Control Lists and asset management.

Network Segmentation

All sensitive networks should be segmented and should avoid the use of Wi-Fi. The segmentation of networks would confine an attack/breach to a specific network and prevent any lateral movement within the organisation.

Encryption

All Wi-Fi networks must secure any data in transit and unauthorised access by using strong encryption and the latest protocols (WPA3).

Endpoint Security

Endpoint security policies and Security Operating Procedures should be implemented for end-user devices connected to networks. This includes updates, password complexity, acceptable use and antivirus policy.

Audit and Inspection

A regime of security audits and vulnerability assessments is to be implemented to enable the detection and mitigation of vulnerabilities.

Monitoring

Network monitoring should be implemented to detect any suspicious behaviour. Any atypical traffic or activity can be a key indicator of a security breach.

Incident Management

Incident Management Plans should be implemented and tested along with the Business Continuity Plan to understand the implications of; and the actions required as a result of a drone-enabled cyber-security attack.

Education and Training

All staff should receive regular updates and appropriate training related to the ever-changing threat landscape. Awareness of potential drone-enabled cyber-attacks should be increased and a process for reporting suspicious drone activity instigated. The training should include an understanding of the security measures and restrictions implemented and the actions to take in the event of an incident.

- Drone Specific Measures

Drone Detection and Countermeasures

The use of radar, RF detection, and visual capabilities to identify and track unauthorised drones. Utilise counter-drone technologies such as jamming, spoofing, or kinetic interception.

Physical Security

Secure all physical perimeters of Critical National infrastructure and key Military Installations and implement robust access control measures to prevent unauthorised access by drones.

Incident Response

Develop and regularly update incident response plans with regards to unauthorised drone activity/incursions.

Public Awareness and Engagement

Increase awareness of the threat from drone-initiated cyber threats among the public and private sectors and encourage reporting of any suspicious drone activity.

International Collaboration

Collaboration with international allies and partners to share intelligence and best practices. Engage in joint exercises to improve coordinated responses to both drone and cyber threats.

Counter-Drone Technology

The UK already employ counter-drone technology at the UK's major airports. There has been substantial investment across the globe in this field with solutions using a combination of radar and sensors. Some of the high-profile solutions are Lockheed Martin's Morfius which is a system that counter-s drone swarms, Qinetiq's Obsidian, and Leonardo-developed FalconShield. The UK Government should extend the use of this capability to those sites highlighted within this paper, including:

- PJHQ
- MoD Main Building
- Key CNI hubs
- GCHQ
- MoD Corsham
- DSTL
- HMNB at Clyde and Portsmouth
- RAF Brize Norton
- Key nuclear facilities at Faslane and Aldermaston

5.4 Conclusions and Future Study

The results of the analysis show different variables, but the key indication is that it is likely that there will be increased hostilities and proxy conflicts between the two nations, resulting in an increased level of cyber-attacks and an increased likelihood of a drone-initiated cyber-attack against key UK MoD and national assets. However, this will likely change in line with any increase in UK or NATO involvement in the Russia-Ukraine conflict. Furthermore, the impacts of such an attack will also increase proportionately.

The analysis indicates both the likelihood of an attack and prioritises which assets would be attacked, highlighting the potential impact. The attacks would target UK MoD Command and Control centres (PJHQ and MoD Main Building), communications networks and Critical National Infrastructure. The attacks would also target GCHQ and MoD Corsham to gather intelligence and cause disruptions in the availability of key networks and systems. The attacks would utilise drones to jam/spoof communication networks and to cause disruptions in the availability of radar and air defence systems. Drones would be used to gather intelligence and disrupt operations at key sensitive sites including DSTL, HMNB at Clyde and Portsmouth, RAF Brize Norton and the key nuclear facilities at Faslane and Aldermaston.

The key mitigation centres on the strengthening of cyber defences across all MoD networks and Critical National Infrastructure and increasing international collaboration with NATO and partner cyber defence agencies.

Future studies in this area would be focused on the use of artificial intelligence and automation algorithms to counter drone swarm-initiated cyber-attacks and to develop an automated response capability to intercept and either deceive, deny or destroy the threat.

References

1. Alshar'e M (2023) Cyber security framework selection: comparison of NIST and ISO27001. Appl Comput J 245–255
2. Amin B, Taghizadeh S, Rahman MS, Hossain MJ, Varadharajan V, Chen Z (2020) Cyber attacks in smart grid—dynamic impacts, analyses and recommendations. IET Cyber-Phys Syst Theory Appl 5:321–329
3. Arsenault AC, Kreps SE, Snider KLG, Canetti D (2024) Cyber scares and prophylactic policies: crossnational evidence on the effect of cyberattacks on public support for surveillance. J Peace Res 61:413–428
4. Balta EC, Pease M, Moyne J, Barton K, Tilbury DM (2024) Digital twin-based cyber-attack detection framework for cyber-physical manufacturing systems. IEEE Trans Autom Sci Eng 21:1695–1712
5. Bennett C, Jahangir M, Fioranelli F, Ahmad BI, Kernec JL (2020) Use of symmetrical peak extraction in drone micro-Doppler classification for staring radar
6. Boakye-Boateng K (2023) Implementation of a trust-based framework for substation defense in the smart grid. Smart Cities 7:99–140
7. Caputo AC (2023) Scenario analysis of offshore wind-power systems under uncertainty. Sustainability 15:16912
8. Carey I, Casimir C (2013) NATO's options for defensive cyber against non-state actors
9. Cecilia A, Sahoo S, Dragicevic T, Costa-Castelló R, Blaabjerg F (2021) Detection and mitigation of false data in cooperative DC microgrids with unknown constant power loads. IEEE Trans Power Electron 36:9565–9577
10. Çetin E, Barrado C, Pastor E (2021) Improving real-time drone detection for counter-drone systems. Aeronaut J 125:1871–1896
11. Çetin E, Barrado C, Pastor E (2022) Countering a drone in a 3D space: analyzing deep reinforcement learning methods. Sensors 22:8863
12. Chen L (2024) Dynamics of cascading failure in cyber-physical power systems from cyber attack. Phys Scr 99:035243
13. Chuter A (2013) British defense firms, gov't team to fight cyberattacks. Defense News (Springfield, Va.) 28:7
14. Claburn T (2022) How Wi-Fi spy drones snooped on financial firm [Online]. The Register. https://www.theregister.com/2022/10/12/drone-roof-attack/. Accessed 26 March 2024
15. Cornish P, Hughes R, Livingstone D (2009) Cyberspace and the national security of the United Kingdom. Threats and responses, vol 1. Chatham House, London, pp 1–46
16. Dibaji SM, Pirani M, Flamholz DB, Annaswamy AM, Johansson KH, Chakrabortty A (2019) A systems and control perspective of CPS security. Annu Rev Control 47:394–411
17. Dov Bachmann SD, Putter D, Duczynski G (2023) Hybrid warfare and disinformation: a Ukraine war perspective. Global Policy 14:858–869
18. Erola A, Axon L, Rensburg AJV, Agrafiotis I, Goldsmith M, Creese S (2021) Control effectiveness: a capture-the-flag study
19. Fadli ZM, Yong SS, Kee LK, Ching GH (2022) Cyber attack awareness and prevention in network security. Int J Inform Commun Technol (IJ-ICT) 11:105
20. Ferenţ D-A, Preja C (2023) NATO's involvement in cyber defence. Intell Info 2
21. Forster A, Wallace W (2001) What is NATO for? Survival 43:107–122
22. Gligorević K (2024) Overview of the possibility application of some nano drone technologies in modern agriculture. Poljoprivredna Tehnika 49:75–96
23. Gonzalez-Jorge H (2024) Counter drone technology: a review
24. Gov.uk (2022) Digital market place—SbD [Online]. https://www.digitalmarketplace.service.gov.uk/digital-outcomes-and-specialists/opportunities/17363. Accessed 08 March 2023
25. Governance I (2023) 27001 adoption [Online]. https://www.itgovernance.co.uk/iso27001. Accessed 03 March 2023
26. Government U (2021) Global Britain in a competitive age: the integrated review of security, defence, development and foreign policy. Her Majesty's Stationary Office

27. Haataja S (2017) The 2007 cyber attacks against Estonia and international law on the use of force: an informational approach. Law Innov Technol 9:159–189
28. Haataja S (2018) Cyber attacks and international law on the use of force
29. Haggit JM, Antonik C, Flach JM, Mcewen TR, Smith MW (2022) Joint activity design evaluation tools for conducting staged-world studies. In: Proceedings of the human factors and ergonomics society annual meeting, vol 66, pp 388–391
30. Hansen L, Nissenbaum H (2009) Digital disaster, cyber security, and the Copenhagen school. Int Stud Quart 53:1155–1175
31. Jackson S (2016) NATO Article 5 and cyber warfare: NATO's ambiguous and outdated procedure for determining when cyber aggression qualifies as an armed attack. The CIP report
32. Kang H, Joung J, Kim J, Kang J, Cho YS (2020) Protect your sky: a survey of counter unmanned aerial vehicle systems. IEEE Access 8:168671–168710
33. Kattel R, Mergel I (2019) Estonia's digital transformation, pp 143–160
34. Knoblauch AM, Rosa SDL, Sherman J, Blauvelt C, Matemba C, Maxim L, Defawe O, Gueye A, Robertson J, Mckinney J, Brew J, Paz EC, Small PM, Tanner M, Rakotosamimanana N, Lapierre S (2019) Bi-directional drones to strengthen healthcare provision: experiences and lessons from Madagascar, Malawi and Senegal. BMJ Glob Health 4:e001541
35. Kotwani B (2023) Red teaming vs. blue teaming: a comparative analysis of cybersecurity strategies in the digital battlefield. Int J Sci Res Eng Manag 07:1–11
36. Kron E (2021) The human problem behind credential theft and reuse. Cyber Secur 4:223–231
37. Kumar MS, Ben-Othman J, Srinivasagan K (2018) An investigation on WannaCry ransomware and its detection. In: 2018 IEEE symposium on computers and communications (ISCC). IEEE, pp 1–6
38. Liu Y, Li Z, Guo G (2022) Interaction network-based resilient consensus of connected vehicles against cyber-attacks. IET Control Theory Appl 17:744–756
39. Lucock X, Westbrooke V (2021) Trusting in the "Eye in the sky"? Farmers' and auditors' perceptions of drone use in environmental auditing. Sustainability 13:13208
40. Madubuike-Ekwe JN (2021) Cyberattack and the use of force in international law. Beijing Law Rev 12:631–649
41. Maghazei O (2022) Drones in manufacturing: opportunities and challenges
42. Marpaung A, Pratomo A, Wicaksono A (2019) Implementation of SWOT analysis in determining competitive strategy on catering and conference services in hotels
43. Mcshane MK, Nguyen T (2020) Time-varying effects of cyberattacks on firm value. Geneva Papers Risk Insur Iss Pract 45:580–615
44. MOD (2021) Red teaming handbook [Online]. www.gov.uk. https://www.gov.uk/government/publications/a-guide-to-red-teaming. Accessed 27 Mar 2024
45. MOD (2023) Defence intelligence—communicating probability [Online]. MOD. https://www.gov.uk/government/news/defence-intelligence-communicating-probability. Accessed 27 March 2024
46. Moşoiu O, Bălăceanu I, Mihai E (2020) Cyber terrorism and the effects of the Russian attacks on democratic states in East Europe. Sci J Silesian Univ Technol Ser Transp 106:131–139
47. Munusamy T, Khodadi T (2023) Building cyber resilience: key factors for enhancing organizational cyber security. J Inform Web Eng 2:59–71
48. NATO (2023) Collective defence and Article 5 [Online]. https://www.nato.int/cps/en/natohq/topics_110496.htm
49. NIST.GOV (2022) NIST RMF [Online]. https://csrc.nist.gov/projects/risk-management/about-rmf. Accessed 16 March 2023
50. Obiuto NC (2024) Reviewing the role of AI in drone technology and applications. Comput Sci Res J 5:741–756

51. Office C (2010) A strong Britain in an age of uncertainty: the national security strategy. The Stationery Office
52. Park D-W (2015) Draft of national cybersecurity act. Int J Secur Appl 9:105–112
53. Park SJ, Kim HT, Lee S, Joo H, Kim H (2021) Survey on anti-drone systems: components, designs, and challenges. IEEE Access 9:42635–42659
54. Pascarella D (2024) Drone intrusion management systems in airports: assessment of ASPRID solution. J Phys: Conf Ser 2716:012070
55. Pasqualetti F, Dörfler F, Bullo F (2013) Attack detection and identification in cyber-physical systems. IEEE Trans Autom Control 58:2715–2729
56. Păunescu D-M (2021) Nato's encounters in the cyber domain. Strateg Xxi Secur Defense Fac 17:153–158
57. Peksa J (2024) A review on the state of the art in copter drones and flight control systems. Sensors 24:3349
58. Radkin AST (2024) UK defence policy in a heightened security landscape. Chatham House
59. Robinson T (2021) Value of Information for argumentation based intelligence analysis
60. Robinson N, Slack C (2019) Co-operation: a key to NATO's cyberspace endeavour. Eur Foreign Aff Rev 24:154–166
61. Samy A, Yu H, Zhang H (2020) Fog-based attack detection framework for internet of things using deep learning. IEEE Access 8:74571–74585
62. Sarker IH, Kayes ASM, Badsha S, Alqahtani H, Watters PA, Ng A (2020) Cybersecurity data science: an overview from machine learning perspective. J Big Data 7
63. Shalamanov V, Bankov B (2022) NATO cyber defence policy and hybrid threats: the way to enhance our resilience
64. Shandler R, Gross ML, Backhaus S, Canetti D (2021) Cyber terrorism and public support for retaliation—a multi-country survey experiment. Br J Polit Sci 52:850–868
65. Shoufan A (2023) Contingency clarification protocols for reliable counter-drone operation. IEEE Trans Aerosp Electron Syst 59:8944–8955
66. Shoufan A, Ruba A (2021) Integrating counter-UAS systems into the UTM system for reliable decision making
67. Sirohi H (2024) A comprehensive review of modern counter-drone technologies: trends, challenges, and future directions. Int J Res Appl Sci Eng Technol 12:4405–4418
68. Snider KLG, Shandler R, Zandani S, Canetti D (2021) Cyberattacks, cyber threats, and attitudes toward cybersecurity policies. J Cybersecur 7
69. Sufi F (2024) A new time series dataset for cyber-threat correlation, regression and neural-network-based forecasting. Information 15:199
70. Szira Z (2023) The development of drone technology and its regulation in the European Union. EU Agrarian Law 12:35–41
71. Taherdoost H (2022) Understanding cybersecurity frameworks and information security standards—a review and comprehensive overview. Electronics 11:2181
72. Taherdoost H, Madanchian M (2021) Determination of business strategies using SWOT analysis; planning and managing the organizational resources to enhance growth and profitability. Macro Manag Public Policies 3:19–22
73. Taiwo JF, Prisca OI, Matthew UO, Onyebuchi NA, Nwamouh UC, Robert UI, Matthew AO (2022) IoT drone technology integration in medical logistics delivery. Sci J Public Health 10:124
74. Tanner NH (2019) Protecting wireless communication. Wiley
75. Tjandra KD, Irawan FG, Nugraha P, Sunindijo RY (2022) Drone readiness in the Indonesian construction industry. Constr Econ Build 22
76. Ukwandu E, Farah MAB, Hindy H, Bureš M, Atkinson R, Tachtatzis C, Bellekens X (2022) Cyber-security challenges in aviation industry: a review of current and future trends. Information 13:146
77. Webber M, Sperling J, Smith MA (2021) What's wrong with NATO and how to fix it. Wiley
78. Xin L (2023) The sensor-actuators stealthy cyber-attacks framework on networked control systems: a data-driven approach. Asian J Control 26:960–973

79. Ye Z (2024) An efficient adjacent frame fusion mechanism for airborne visual object detection. Drones 8:144
80. Zailani MAH, Raja Zahratul Azma Raja S, Rahana AR, Ismail A, Shahnaz IS, Chan K-C, Jamaludin M, Mahdy ZA (2021) Drone versus ambulance for blood products transportation: an economic evaluation study. BMC Health Serv Res 21
81. Zvozdetska O (2018) NATO's new strategic concept in cybersecurity issues in the context of up-to-the date vulnerability and threat information. Mediaforum Anal Forecasts Inf Manag 71–93

Chapter 6
Policy and Compliance: Regulatory Perspective for Autonomous Aerial and Surface Operations

Chisom Sylvia Onunta and Umair B. Chaudhry

Abstract Today, unmanned aerial systems (UAS) are increasingly used by individuals, organizations, government, businesses in the collection of military and civilian applications like transportation, border surveillance scientific research, port security, and more. The need for governance for these unmanned aerial vehicles is vital for safety. Overall use of unmanned aerial systems must be properly regulated to ensure correct use, so it doesn't endanger the public or property. This paper provides a comprehensive review of regulatory frameworks governing unmanned aerial systems, it also discusses the aspect UK government regulation on each of the unmanned aerial systems, risk governance covering safety and cybersecurity and the need for a suitable framework relating to unmanned aerial systems usage. While focusing on the United Kingdom, this research explores the legal implications of commercial, civil UAS use, its challenges and opportunities. It also presents the need for a legal framework and regulation that can accommodate the evolving UAS technology while ensuring public privacy and safety.

Keywords Drones · Governance · GDPR · Regulation · Unmanned aerial vehicles · Unmanned surface vehicles

6.1 Introduction

Unmanned aerial systems (UAS) represent groundbreaking advancements for humanity in the technological era. The progress of unmanned aerial systems has profoundly transformed various industries, presenting both numerous opportunities and challenges. The autonomous mobility of systems aims to mitigate human errors while driving, as 94% of accidents in (autonomous vehicles AVs are attributed to human error [1]. By calculating with high efficiency, the risk of accidents due to

---

C. S. Onunta · U. B. Chaudhry (✉)
Northumbria University London, London, UK
e-mail: umair.chaudhry@northumbria.ac.uk

U. B. Chaudhry and H. Jahankhani, *Autonomous Revolution: Strategies, Threats and Challenges*, Advanced Sciences and Technologies for Security Applications,
https://doi.org/10.1007/978-3-031-96782-5_6

driver negligence is reduced. This has led UAS to rapidly evolve from niche gadgets into essential tools in many industries' technological arsenals.

These benefits hinge on the precision of each vehicle's control system. If accuracy isn't ensured, safety and economic benefits are compromised, and autonomous systems can, in fact, cause serious damage. No matter how advanced a system is, there will be losses, as performing tasks inevitably involve errors—especially with complex systems. This is why countries around the world are cautious and skeptical about the legal regulations surrounding autonomous systems, given that safety and security are critical in UAS operations [5]. Integrating UAS into civilian airspace or on the ground demands strict regulations on the ground to avoid collisions and guarantee the security of both ground and airborne entities. The security and effectiveness of UAS have been greatly enhanced by recent developments in regulations, highlighting the critical role that technology plays in resolving regulatory issues.

The legal consequences of UAS operations stretch beyond security, safety, involving notable privacy concerns. The proliferation of UASs exacerbates concerns regarding intrusive surveillance, prompting ethical and legal debates on balancing technological progress with individual privacy rights [39]. Moreover, the global aspect of UAS regulation [34] brings both hurdles and prospects for alignment. The varied regulatory strategies employed by different nations highlight the necessity for international collaboration to create unified UAV regulations. According to Shakhatreh et al. (2019), these activities are crucial for facilitating cross-border UAV operations and guaranteeing a unified worldwide regulatory framework. The importance of global governance frameworks in UAS regulation is shown by the crucial role played by international organizations like as the International Civil Aviation Organization (ICAO) in spearheading these harmonization efforts.

Additionally, designing a framework that supports new technology without disrupting the existing transportation and traffic regulations is a major challenge. This involves addressing the positioning of UAS within transportation law, determining liability in traffic accidents (whether it falls on the manufacturer or the UAS owner), and ensuring safety protocols for various types of UAS. A report by Lloyd's of London highlighted the potential threats posed by the lack of a robust regulatory framework for UAS, emphasizing the need for coordinated insurance standards and clarity on third-party liability, among other insurance aspects related to its use. "Regulation itself will change, as it always has. But our goals will not change. We want transport to be cleaner, safer, healthier, greener, cheaper, more convenient, and more inclusive. As regulators, we will judge every innovation on whether it serves those ends or undermines them." [22, 23].

In summary, the legal frameworks governing UAS operations must develop concurrently alongside the technological advancements and growing applications of UASs. This research explores the current legal regulations, identify challenges and gaps, and offer recommendations for creating comprehensive, adaptable legal frameworks to accommodate the evolving nature of UAS technology in the United Kingdom. Through a thorough examination of international and domestic regulations, this study aims to contribute to the ongoing dialogue on crucial aspects

UAS governance which includes regulations, international frameworks, national regulations, security and law enforcement, privacy, and data protection.

6.2 Research Methodology

This study employs the method of secondary research from analysis of existing data. Using qualitative descriptive data analysis method to investigate the legal frameworks governing autonomous systems. This methodology was selected to methodically collect, review, and analyze the breadth of academic literature, government publications, legal documents, industry relevant reports and practical studies related to autonomous system regulation. The objective is to assess the present state of autonomous systems legislation, pinpoint regulatory gaps, and evaluate the effectiveness of existing legal structures. Describing the procedures utilized for the critical descriptive analysis and the literature review.

Scope and Sources: A literature review that includes a wide variety of sources guarantees a comprehensive analysis of the subject. To remain relevant to the rapidly developing UAS technology and the constantly changing regulatory landscape, scholarly publications, legal status, government websites, international recommendations, and case studies are critically analyzed, with an emphasis on works published in the recent ten years. Peer-reviewed publications, international aviation authority reports, and legal databases of information are examples of primary sources.

Search Strategy: A structured search strategy was employed using keywords related to UASs ("drones," "unmanned aerial vehicles," "UAS regulations," "USV regulation," "MASS regulations" UK regulations on UAV, USV, AV) across multiple databases, research gate, including library of congress [13], IEEE Xplore, JSTOR, academia.edu, Google Scholar. The search is streamlined by applying filters for relevance, publication date and citation count to prioritize recent and high-impact literature.

Selection Criteria: Documents that focus on UAS regulations or legal issues, studies that address civilian applications of UASs, works that provide a comparative analysis of international UAS laws, the United Kingdom, and three main approaches to unmanned aerial system regulations and policies are the criteria used to choose the literature review. Resources from DVSA (Driver and Vehicle Standards Agency), DVLA (Driver and Vehicle Licensing Agency), DfT (Department for Transport) [8] and its Motoring Agencies, CCVA (Centre for Connected and Connected Vehicles), FAA (federal aviation authority), and CAVPASS (connected and automated vehicles; process for assuring safety and security). Innovate UK, the International Maritime Organization (IMO), the Civil Aviation Authority (CAA), the Law Commissions, the National Cyber Security Centre (NCSC), and COLREGs (international regulations for preventing collision at sea) are among the government departments and their agencies that were chosen for this study.

The results of the literature review are arranged thematically in the critical descriptive analysis. Themes such as safety, privacy, cybersecurity, airspace integration, and international harmonization within each unmanned aerial system are developed based on recurrent problems and difficulties in UAV laws. Every subject is looked at to describe the state of the law today, evaluate regulatory strategies critically, identify best practices, and identify any gaps. The analysis follows multiple evaluation criteria to determine the effectiveness of current UAS regulations in the United Kingdom.

By presenting a summary of the United Kingdom governments actions, its challenges and opportunities on each sector automotive (AV) autonomous vehicles on public roads; maritime (USV) unmanned surface vehicles and aerial (UAV) unmanned aerial vehicles while offering guidance for further exploration in this dynamic field.

6.3 Literature Review

International civil aviation organization (2019) created a compatible International Aviation Trust Framework (IATF) drone regulations offering global guidelines for UAS regulations with an emphasis on safety and environmental considerations. Setting international aviation rules, particularly those pertaining to UAS operations, is a major responsibility of the International Civil Aviation Organization (ICAO). As UAS technology develops, ICAO has worked to offer guidance, policies, and recommendations to guarantee their safe, effective, and secure integration into international airspace. However, the rapid advancements and diverse applications of UAS pose challenges to the timely updating and implementation of international standards [31].

ANSI (American National Standards Institute) with Consumer Technology Association (CTA) issued the ANSI/CTA-2088 (2020), which outlined some simple cybersecurity instructions for user devices like internet of things (IoT). This standard emphasizes the importance of device identifier security, secure access, protection of data at rest and in transit, utilization of industry-standard communication and cryptography protocols, data validation, event logging, the capability to patch vulnerabilities, and reprovisioning features. In 2022, ANSI/CTA introduced ANSI/CTA-2088.1 specifically for drones, outlining the cybersecurity requirements and recommendations modified to the specific uses, capabilities and applications of small UAS.

European Union [16] presents the EU drone regulations where the EU established an extensive regulatory framework that encompasses guidelines for UAS design, production, and operation. The European Union's strategy for UAS regulation, especially regarding the potential for final stage delivery, serves as a model for striking a balance between innovation with privacy and safety concerns. This regulatory framework underscores the importance of technology in improving UAS operational efficiency while safeguarding citizens' rights and safety.

The "Standard of Excellence in Cybersecurity" (2020) was released by the European Air Traffic Management system industry consortium, the Civil Air Navigation Services Organization (CANSO). High-level UAS security recommendations are included in this paper, including information exchange, risk assessments, protective technologies, response planning, asset management, supply chain risk management, and more.

Federal aviation Administration FAA (2020) part 107 Regulations. The FAA created rules for UAS operation in United Kingdom, including operational limitations registration requirements and airspace restrictions. FAA (Federal Aviation Authority) created comprehensive regulations governing UAS operations, emphasizing safety, privacy, and airspace integration. On the other hand, emerging economies may have more fundamental regulations, reflecting the evolving stage of UAS integration into their economies and societies.

According to [29] and UAS cyber security threats (European Organization for the Safety of the Air Navigation 2023), The cybersecurity of Unmanned aerial Systems (UAS) is a significant issue because of the rising threats of cyberattacks. These attacks can target the confidentiality, availability and integrity of UASs, potentially leading to serious outcomes like the theft or exposure of sensitive data (Hamza et al.). Directives (EU) 2022/2555 [EU, 2022/2555] on strategies for a high common level of cybersecurity throughout the Union and (EU) 2022/2557 [EU, 2022/2557] on the flexibility of essential entities were released by the European Union on December 14, 2022 [4].

Cybersecurity and Infrastructure Security Agency (CISA), as a U.S. federal agency under the Department of Homeland Security (DHS), issued "Cybersecurity Best Practices for Operating Commercial Unmanned Aircraft Systems" (2019). This concise document addresses the secure utilization of UAS software and firmware, the safe execution of UAS operations, the protection of data transfer and storage, and the sharing of information and vulnerabilities.

UK regulation ((EU) 2020/945) The UK Drone Delivery Group released a White Paper outlining essential measures for advancing drone development beyond visual line of sight (BVLoS) in the United Kingdom, which would significantly boost the country's drone industry. It noted that regulatory obstacles make testing and technological advancement costly and extremely slow, and it cautioned that standards are being set without considering a large portion of the existing industry knowledge base.

Rules on safe use of automated vehicles on GB roads [28] document outlines proposed regulatory changes to ensure their safe operation of AV on British roads. It also affirms the UK's plans to pursue more ambitious applications of this technology, supported by the necessary research and development to ensure safety. CORELGs (International Regulations for Preventing Collisions at sea) established by Maritime International Organization published a guide and rules for executing standard avoidance maneuvers. The DFT (Department for transport) in 2016 initiated a study on the impacts of piloted drone collisions with manned aircraft, aiming to enable safe drone operations through a "detect and avoid" system that ensures drones do not conflict with manned aircraft.

6.3.1 Autonomous Vehicles

Automated vehicles (AVs), often called self-driving, robotic vehicles or driverless vehicles, present significant possibilities in the transport field. These zero emissions shared and connected AVs are anticipated to lower accidents, fatalities, emissions, injuries and reduce traffic congestion on the roads, while promoting a more inclusive transportation system [7]. However, AVs also introduce several challenges, including new types of accidents from system failures, privacy concerns, and potential job losses for drivers.

In the "Future of Mobility" report, the UK government identifies significant prospects for new mobility solutions to aid in the transition to net zero emissions and solve congestion problems. Shared, electric, and connected AVs could lead to seamless mobility, characterized by availability, sustainability, affordability, convenience and efficiency. However, the future of mobility could evolve in various ways. For instance, as noted by Wadud et al. [38], continuous operation of autonomous taxis could increase energy consumption. Additionally, while AVs might enhance accessibility for disables or older individuals [11], this could also lead to higher traffic density due to increased usage. Although fully autonomous cars are meant to be safer than non-autonomous vehicles, this is difficult to verify. Developers need new methods or standards to measure safety, and lawmakers should adopt flexible regulations that evolve with the technology [17].

Regulation: The Automated Vehicles (AV) Act, implemented on 20 May 2024, paves way for self-driving vehicles to be used on the roads by 2026 in the United Kingdom [35]. The AV Act is a pivotal element of this legislation, designed to advance the autonomous vehicle industry. It primarily aims to enhance road safety by minimizing human error. Three major developments towards regulations have been made. First, it gives a guide to a comprehensive framework which specifies liability of a self-driving vehicle when a listed vehicle's "when driving itself" causes an insurer to be held accountable for accidents (Section 2(1) of AEV Act 2018) which marks a significant step in building public trust and acceptance of autonomous vehicle. Secondly, the description of a self-driving car was provided. Section 8(1) AEV Act 2018 defines "driving itself" as "operating in a mode in which it is not being controlled, and does not need to be monitored, by an individual".

A regulatory framework for automated vehicles introduced the ALKS (automated lane keeping system), which enables drivers to handover control to the vehicle. ALKS was designed for use at low speeds, up to 37 miles an hour, on motorways [3].

With ALKS, automakers will create, design, and produce automobiles for international markets. Therefore, to prevent a patchwork of disparate national legislation, it is crucial that contracting parties to adhere to and implement international regulations.

Safety: The evidence clearly shows that the potential of self-driving vehicles can only be achieved if we also change the way we ensure their safety. Therefore, the United Kingdom government has developed a new safety framework for self-driving vehicles to meet the needs of both society and industry, ensuring their safety when they are first deployed on UK roads and throughout their use. This framework will be guided by an overarching safety ambition for self-driving vehicles, which will be elaborated in the National Safety Principles. The driving systems will be assessed against safety criteria, usually through authorization to international or national technical standards.

Under the introduced legislative framework, any fully autonomous vehicle must pass through two stages before being permitted to operate on UK roads. Initially, it must comply with relevant technical safety standards, such as UNECE Regulations or GB vehicle type approval standards for automated vehicle technologies. Subsequently, it must receive government authorization to drive autonomously.

Cybersecurity: Tesla in 2016 introduced the first self-driving car, but within a year, it was compromised when hackers remotely took control of a Tesla Model S from twelve miles away. They accessed the brakes, door locks, dashboard computer screen, and other electronic components. These cybersecurity issues pose risks of various accidents.

In response, the government issued key cybersecurity principles for connected and automated vehicles [14, 15]. Jesse Norman, Future of Mobility Minister, highlighted that as vehicles become intelligent, the opportunity for future mobility grows, but so does the risk of data theft and hacking. On December 19, 2021, The British Standards Institute issued a cybersecurity standard, to enhance the flexibility and preparedness ensuring that UK remain at the cutting edge of transport technology advancements.

This regulation closely aligns with ISO 21434—Road Vehicles—Cyber Security, a standard framework that includes defining cybersecurity policies, managing cyber risks, and fostering a cybersecurity culture. The concept, product development, manufacturing, operation, maintenance, and decommissioning of electrical and electronic systems in road vehicles, including their interfaces and component parts, are all included in the cybersecurity risk management process. Companies must create autonomous and connected cars to avoid cyber threats and handle possible cyberattacks, according to these non-binding guidelines.

Despite these regulations, the UK Parliament [36, 37] still faces several key questions:

What regulations could ensure that the potential benefits of AVs are achieved, such as contributing to net zero targets and enhancing road safety? What regulations could address concerns like safety, cyber-attacks, increased congestion, and negative environmental impacts? How will UK regulations align with international frameworks, like those being developed by the United Nations?

6.3.2 Unmanned Autonomous Aerial Vehicles

UAVs, sometimes known as "drones," come in a range of sizes, from tiny to massive planes. They can be fully autonomous or controlled by a remote pilot [10]. At first, the military mostly used drones for surveillance. These days, the public primarily uses them for photography, videography, and other uses [9. Drones with limited or partial automation are typically seen in the UK, where a pilot maintains control. The exception is military drone swarms, which are still managed from a central ground station despite having a greater potential for automation. Royal Navy, 2021. In the UK, business operators who have received special approval from the Civil Aviation Authority (CAA) are the primary users of drones, aside from recreational users. For instance, Amazon Prime Air received special authorization to develop and test drone deliveries, but the timeline for the technology's rollout remains uncertain.

Regulations: The 2024 UK drone regulations are founded on the principle that "UAVs operating in the UK must adhere to the same safety and operational standards as manned aircraft when performing similar operations in the same airspace" (CAP 722—Unmanned Aircraft System Operations in UK Airspace—Guidance).

Although this principle seems straightforward, most drones lack the inherent safety features of manned aircraft and are usually not operated by highly trained pilots. As drone usage becomes more widespread, the UK Civil Aviation Authority (CAA) faces the challenging task of developing regulations that ensure public safety and security while supporting the industry's need to scale operations significantly. The key regulations and developments from CAA include

- It is compulsory for all small drone operators between 20 and 250 kg to enrol themselves by taking the online competency test.
- It is forbidden to fly any drone above 400 ft or 1 km near of an airport [33], this was put into action after Gatwick incident in 2018.
- The application known as Drone assist was developed by the CAA and NATS (National apprentice training scheme) to help people understand safety regulation and rules.
- Regarding liability, the aircraft owner would be responsible for accidents involving fully autonomous drones, as any drone over 20 kg is required to have insurance.

The Air Navigation Order 2016 (as modified, under the Civil Aviation Act 1982) and the Regulations under UK Regulation (EU) 2018/1139 (the Basic regulation) are the two separate frameworks that govern UAS activities in the UK. The drone regulations in the UK are based on the drone's size and weight, closeness to other persons.

McLachlan et al. [27] shows that the drone regulations are structured around risk levels and categorized into three groups: Open, Specific, and Certified. The Open category is categorized for low-risk drone flights, for instance when you are flying a lightweight drone or within the countryside. The Specific category is designated for higher-risk flights, such as operating a heavier drone over urban areas. To fly within this category, obtaining an operating approval is required. The Certified category is for larger drones which must meet a specific safety certification like that of an aircraft.

To ensure consistent, effective and standardized communication, numerous additional regulatory bodies have specified distinct UAV requirements.

- The Institute of Electrical and Electronics Engineers (IEEE) created the Drones Working Group (DWG) in 2015 whose primary goal was to produce a hierarchy for consumer drones to prioritize privacy and security issues. To achieve this, the DWG created techniques evaluations, systems, testing and specifications for consumer drones to protect the privacy and security of the public and their properties. Similarly, the IEEE Conformity Assessment Program (ICAP) and OM/AerCom SC are working together to create a conformity assessment program for UAVs to demonstrate compliance with IEEE standards. The IEEE drone application certificate program, involving industry stakeholders, has also been introduced to support regulators, solution providers, system integrators, manufacturers, users, and other potential beneficiaries. This program includes the academic community and testing laboratories to show buyers that UAV systems or products meet IEEE standards. The European Telecommunications Standards Institute (ETSI) examines UAV operations, use cases, and the upcoming Internet Protocol (IP) suite framework, as well as the spectrum regulations necessary to integrate UAVs into existing LTE networks. ETSI has also created European standards at the request of EC regulations to facilitate the widespread use of services such as Data Link Services (DLS), Advanced Surface Movement Guidance and Control, and the Airport Collaborative Decision-Making procedure. Additionally, the Radio Equipment Directive has been established for UAVs to operate in line with spectrum management rules, and ETSI standards can be used to demonstrate compliance with specific aspects of this directive.

- In the United Kingdom, civil aircraft activities in a national space (NAS) are overseen by the Federal Aviation Authority (FAA). The FAA published an updated UTM concept in March 2020 to specify the operational and technical criteria for UTM ecosystem airspaces, with the goal of providing an operational and regulatory framework for unmanned traffic management (UTM) for data or information exchange. Additionally, the FAA has set a weight limit of fewer than 55 pounds for UAS. The FAA separates the current airspace at A to G. Class A-G airspace is controlled, whereas G airspace is unregulated. It indicates that class G will be supervised by (ATC) entities rather than air control. FAA regulations allow UTM operations up to 400 feet above the ground. Remote pilots are not required to obtain authorisation from authorities or approved training centres (ATCs) when operating in uncontrolled airspace. However, they must notify ATC of their intention when they enter regulated airspace. Additionally, the UAV operator should ensure that the anticipated UTM, operations comply with specific regulations (such as e-registration of the operator and UAV) and meet safety standards such drone airworthiness and communication capabilities.
- The European Aviation Safety Agency (EASA) has also published regulations for UAV operations. EASA's concepts are based on an operation-centric and proportional strategy, addressing operational conditions and methods rather than solely focusing on UAV parameters. These efforts are guided by the European roadmap for integrating civil remotely piloted aircraft into the European aviation system. The paper outlines various levels of integration and harmonization which includes three annexes with clear strategies to implement and enhance regulations. It also considers future research initiatives and the societal impacts of UAV services

Safety and privacy: One major concern is that drones can intentionally or unintentionally collect data on individuals without their knowledge or consent. Safety issues are categorized into risks to air users, ground risks, and the malicious use of drones. The Information Commissioner's Office (ICO) has declared that the Data Protection Act 2018 and GDPR, which also apply to drone use, protect personal information, including location and personal photos, for commercial drone operators [20]. According to UK drone law, most drones or model aircraft must be registered before being flown outdoors. This registration procedure entails registering for an operator ID if in charge of a drone or model aircraft, as well as completing a theoretical test to receive a flyer ID. This guarantees that drone owners can fly their drones safely and responsibly and have a firm grasp of safety requirements.

In addition to registration, drone operators must adhere to the Civil Aviation Authority (CAA) regulations for drone usage in the UK. These rules include flying drones within the line of sight, keeping a safe distance from people and property, and not flying drones above 400 feet or in restricted airspace. The amendments to the Air Navigation Order 2016 have introduced significant changes, including moving many drone regulations to the "Unmanned Aircraft Regulation." This regulation, initially from the EU (UK Regulation (EU) 2019/947), was retained after the UK's departure from the EU.

Cybersecurity: Because industry systems are becoming more interconnected, the aviation sector needs to be vigilant against direct and indirect cyberattacks, which are often the result of careless or reckless activity. The risk profile is constantly shifting because hackers—whether its humans, AI systems, or self-replicating viruses—are constantly searching for ways to take advantage of weaknesses and can swiftly create new methods to breach any system's cyber security. Because of this, the aviation sector must have strong security measures, and the CAA's cyber strategy must be regularly evaluated and updated to reflect the latest developments in cyber vulnerability and intent. To establish a cyber-resilient aviation system, UK aviation created a framework for cyber regulations and policies regarding UAVs that:

- Outlines the CAA's responsibilities for cyber security under current international, EU, and UK rules.
- Establishes process for ongoing evaluations of new cyber standards, assurance, and cyber management initiatives for aviation entities that fall under its purview.
- Recognises the gaps and how these standards, assurance, and cyber management activities relate to the legislative requirements outlined in strategic priority one.
- Identifies the main structural cyber vulnerabilities in the aviation system of the United Kingdom, namely in relation to interactions between various entities.
- Combines with the regulatory regime, which includes the Safety Management System (SMS), Performance Based Oversight (PBO), and the Regulatory Safety Management System (RSMS), and cultivates industry partnerships to guide our efforts. Additionally, it outlines necessary training and establishes a decision-making process focused on compliance and oversight activities.

Case Study (Amazon)

Homeowner 1 is located in Manhattan, Kansas, at 1814 Todd Road. Homeowner 2 is located at 1812 Todd Road, next door, and has a "no trespassing" sign up. Homeowner 1 has a Prime-air subscription and is an Amazon consumer. Homeowner 1 places an Amazon order for peanuts, which will be delivered to their house via Prime Air in 30 min. The operator looks at aerial and ground photos to plan the UAV route and spots trees that could block delivery. The pilot finds a way to Homeowner 1's house along Todd Road amid the woods. The UAV hovers over Homeowner 2's front porch.

Arriving to deliver to 1814 Todd Road is the Amazon UAV. However, the UAV delivers its delivery to homeowner 2's house at 1812 Todd Road because of an unforeseen GPS software error that causes the number 4 after the number 1 to be incorrectly perceived as a 2.

A relative of homeowner number two is seeking to move because of a severe peanut allergy. The EpiPen could have prevented anaphylactic shock, but this person forgot it. As fate would have it, no one was home when the UAV delivers its package. The relative, who is unfamiliar with this contemporary technology, goes outside to retrieve the delivery. Regretfully, there is peanut residue on the box of the Amazon product. The relative has a severe allergic reaction right away and died from airway blockage without their EpiPen.

Some of the Legal Issues Raised

- Did the Amazon UAV unintentionally land on the property, thereby breaching Kansas Statutes Annotated (KSA) 21-3721, committing criminal trespass, or engaging in the invasion of privacy?
- Was Homeowner 1 contributorily negligent in subscribing to Amazon Prime-air delivery, given that a reasonable person might anticipate the delivery landing at his neighbour's property, which had a posted no-trespassing sign?
- Did the incidents that transpired because that of the Amazon UAV pilot choosing to fly over homeowner 2's airspace constitute a breach of the COA or another certification?
- Did homeowner 2 commit criminal trespass as described by KSA 21-3721 when they flew through their house?
- When Amazon sells signage on its marketplace that clearly advise of KSA-3721, might it claim that it was not aware of the statute's requirements?
- Given that it was foreseeable that the UAV may carry peanuts—which are known to induce severe allergic responses in many—to the incorrect address, was Amazon held criminally and/or civilly liable for the death of homeowner 2's relative?
- Do the "Terms of Service" agreed to homeowner 1 and Amazon have any effect against legal or criminal claims brought by homeowner 2, the relative, or the survivors?
- Should the FAA use the National Transportation Safety Board (NTSB) to investigate if Amazon and the parties file a lawsuit to cancel some or all their COAs or permits for breaking laws or customs?

Some laws that may be implicated in Amazon UAV misadventure in delivering the parcel include:

Federal Aviation Administration (FAA) regulations: These govern the operation of UAVs and may impose penalties for unauthorized or unsafe operations.

State trespass laws: These could hold amazon liable for trespassing onto private property with his drone.

Negligence laws: If drone actions were deemed negligent and caused damage, who could be held liable for that damage.

Other parties that might be subject to claims include:

Amazon: If it was found that Amazon's systems or procedures contributed to the incident.

Homeowners: If they suffered damages due to the misconception of the drone.

The government's difficulty in determining who will be the legislators, regulators, and adjudicators of UAS legislation in the future is to balance excessive regulation with just enough laws to guarantee public safety. While too little monitoring and safety could be compromised by market competition and profit, too much oversight and safety could also have a detrimental effect by impeding the advancement of novel technology that could improve UAS safety in the future.

6.3.3 Unmanned Autonomous Surface Vehicles

According to business analysts, automation in the maritime sector represents the future of shipping, ports, and vessels, also known as unmanned surface vehicles (USVs) or maritime autonomous ship systems (MASS). USVs are defined as ships that can operate independently of human interaction to varying degrees. They have extremely nonlinear dynamics and carry out tasks in congested surroundings without human assistance. Maritime autonomy encompasses both completely autonomous and remotely operated shipping technology, both of which are advancing quickly. Future ships will be able to make critical operational decisions by effectively assessing their environment and health.

While technological advancements and market demand are essential for innovation, regulatory aspects become a significant consideration [30]. This is especially true for autonomous vessels, where key developments are advancing the field. Despite the rapid advancements in maritime science and technology, autonomous vessels must adhere to international regulations to operate safely across nations and even in seabed areas beyond national jurisdiction. Some regulatory aspects of manned vessels, such as certain clauses of the International Safety Management (ISM) Code, may apply to unmanned vessels. However, there is need for a particular international regulation that account for the unique characteristics of unmanned vessels.

Initial attempts to create software and algorithms enabling unmanned vessels to adhere to the COLREGs (Convention on the International Regulations for Preventing Collisions at Sea) started about 2000 and continued until 2010. Additional policy developments were sparked by the establishment of the ASTM (American society for testing and materials) Committee, which was tasked with developing technical standards for autonomous maritime vehicles and included a subcommittee on regulatory matters. Through their Maritime Advocacy Committee, the Association for Unmanned Systems International (AUVSI) tackled the problem in 2011, with a special emphasis on educating and interacting with the U.S. Navigation Safety Advisory Council (NAVSAC). The U.S. Coast Guard, the appropriate authority for U.S. waters, receives advice from this council. Following several meetings, this work was completed in late 2012 with a resolution that included suggestions for both legislative changes, such as amending specific COLREGs, and technological solutions, including the automated identification system (AIS).

An initiative to create voluntary best practices for unmanned vessels, known as marine autonomous systems (MAS), was started by marine UK, the UK industry association. The UK Maritime Autonomous Systems Regulatory Working Group (MASRWG) published the first edition of the UK Industry Code of Practice in 2017, with a primary focus on the technical aspects of MAS design and construction. The original guidance placed a strong emphasis on design and manufacture, even if it also addressed construction and operation. The MASRWG revised the Code of Practice to give USV operations more attention considering the notable expansion of autonomous systems. This included recommendations for platform registration and skills training.

Regulations: A specific industry principle for training and development is outlined in the UK Maritime Autonomous Systems (Surface) (MAS(S)) Code of Conduct: The industry will guarantee that all USV operational workers receive the proper amount of training and certification to comply with their Safe Operating Plan. The training and certification of MASS operational staff should be at least as high as that required for a similar manned vessel, and it should be in accordance with the same recognized Flag State Authority or national equivalent requirements. Staff development procedures will be in place at companies in the industry to track and advance skill development. It is crucial that the training be delivered by a regulated authorized organization or other qualified authority in accordance with the relevant, accepted standards.

COLREGs in Collision Avoidance: COLREGs are required to operate marine vessels. Developing avoidance techniques for USVs based on COLREGs is essential to ensuring maritime safety. For USV collision avoidance [6], the following guidelines are applied for different COLREG encounter scenarios:

- Rule 13 (overtaking) states that when the OS (own ship) approaches the target ship (TS) from a direction that is more than 22.5° behind its beam, it is considered to be overtaking. In this case, either the port or the starboard side of the TS will allow the OS to pass it.

- Rule 14 (head-on): states that both the OS and the TS (target ship) must change their courses to starboard when they cross paths that are reciprocal or almost reciprocal.
- Rule 15 (crossing): When two vessels meet one another between 15° 15° and 112.5° 112.5° (port and starboard), this is referred to as crossing. Keeping out of the way is required of the vessel with the other on its starboard side.

The UK Maritime and Coastguard Agency (MCA) released Marine Guidance Notice MGN 702 (M), which outlines specific regulations for Maritime Autonomous Surface Ships (MASS) that are less than 2.5 m in length overall. Published on August 5, 2024, the notice introduces a General Exemption that allows these small, unmanned vessels to operate without the need for certification traditionally required by the MCA or its delegated bodies. Although this regulation has some clauses which involves meeting all the specified conditions.

Safety: The UK MASS code of practice states that to guarantee maritime safety, human injury or death must be prevented, and environmental, property and marine environment damage must be avoided. The successful implementation of a Safety Management System (SMS) as part of the management and operation of MASS would be made possible by the implementation and maintenance of a safety management system that meets the following functional requirements Established channels of authority and communication between MASS and shore personnel; procedures for reporting accidents and infractions of the Code's regulations; guidelines and protocols to ensure the safe operation of MASS and environmental preservation in compliance with applicable international and Flag State laws; protocols for anticipating and handling emergencies; protocols for management reviews and internal audits; To capture mission data for subsequent study, a voyage data recorder can be needed.

The International Safety Management Code (ISM Code) may have a requirement for Document of Compliance and Safety Management Certificate. The health and safety of all employees on or near the USV are the responsibility of the MASS's owner or master. The Merchant Shipping and Fishing Vessels (Health and Safety at Work) Regulations 1997 (SI 1997 No. 2962) and other relevant regulations are applicable when hiring support crew. Under Regulations 3 and 4 of the International Safety Management Code (ISM Code), USVs are required to possess a certain degree of safety certification, which includes a Safety Management Certificate and a Document of Compliance.

Cybersecurity: All hardware and software in USVs must be designed and manufactured with cyber safety and security in mind. Throughout the design phase, this integration should be considered. Future patches and updates could have unanticipated negative consequences on the security integrity and functionality of the system. To the extent that it is practical and required, USVs should have cybersecurity protections in place to safeguard sensors and control systems. The command link, steering generation and transmission, waypoints, and the capacity to execute an Emergency Stop are just a few of the systems and the vessel's performance that should be examined for security flaws through a Cyber Security Analysis. According to the MOSS UK Code of Practice, the Cyber Security Analysis should be evaluated and implemented similarly to a Risk Assessment, determining the necessity of corrective actions to lower risk to a manageable level.

6.4 Opportunities and Challenges

6.4.1 The Gain Through Autonomous Vehicles

Autonomous vehicles (AVs) have the potential to bring substantial economic benefits to the UK, including attracting international investment and boosting the UK's reputation as a global leader in science. By 2035, the UK market for AVs could be valued at up to £42 billion, representing around 6% of the £650 billion global market, and creating up to 38,000 new jobs in the sector. In response to the government's 2021 call for evidence on the 'Future of connected and automated mobility in the UK', respondents highlighted that the emerging CAV opportunities could not only ensure the future stability of the UK's automotive, logistics, and mobility sectors but also open new markets for other emerging technologies. These include artificial intelligence, machine learning, cybersecurity, Mobility as a Service (MaaS), and innovative insurance products. Such advancements could align with broader government priorities, such as Levelling Up, achieving Net Zero, and the 'Global Britain' strategy outlined in the 2021 publication, 'Build Back Better: Our Plan for Growth'.

The global race to advance and implement autonomous vehicle (AV) technologies presents numerous challenges beyond just technological development. Building public trust, addressing regulatory hurdles, and integrating these services into current and future transport networks have raised questions for both companies and governments aiming to bring self-driving vehicles to market. Consequently, the industry and government have had to develop technology, policy, and regulations in concurrent, fostering transparent analysis of the obstacles to introducing self-driving vehicles, as outlined in Zenzic's 'UK CAM Roadmap to 2030' (2019). Similarly, rapid technological advancements in other transport modes also pose regulatory challenges. For example, on behalf of the Centre for Connected and Autonomous Vehicles (CCAV), the Law Commission is presently examining the regulatory framework for the safe deployment of automated vehicles in the UK. Also, there are recent complaints that AV obstruct emergency response, entrench transportation networks and block traffic.

6.4.2 Assistance Through Aerial Vehicle

UAVs can help with data collecting, debris navigation, and the requirement for prompt responses. Drone technology is increasingly being used by armatures to gather high quality imaging data of remote locations, like as mountain tops, coastlines, and islands. It helps rescue teams, evaluate damage, launch urgent discovery, and dispatch supplies, including medical kits and first aid. UAVs are regarded as essential in fields such as public security, rescue operations, and catastrophe risk management. UAVs are also used for monitoring of overhead power lines, space exploration, emergency and medical services, maritime communication and surveillance, real time monitoring of road traffic, infrastructure and construction inspection etc.

Due to past UAV-related incidents in UK airspace, ensuring that drones operate securely alongside human aircraft and beyond visual line of sight (BVLoS) systems is one of the largest problems. Urban air mobility (UAM) and flying taxis are being developed by various aircraft manufacturers [19]. Noise is one of the problems here. Martinez and Li [26] observed that the sound produced by smaller drones differs from that of normal planes. According to them, drone noise is viewed as less acceptable than that of civil aircraft. This is because drones emit a high-pitched noise while flying at lower altitudes (in places that aren't subjected to aeroplane noise). Ensuring the security of sensitive data from drones or UAVs, such as position, location, etc., is another crucial concern. Since UAVs lack encryption, they run the risk of being taken over. Using UAVs raises serious concerns about hacking and cyber liability. UAVs are susceptible to possible data leak concerns during military operations. Hackers may take full control of UAVs to steal information, violate privacy, or engage in any unlawful action, including smuggling.

Notwithstanding the widespread use of UAVs, regulatory agencies urgently need to develop standards for how they should operate in various nations. The ambiguity or absence of important standards and rules for UAV operations, permitted airspace, permitted weight and size, permitted height, privacy or secrecy considerations, and safety needs and features are major obstacles to the widespread deployment of UAVs. It is evident that government regulations pertaining to the use of UAVs are not very diverse. Commercial aircraft navigation may be impacted by UAVs. Therefore, nations should enact laws and guidelines to ensure that UAVs operate properly. For example, the FAA oversees issuing certifications and air traffic regulations for UAVs in the United States. Because different nations have distinct laws and standards, international cooperation or coordination can also help to establish global rules. In summary, there are several issues with UAV regulations, including inadequately detailed legal procedures and lengthy licensing processes that restrict the technology's broad application, accessibility, and adaptability. In this sense, artificial intelligence (AI) algorithms for standard compliance can be a practical solution that facilitates risk reduction, improved security of sensitive data, and prompt reaction to new standardisation demands. In this manner, we can observe AI-powered UAVs that are approved for domestic use in different nations, fulfilling the operational requirements of each distinct region.

6.4.3 *Limitations to Surface Vehicles*

The efficient operation of autonomous technology demands compliance with and introduction of a set of agreed standards for the safe operation of equipment, at both international domestic levels. "Regulatory challenges have been identified by the industry as obstacles to the growth of maritime autonomous systems. The expansion of maritime autonomy will have extensive regulatory implications, including those related to land-based infrastructure, data collection and usage, cybersecurity, and the future maritime workforce. Xu and Zhang [40] assessed the oversight issues posed by the application of USVs to maritime authorities [24, 25] and noted that legal regulations for USVs are hindering the growth rate of USV technology. The impact of automation on maritime employment has become a significant concern [18] (Pribyl and Weigel 2018) [2], further aggravating existing seafarer shortages, which has also been recognized as another challenge."

6.5 Conclusions and Future Directions

Current Unmanned Aerial Systems (UAS) regulations primarily focus on safe drone operation. "Compared to the regulation for autonomous vehicles, there has been minimal focus on integrating other UASs into airspace and establishing a liability framework for self-flying aerial vehicles." Regulations should include all unmanned serial systems.

Urgent Need for UAV Cybersecurity Regulations, the widespread adoption of Unmanned Aerial Vehicles (UAVs) in critical infrastructure demands immediate cybersecurity attention. Despite their increasing use, particularly in monitoring and operations, there is a striking absence of modern regulations and standards addressing UAV cybersecurity risks [12]. Implementing robust cybersecurity guidance and standards is paramount to safeguarding vulnerable systems. Data Governance for UAS Data, implementing manufacturer-wide policies and regulations for data management can streamline data governance for Unmanned Aircraft Systems (UAS) data, preventing issues associated with large-scale data collection.

The difference in international standards is one of the most notable obstacles to UAS regulation. The structure of airspace, technological infrastructure, and security concerns are only a few of the many variables that affect national rules. International organizations like ICAO struggle to strike a balance between the need for efficiency and international security and the sovereignty of individual nations over their airspace, land, and sea [31, 32]. The regulatory agency responsible should standardize regulations for every UAS crossing countries or airspace for security, safety, and privacy purposes. Restricting the risks emerging from UAS will prevent the industry from growing, its vital to create a framework to share liability for damages between UAS manufacturers and owners because attributing complete liability to the owner will dissuade consumers from buying, whereas assigning it fully to the manufacturer will subsequently reduce research and creation of UAVs.

For road authorities to fully comprehend the ramifications of self-driving cars, further research is required. National Highways will investigate the operational effects of autonomous vehicles on the SRN (strategic road network) soon. The effects of ALKS and other evolving types of self-driving technology on operational processes, like incident traffic management, will be covered. To accommodate the different use cases for self-driving cars, National Highways will also be investigating road layout, lane markings, roadworks design, digitization, and implementation standardization. External interference, like waves, wind, and current, always affects a USV and will lead it to go off course. Therefore, developing a collision avoidance strategy as well as regulation for a USV under external interference and reducing the impact of external interference will be vital to future study.

In summary, the success of autonomous technologies and systems in the UK depends on a robust regulatory framework that assures innovators and investors, while instilling public with confidence in the security, safety, and societal benefits of these technologies. Addressing the regulatory challenges posed by UASs requires a versatile approach that balances privacy, safety, and innovation. Legislators must continue to be flexible in revising existing laws to reflect new operational procedures and technological developments. Cooperation among industry stakeholders, regulatory bodies, and technology developers is essential for creating regulations that ensure the beneficial and safe use of UAS technology. As UASs continues to penetrate various sectors, constructing detailed, adaptable and evolving regulatory frameworks will be crucial for fully harnessing their potential while reducing associated risks.

References

1. 2020 Commission Work Programme – key documents (n.d.) European Commission. https://commission.europa.eu/publications/2020-commission-work-programme-key-documents_en#details
2. Autonomous Vessels: How an emerging disruptive technology is poised to impact the maritime industry much sooner than anticipated. (n.d.) vLex. https://law-journals-books.vlex.com/vid/autonomous-vessels-how-an-936457201
3. Aviation autonomy – Law Commission. (n.d.) https://lawcom.gov.uk/project/aviation-autonomy/
4. Barrado C, Boyero M, Brucculeri L, Ferrara G, Hately A, Hullah P, Martin-Marrero D, Pastor E, Rushton AP, Volkert A (2020) U-Space Concept of Operations: a key enabler for opening airspace to emerging Low-Altitude operations. Aerospace 7(3):24. https://doi.org/10.3390/aerospace7030024
5. Birdsall A (2022) New technologies and legal justification: the United Kingdom's use of drones in self-defence. Glob Const 11(2):197–216
6. Deng F, Jin L, Hou X, Wang L, Li B, Yang H (2021) COLREGs: compliant dynamic obstacle avoidance of USVs based on the dynamic navigation ship domain. J Mar Sci Eng 9(8):837. https://doi.org/10.3390/jmse9080837

7. Department for Transport (2019) https://assets.publishing.service.gov.uk/government/uploads/system/uploads/attachment_data/file/877630/technology-innovation-route-map-document.pdf
8. Department for Transport (2022) Rules on safe use of automated vehicles on GB roads. https://www.gov.uk/government/consultations/safe-use-rules-for-automated-vehicles-av/rules-on-safe-use-of-automated-vehicles-on-gb-roads
9. Departmental Overview 2015–16 Department for Transport – NAO overview. (2022, August 22). National Audit Office (NAO). https://www.nao.org.uk/overviews/departmental-overview-2015-16-department-for-transport/
10. Exploring urban drone integration. (n.d.) Nesta. https://www.nesta.org.uk/report/flying-high-challenge-future-of-drone-technology-in-uk-cities/exploring-urban-drone-integration/
11. Fagnant DJ, Kockelman K (2015) Preparing a nation for autonomous vehicles: opportunities, barriers and policy recommendations. Transp Res Part Pol Pract 77:167–181. https://doi.org/10.1016/j.tra.2015.04.003
12. Flight Critical. (2024) Cyber Security Cooperative Research Centre. https://cybersecuritycrc.org.au/flight-critical/index.html
13. Gesley J, Feikert-Ahalt C, The Law Library of Congress, Global Legal Research Directorate (2023) Regulation of advanced air mobility: European Union, United Kingdom. The Law Library of Congress, Global Legal Research Directorate. Washington, DC. Retrieved from the Library of Congress. https://www.loc.gov/item/2023555903/
14. HM Government (2022) Connected and automated mobility 2025: realizing the benefits of self-driving vehicles in the UK. (https://assets.publishing.service.gov.uk/government/uploads/system/uploads/attachment_data/file/1099173/cam-2025-realising-benefits-self-driving-vehicles.pdf
15. HM Government (2022) Road vehicles, functional safety (2019). https://www.iso.org/obp/ui/en/#iso:std:iso:26262:-3:ed-2:en
16. Implementing regulation – 2019/947 – EN – EUR-Lex (n.d.) https://eur-lex.europa.eu/eli/reg_impl/2019/947/oj/eng
17. Kalra N, Paddock SM (2016) Driving to safety: How many miles of driving would it take to demonstrate autonomous vehicle reliability? Transp Res Part Pol Pract 94:182–193. https://doi.org/10.1016/j.tra.2016.09.010
18. Kim M, Joung TH, Jeong B, Park HS (2020) Autonomous shipping and its impact on regulations, technologies, and industries. J Int Marit Saf Environ Affairs Shipp 4(2):17–25. https://doi.org/10.1080/25725084.2020.1779427
19. Kinjo H, Mitsui & Co. Global Strategic Studies Institute (2018) Development trends and prospects for evtol: a new mode of air mobility. In Mitsui & Co. Global Strategic Studies Institute Monthly Report. https://www.mitsui.com/mgssci/en/report/detail/__icsFiles/afieldfile/2019/07/18/1906m_kinjo_e.pdf
20. Kumar A (2021) Commercial drone use in the context of governance, ethics, and privacy: a techno ethical review. ACADEMICIA: Int Multidiscipl Res J 11(11):171–177
21. Lee D, Hess DJ, Heldeweg MA (2022) Safety and privacy regulations for unmanned aerial vehicles: A multiple comparative analysis. Technol Soc 71:102079. https://doi.org/10.1016/j.techsoc.2022.102079
22. Lisinska, J. (2021). Autonomous vehicles: automotive, maritime and aerial A policy landscape review. https://doi.org/10.18742/pub01-064
23. Lisinska J (2021) Autonomous vehicles on public roads in maritime and aerial—a policy landscape review. https://doi.org/10.18742/pub01-06
24. Maritime UK (2018) Maritime autonomous surface ships—UK code of practice. https://www.maritimeuk.org/media-centre/publications/maritime-autonomous-surface-ships-uk-code-practice/
25. Maritime UK (2020) Maritime Autonomous Ship Systems (MASS) UK industry conduct principles and code of practice. Version 4
26. Martinez AT, Li Z (2020) Metrics for assessing the perception of drone noise. HAL (Le Centre Pour La Communication Scientifique Directe). https://doi.org/10.48465/fa.2020.0018

27. McLachlan S, Dube K, Schafer B, Gillespie A, Fenton N (2022) The chaotic state of UK drone regulation. arXiv:2205.01041
28. Ministry of Defence. (2017, September 28) Drones: how to fly them safely and legally. GOV.UK. https://www.gov.uk/government/news/drones-are-you-flying-yours-safely-and-legally
29. Nichols RK, Mumm HC, Lonstein WD, Ryan JJCH, Carter C, Hood J-P (2019) Unmanned aircraft systems in the cyber domain. NPP eBooks 27. https://newprairiepress.org/ebooks/27
30. Ringbom H (2019) Regulating autonomous ships—concepts, challenges and precedents. Ocean Dev Int Law 50(2–3):141–169. https://doi.org/10.1080/00908320.2019.1582593
31. Shrestha R, Zevenbergen J, Panday US, Awasthi B, Karki S (2019) Revisiting the current uav regulations in nepal: a step towards legal dimension for uavs efficient application. the International Archives of the Photogrammetry, Remote Sensing and Spatial Information Sciences/ International Archives of the Photogrammetry, Remote Sensing and Spatial Information Sciences XLII-5/W3:107–114. https://doi.org/10.5194/isprs-archives-xlii-5-w3-107-2019
32. Stöcker C, Bennett R, Nex F, Gerke M, Zevenbergen J (2017) Review of the current state of UAV Regulations. Remote Sensing 9(5):459. https://doi.org/10.3390/rs9050459
33. The Drone and Model Aircraft Code | UK Civil Aviation Authority (n.d.) https://register-drones.caa.co.uk/drone-code
34. UAS Regulation (UK regulation) Published by the Civil Aviation Authority (2024). https://regulatorylibrary.caa.co.uk/2019-947-pdf/PDF.pdf
35. UK Government (2024) Self-driving vehicles set to be on roads by 2026 as Automated Vehicles Act becomes law. GOV.UK
36. UK Parliament (2023) Self driving transport committee. https://publications.parliament.uk/pa/cm5803/cmselect/cmtrans/519/report.html
37. UK Parliament (2024) Automated vehicles are in trial and development stages across the UK. What are the opportunities and challenges of them being on UK roads? https://post.parliament.uk/autonomous-transport/#heading-2
38. Wadud Z, MacKenzie D, Leiby P (2016) Help or hindrance? The travel, energy and carbon impacts of highly automated vehicles. Transp Res Part Pol Pract 86:1–18. https://doi.org/10.1016/j.tra.2015.12.001
39. West JP, Bowman JS (2016) The Domestic Use of Drones: An Ethical Analysis of Surveillance issues. Pub Admin Rev 76(4):649–659. https://doi.org/10.1111/puar.12506
40. Xu Chunsong, Zhang Renping (2017) Thoughts on the development of unmanned ships and the revision of maritime conventions. World shipping 40(03):20–22. https://doi.org/10.16176/j.cnki.21-1284.2017.03.004
41. Zhuang J, Luo J, Liu Y, Bucknall R, Sun H, Huang C (2019) Collision avoidance for unmanned surface vehicles based on COLREGS. In: 2019 5th international conference on transportation information and safety (ICTIS), Liverpool, UK, pp 1418–1425. https://doi.org/10.1109/ICTIS.2019.8883829

Chapter 7
How Can Existing Cyber Frameworks Be Better Implemented to Ensure Operational Resilience Within the UK Aviation Industry in the Event of a Cyber-Attack Against Its Mission Critical, Satellite-Based Services?

Rob Patton and Hamid Jahankhani

Abstract The Global Navigation Satellite Systems (GNSS) is used for a large number of essential services worldwide, including the critical communications, navigation and surveillance functions required by commercial and military aircraft to conduct normal operations. A vast increase in GNSS jamming and spoofing attacks, concentrated in and around conflict zones, has forced a review of the UK aviation industry's resilience to this threat due to the indirect flight safety impact on commercial air traffic operating in affected regions. Numerous cyber security frameworks exist that attempt to offer mitigating controls and promote resilience to prevailing cyber threats, but very few are specific or comprehensive enough to cover the unique nature and complexity of aircraft systems and services. International and domestic aviation bodies offer guidance and discourse about understanding the risk and the need to work collaboratively to combat its effect. A review of existing UK aviation cyber security policy has been conducted within this research, looking specifically at the UK Civil Aviation Authority's (CAA) Cyber Security Oversight Strategy to identify its strength in threat identification, but also its weakness in mandating mitigating controls, or offering an adequately detailed self-assessment and audit process to bring GNSS system vulnerabilities into scope. The proposed revisions, validated against two separate spoofing case studies, include a requirement to specifically assess GNSS resilience, the incorporation of jamming and spoofing response into pilot simulator training, the improvement of threat intelligence and production of operational notifications, the appraisal of conventional analogue technologies for system redundancy and the enforcement of an audit against the National Cyber Security Centre's Cyber Assessment Framework (CAF), repeated every 2 years. An aviation CAF overlay, offering specific guidance on successful completion, should also be produced.

R. Patton · H. Jahankhani (✉)
Northumbria University London, London, UK
e-mail: Hamid.jahankhani@northumbria.ac.uk

U. B. Chaudhry and H. Jahankhani, *Autonomous Revolution: Strategies, Threats and Challenges*, Advanced Sciences and Technologies for Security Applications,
https://doi.org/10.1007/978-3-031-96782-5_7

Keywords Jamming and spoofing attack · National Cyber Security Centre's Cyber Assessment Framework (CAF) · Civil Aviation Authority's (CAA) · The Global Navigation Satellite Systems (GNSS) · Cybersecurity · Satellite security · LEO · MEO

7.1 Introduction

The aviation industry is heavily reliant upon international satellite infrastructure, namely the GNSS, to communicate, navigate and monitor global flight operations.

There has been a significant rise in the number of reported instances of GNSS jamming and spoofing, which disrupt, deny or falsify data signals, resulting in reduced system reliability and the degradation of flight safety [86].

The Position, Navigation and Timing (PNT) signals transmitted from the GNSS are critical to flight operations, but also to wider critical national infrastructure (CNI), and their integrity must be assured [41]. In conjunction with secure system design and defence should be policy-based resilience, however, global aviation cyber security policy and guidance is in its infancy and inadequate to protect the complex aerospace systems used today.

Existing policy needs to be reviewed and revised so that the threat to GNSS services can be adequately acknowledged, understood and addressed. This requires prescriptive guidance and the inclusion of relevant system hardware and software within the scope of cyber audits incorporated within the UK aviation regulator's cyber security oversight strategy. The current approach of generic cyber security advice and a call-to-arms needs to be reconfigured to be fit for purpose, keep pace with the changing threat landscape and to ensure sufficient and appropriate resilience within the UK aviation industry.

7.2 Literature Review

This chapter will provide the context of the research by initially introducing the space environment and the satellites and space-based assets that operate within it. This important background information will inform the utilisation of many of the satellite services they provide and convey the considerable dependency that many industries place on them, paying particular attention to the airline industry. Several threats faced by the industry to the correct and reliable functioning of these systems are identified, leading to the critical discussion of satellite security and the cyber security frameworks, risk management processes and remediation guidelines that are currently being used, so that specific areas of improvement may be identified.

7.2.1 Satellite Infrastructure and GNSS

As of 28th May 2024, the active satellite tracking website, "Orbiting Now" reports that there are more than 10,000 satellites currently in orbit, offering a range of services that include positioning, navigation, communications, Earth observation and space science, moving in various trajectories around the Earth [87].

Satellites vary in size, mass and capability, dependent upon their payload, power requirements and the size of their antenna and solar array, all balanced against their production and launch budget [76]. For example, ViaSat-1, a broadband internet communications satellite, had a launch mass of close to 7000 kg [73], yet technological advancements and more efficient communications has allowed for the design of CubeSats and other nanosatellites that can measure 10 cubic centimetres and weigh around 1 kg [74, 112], saving weight and therefore cost, or saving space aboard launch vehicles, allowing multiple satellites to be launched at the same time.

Around 80% of operational satellites are in Low Earth Orbit (LEO) at an altitude less than 2000 km above the Earth's surface, orbiting once every 90–120 min, whilst a Geostationary Earth Orbit (GEO), at 35,786 km, fixed above a particular point on Earth [87] provides the second most popular orbital regime.

As their name suggests, LEO satellites orbit much closer to the Earth. Their relative proximity naturally offers increased signal strength and lower latency making them ideal for real-time communications, whilst also increasing their relative velocity and rate of geometric change, which improves the accuracy of position determination due to the reduction in convergence time, a measure of the time taken to calculate an accurate and stable solution [39]. Satellites in a higher orbit benefit from a larger field of view, enabling navigation, communications and imagery using fewer satellites, whilst leveraging their longer dwell time over a particular location to maximise image detail and fidelity, contributing to increased effectiveness of aerospace and maritime navigation, strategic reconnaissance and humanitarian assistance [61].

Major satellite operators include government agencies such as the National Aeronautics and Space Administration (NASA), the European Space Agency (ESA) and Roscosmos, the Russian Federal Space Agency, alongside a growing number of private companies including Inmarsat, Iridium and SpaceX.

Beyond strategic communications and observation, military use of satellites incorporates the early detection and tracking of missile threats, with space-based assets providing wider coverage of infra-red heat signatures than conventional ground-based detection [16]. Meanwhile, private sector utilisation includes extensive imagery to monitor and perform trend analysis on weather patterns, tidal flows, deforestation and atmospheric composition [1, 119], as well as being used for mapping as in the case of Google Earth [72], whilst around 50–60% of space-based platforms are utilised for television, telephone, internet and radio communications [63, 106].

It is estimated that there are hundreds of thousands of fragments of space debris currently in orbit, resulting from fuel and battery explosions or satellite collisions, which create a significant hazard to the operation of other platforms [26]. Satellite imagery is currently used to track debris and communicate potential collision risks, whilst missions such as RemoveDEBRIS from the University of Surrey are currently trialling space-based methods to remove it [33]. Ironically, the existence of multiple platforms in space further contributes to the issue of space debris.

Satellite communications networks are configured with a space, ground and user segment, where the space segment represents the space-based communications platform(s) or constellation, the ground segment comprises the signal processing infrastructure and orbit update and management systems, and the user is any end device that receives and utilises the transmitted satellite signal [15].

Legacy satellites provided a low-power, point-to-point, line-of-sight, microwave communications relay between ground-based stations that required large and costly antennas due to limited satellite performance, in what is known as a 'bent-pipe system'. Advancements in communications technology and the introduction of optical communications capable of higher data rates and bandwidth, enabled multi-point data transmissions. Together with increased power and efficiency in launch vehicles that allowed for an increased payload and improvements in power supply technology extending satellite lifetime, overall development cost reduced whilst ground stations and their associated antennas shrunk in size [66, 103].

Modern satellites also employ Inter Satellite Link (ISL) allowing satellite-to-satellite communication that increases platform autonomy whilst reducing dependence on ground infrastructure [42]. This maintains perpetual satellite visibility and integrity for critical systems including PNT services, provided extensively by the GNSS [71] (Fig. 7.1).

GNSS satellites are mostly found in medium Earth orbit (MEO) due to its stability and global coverage [82], with the most ubiquitous and familiar of the GNSS constellations being the Global Positioning System (GPS), first launched experimentally in the late 1970s [91]. It was originally managed and maintained by the United States Air Force, however control has now been passed to the United States Space Force for satellite launch, oversight, orbital control and maintenance [99].

The GPS constellation consists of 31 operational satellites, allowing for redundancy for the nominal 24 satellites that the United States government has committed

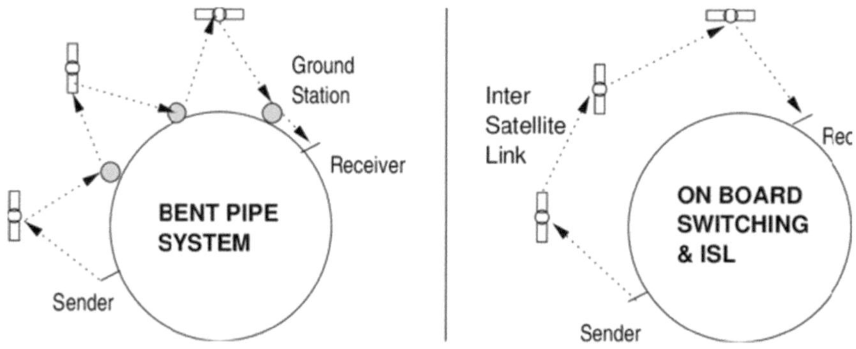

Fig. 7.1 Traditional bent pipe system infrastructure versus inter satellite link [103]

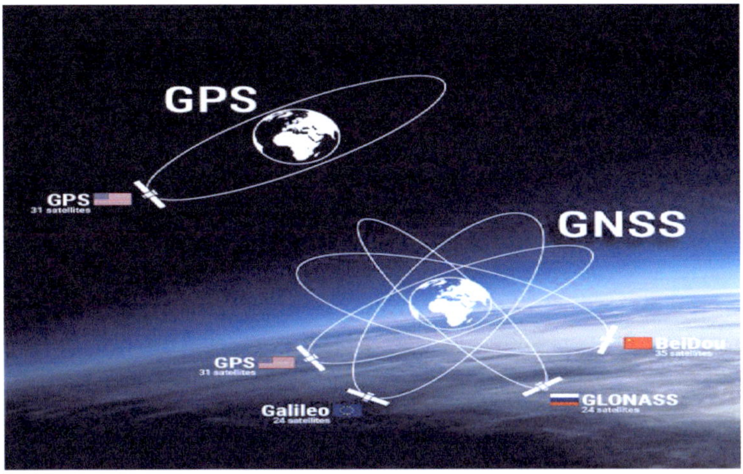

Fig. 7.2 GPS versus GNSS visualisation [103]

to maintaining. They follow 6 orbital planes at approximately 20,200 km above the Earth. This arrangement ensures visibility of at least 4 satellites at any time, enabling precise timing and a three-dimensional position, accurate to a mandated performance standard of better than 2 m, subject to atmospheric conditions, interference from buildings, and user device constraints [75].

Whilst GPS is used extensively, it is just one of four satellite constellations, at present, that make up the GNSS. The European Union, Russia and China also have their own interoperable constellations, approved by the International Civil Aviation Authority (ICAO), that utilise similar infrastructure [52] (Fig. 7.2).

The Galileo constellation, made up of 30 MEO satellites, including 3 spares, each with a 700 kg launch mass [27], was developed by the European Union and ESA in 1998, providing complimentary services to GPS whilst also providing an enhanced service to safety–critical users that requires signal authentication alongside integrity monitoring and user notification of system degradation [60]. Galileo has been described as the world's most precise satellite navigation system meeting the strict ICAO navigation specifications, and is progressively being incorporated into aviation operations [28].

GLONASS is the Russian counterpart to GPS, which was deployed in 1982 and consists of a nominal 24 operational satellites, orbiting at 19,100 km above the Earth [92]. The constellation has been integrated within the civil aviation sphere since 1996 [48], whilst the Chinese BeiDou System (BDS) is the most recent addition to the GNSS, having become fully operational in 2020, offering a service independent of the US GPS [6].

Multi-GNSS constellations have been proven to increase the accuracy of position and time [67], consequently it would follow that GNSS systems would collectively offer increased availability, reliability and accuracy, coupled with wider coverage, allowing for continuous operations and improved performance. However Mi et al. [70] argues that inter-system biases exist that need to be controlled and rectified before full interoperability can be achieved [70].

PNT data transmitted by these constellations, alongside communications and surveillance architecture, provides input to the component constellations of GNSS to support space, aviation, maritime and military operations, as well as to provide services to the automotive, logistics, utilities and finance industries, amongst others [64]. Timing is an important feature of these services as without an accurate time reference, many services would fail to function correctly. Financial transactions need to be timestamped, whilst the utilities and telecommunications sectors require precise time synchronisation to ensure efficiency, reliability and stability within power grids and mobile phone networks respectively. GNSS satellites have atomic clocks that keep UTC time, which can also be used by ground stations to calculate an accurate position fix, having deduced the latitude, longitude and altitude, knowing that radio waves transmitted from satellites travel at the speed of light. Position fixes are crucial for rapid resource deployment and coordination of emergency response and law enforcement as well as real-time traffic management, navigation and collision avoidance for road, rail, maritime and aviation operations [41].

These sectors are interconnected and support a multitude of essential services that are crucial for economic stability and public safety, with the disruption or denial of these services potentially threatening national security and defence. Most major political players, including the United Kingdom, the United States, the European Union, Russia and China have identified communications, transportation and space as integral pillars of their critical infrastructure and have implemented legislation and controls to combat external threats and increase sector resilience.

7.2.2 Aviation Applications of GNSS

As aircraft manufacturers strive to achieve the best in flight safety, fuel and operational efficiency, airframe reliability and passenger comfort, digitalisation has become pervasive throughout aircraft system design, from fly-by-wire flight controls with no physical connection, only electronic interfaces and control loops, to digital navigation and communications infrastructure [68, 110].

Many of these onboard systems rely on GNSS to transmit and receive critical data, with arguably the most important being those which support the ability to navigate accurately and efficiently for many thousands of miles, for significant periods of time, over remote and inhospitable terrain. Legacy ground-based analogue communication, navigation and surveillance (CNS) infrastructure requires line-of-sight and cannot maintain the same level of accuracy, capacity and reliability that more modern space-based, digital systems provide [25].

As air traffic increases and rising operational costs demand increased efficiency in fuel, emissions, route and traffic management, airspace design and navigation procedures have developed such that routes and trajectories have become optimised and separation between aircraft has reduced, allowing for a significant increase in aircraft capacity on airways [54]. This necessitates a far higher accuracy in PNT than has previously been provided by conventional navigation aids, particularly in polar and oceanic regions. Performance-Based Navigation (PBN) is a navigation concept that prescribes a required level of performance from aircraft, aircrew and systems, mandating requirements for accuracy, integrity, functionality and continuity. With the implementation of PBN, and the increasing accuracy of GNSS, lateral separation can be substantially reduced. The availability of certain airways and airspace is dictated by an aircraft's ability to maintain a required navigation performance (RNP), with the specifications mandated within the relevant authority's airspace management documentation and can vary dependent upon location, the time of day, aircraft altitude or the expected level of air traffic.

Failure of onboard systems or errors in data processing would require priority flight monitoring, increased separation minima, a reroute or reallocation of flight level, all in the interest of flight safety. The same would occur should an issue with GNSS transmissions become evident, with the required accuracy no longer being achieved. It is important, therefore, that anomalous signals are quickly detected and reported to air traffic control (ATC). This is currently enabled through onboard automatic performance monitoring and the display of the Estimated Position Uncertainty (EPU) along with the number of visible satellites [53].

Rectification of these errors can be achieved through satellite signal augmentation, a process that utilises PNT data adjustments calculated by ground reference stations and relayed by GEO or LEO satellites, to correct for ephemeris or orbital errors, atomic clock inaccuracies or atmospheric signal delays, thus enhancing the accuracy and availability of GNSS signals and assuring compliance with RNP specifications, such is the process employed by satellite-based augmentation systems (SBAS), such as the Wide Area Augmentation System (WAAS) in the United States, or the European Geostationary Navigation Overlay Service (EGNOS) [120].

Augmentation systems can also be aircraft-based (ABAS), such as Receiver Autonomous Integrity Monitoring (RAIM) that provides fault detection and exclusion (FDE) whereby GNSS signals that exceed given tolerances can be identified and eliminated from the position calculation, allowing for uninterrupted navigation. Finally, Ground-based Augmentation Systems (GBAS) operate in the terminal area, either on or in the vicinity of an airport, where it is used to augment GNSS PNT data to calculate vertical and lateral deviations for precision approaches [54] (Fig. 7.3).

The increased navigation capability brought about by GNSS and signal augmentation, coupled with the benefit of no longer being restricted to using conventional ground-based navigational aids, permits an aircraft to fly optimised routes and curved tracks, taking fuel efficiency, time, environmental impact, terrain avoidance and noise into account, ultimately equating to considerable cost savings for an aircraft operator and opening up potential routes and airports that were previously inaccessible [60], with imminent SBAS approach ubiquity helped by the European Union mandating their use from 2030 [3].

From a communications perspective, SATCOM (Satellite Communications) is a vital aviation technology permitting long-range communication that enables and

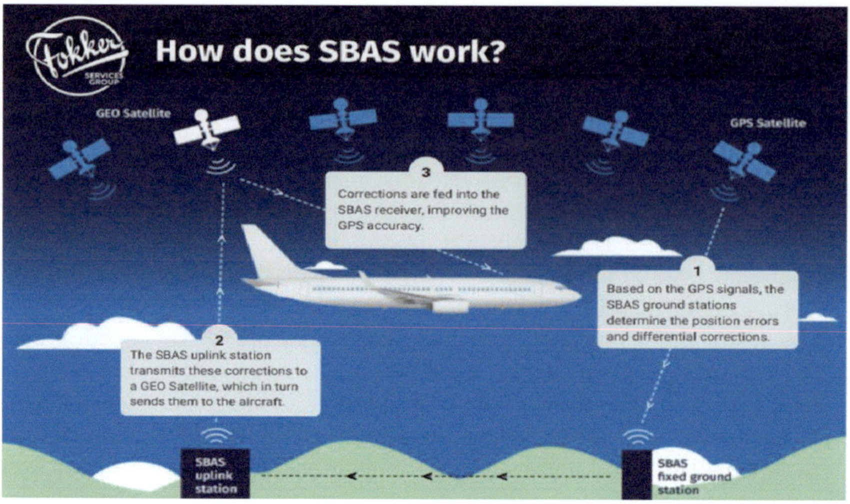

Fig. 7.3 Satellite based augmentation system

supports Air Traffic Management (ATM) and efficient communication with ATC and Airline Operations Centres (AOC) beyond the range of conventional radar and Very High Frequency (VHF) radio, typically in oceanic and polar regions [5, 56]. SATCOM, and the underlying GNSS infrastructure that enables it as part of the Aeronautical Telecommunication Network (ATN), provides a high speed, air-to-air and air-to-ground satellite relay for voice and data transmissions, including onboard Wi-Fi as part of the passenger flight experience [97].

Another component of the ATN is the Aircraft Communication Addressing and Reporting System (ACARS) that is a datalink and SATCOM messaging service used for acquiring departure and oceanic clearances from ATC, aircraft engine and system remote monitoring through the Aircraft Condition Monitoring System (ACMS) and for administrative communications [65, 94]. Two-way text communication between an aircraft and ATC is enabled by Controller Pilot Data Link Communications (CPDLC), sending non-urgent messages and instructions in written form to avoid misinterpretation and congestion of VHF/HF frequencies [2].

GNSS and the ATN is also an enabler of aircraft surveillance services, most notably Automatic Dependent Surveillance-Broadcast (ADS-B), an automated datalink broadcast of an aircraft's or aerodrome vehicle's information such as identification and position [47], that can be used as input for other information systems to generate aircraft symbology on an aircraft navigation display for situational awareness and collision avoidance, as well as providing precise data to ATC for efficient airspace management and aircraft separation [29] (Fig. 7.4).

This subsection has begun to highlight the level of GNSS dependency within aviation yet does not mention the numerous subsystems that also rely on accurate GNSS transmissions. The failure of any one of these may have severe consequences.

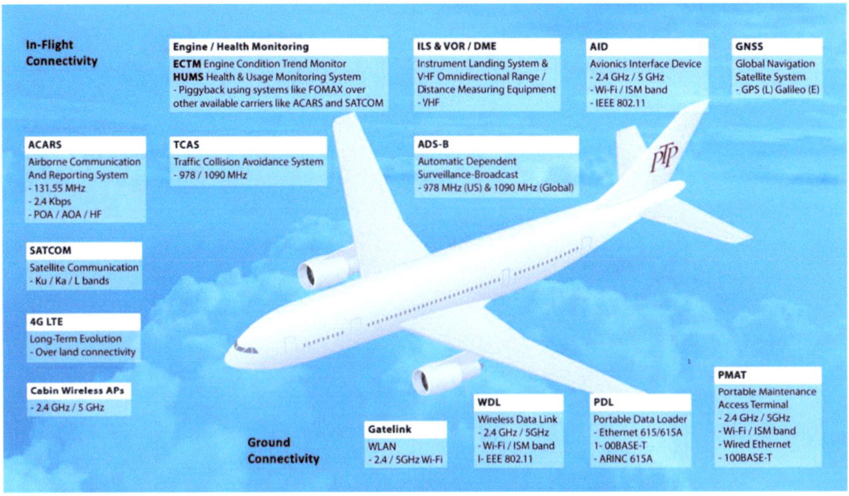

Fig. 7.4 Examples of in-flight connectivity and GNSS-dependent systems [88]

7.2.3 Current Threat to GNSS

The ubiquity of GPS-enabled applications and the increasing reliance on GNSS for critical aviation functions, amongst the many other dependencies discussed in Sect. 7.2.1, creates an ever-increasing attack surface that exposes society and industry to threats and vulnerabilities that must be addressed to ensure reliable and safe operations, particularly as their attractiveness to malicious actors is increasing [59].

Due to the operational distance of the space-based transmitter from the aircraft or ground-based antenna, GNSS signals are weak and are travelling through areas with crowded frequencies. This weakness and congestion make GNSS signals susceptible to RFI, or jamming, which can be unintentional and a result of the environment or a nearby disruptive electromagnetic signal, or intentional and caused by deliberate or malicious acts.

Satellites operate in a hostile environment where they need to overcome challenges posed by high levels of solar radiation, extremes of temperature and the risk of collision with space debris, all whilst maintaining a speed of around 7000 miles per hour to fight against the Earth's gravitational pull and hold its orbit [83]. To that end, unintentional signal disruptions and denial should be expected, and even catastrophic failure of the asset should be planned for.

GNSS jamming blinds a receiver, forcing temporary or permanent degradation of the signal, affecting satellite signal lock, position accuracy and system reliability and can affect an entire area or be manipulated to selectively target specific GNSS frequencies [117].

Spoofing is another common threat to GNSS and can only be intentional as it involves the transmission or injection of false signals or data to force end-user systems

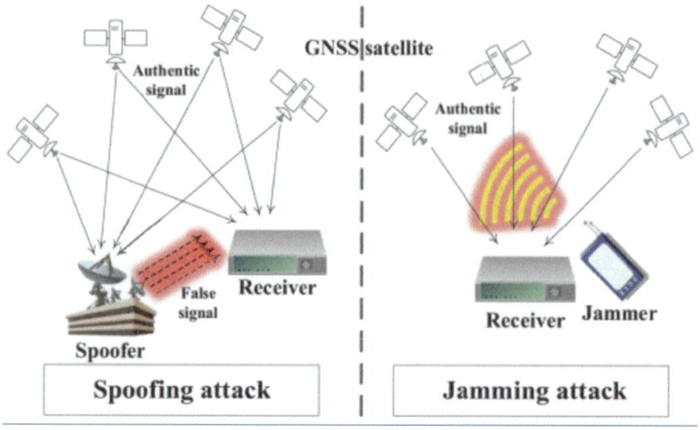

Fig. 7.5 GNSS spoofing and jamming [118]

to believe they are positioned somewhere other than their true position, without disrupting the GNSS operation, in contrast to jamming [21]. The erroneous PNT information calculated during a spoofing attack could potentially lead to dangerous and often unnoticed navigation errors that could jeopardise flight safety [121], particularly within heavily congested airspace or during critical phases of flight such as departure and landing (Fig. 7.5).

Jamming is far simpler to execute than spoofing as it requires less technical knowledge or understanding of satellite engineering and protocols. It simply requires the introduction of an obstructive signal transmitted at the same frequency, which decreases the signal-to-noise-ratio (SNR), the ratio of signal power to background noise, taking the SNR below the threshold required for adequate functionality and performance [57]. Spoofing can be conducted using a software-defined radio (SDR), a communication device that implements traditional radio hardware, signal processing, amplification, and modulation all through computer software, introducing flexibility of reconfiguration, multiple frequency and protocol management, and ease of system upgrade to its operation [38].

What previously cost tens of thousands of pounds to execute and required relatively powerful computing infrastructure can now be accomplished at a fraction of the cost on little more than a laptop with a spoofing application installed. This increased availability and ease of concealment is complicating the problem and exacerbating the threat to GNSS [37, 62].

Low-cost jamming devices are also problematic, including simple homemade devices or personal jammers purchased and used by many employees to hide their location from their employers who track their location whilst in company vehicles. Whilst effective in their aim, these cheap, yet illegal devices, can also block other GNSS signals, exemplified in a 2013 incident at Newark Liberty Airport, in which ATC flight efficiency operations were accidentally disrupted [14, 21].

Technical weaknesses further amplify vulnerabilities, making exploitation easier for an attacker. A study conducted by the Ruhr University Bochum discovered numerous basic security flaws within satellite infrastructure including a lack of encryption, effective access controls and software coding vulnerabilities [115]. These have all been referenced within the Space Attack Research and Tactic Analysis (SPARTA) exploitation matrix from the Aerospace Corporation, which has identified numerous cyber threats and avenues of attack, akin to the more familiar MITRE ATT&CK framework for enterprise systems, yet specific to assets within the space domain. It covers sub-techniques including spacecraft communications reconnaissance, launch operations compromise and the threat of lateral movement through constellation hopping, but with SPARTA v2.0 in April 2023, additional controls for jamming and spoofing under the execution technique were added, highlighting the pre-eminence of the threat to GNSS [101].

A further threat acknowledged by SPARTA is that of the comprise of ground stations, considered by Varadharajan and Suri [108] to be the easiest means of targeting space systems due to the exposure to orbital control and tracking infrastructure via traditional IT systems and protocols, which are equally vulnerable to malware and exploitation through social engineering if security misconfigurations are not remedied and the correct controls are not in place [108].

Successful exploitation of a ground station and direct or indirect access to a GNSS satellite could allow a malicious actor to modify data to alter its orbit, potentially commanding a trajectory into a debris field, a collision with another satellite or space vehicle, or forcing a deorbit and burn up of the satellite on re-entry into the Earth's atmosphere, whilst potentially locking out legitimate users to prevent recovery. A degraded signal could significantly affect PNT data without being obvious to a user, whereas denial of GNSS signals and the subsequent loss of satellite lock would immediately affect system status and reliability [7], necessitating a reconfiguration or analogue fallback, if available [83]. If the effects were prolonged, the widespread disruption could have significant consequences including the collapse of the financial system and breakdown of power grids, which has the potential to endanger life [107].

Aircraft subjected to jamming or spoofing attacks would lose the ability to efficiently navigate within required performance specifications and pilots would lose critical systems for situational awareness, communication and surveillance, creating a sustained period of disorientation, exacerbated by the numerous visual and aural warnings and alerts triggered in the cockpit. Unfavourable weather conditions, geography and visibility, could worsen the impact and lead to the unintentional exceedance of lateral or vertical limits, the encroachment into restricted areas, or even controlled flight into terrain (CFIT), however, system reconfiguration to navigational aids independent of satellite technology, and mode reversion by the Flight Management System (FMS) should be prompt [34]. Mitigation through signal augmentation such as RAIM, described in Sect. 7.2.2, would be ineffective against widespread jamming as there would be no single bad signal to exclude, and the effect is minimal against modern spoofing techniques [31].

There are also the financial implications of an attack to consider, which could include compensation for service interruptions and the upgrade or repair of a satellite, potentially costing millions of pounds [58].

7.2.4 Satellite Security and Cyber Frameworks

It is essential that robust security measures are implemented to mitigate existing and future threats to GNSS and flight operations. Research into modern technical mitigations include anti-spoofing algorithms for signal processing, detection and optimisation that monitor abnormal changes in PNT data and note changes in the direction of arrival of a transmission, whilst cryptographic signal authentication includes digital signatures within navigation transmissions, combating the notorious lack of encryption in satellite communications [69, 118]. Further integration and formal approval of other GNSS constellations into normal flight operations will increase availability and integrity through automatic failover to another system, possibly combined with increased satellite networking through ISL and adjustments to orbits to enable a concentration of satellite signal redundancy over a particular geographical region [118].

Of particular note for aviation dependencies is the implementation of adaptive frequency filtering. This technique utilises Kalman filtering that blends Inertial Navigation System (INS) and GNSS data that aims to provide an accurate and updated position even under RFI conditions, an approach being trialled within advanced RAIM (ARAIM) [36]. Kalman filtering highlights the importance of the deployment of redundant PNT systems and analogue fallback.

Retrofitting space-based platforms already in orbit is extremely difficult to achieve considering their environment and velocity but it is also prohibitively expensive, and system downtime can significantly impact operations. Instead, an approach akin to the 'Secure by Design' principle promoted by and embedded within UK government departments should be adopted, in which risk-driven controls and assurance should be designed and built into new technologies from the ground up, with security in mind from the outset, ensuring defence in depth [13]. A defence in depth approach is fundamental to an effective and comprehensive security program that is policy-led and greater than the sum of its parts.

Whilst CNS resilience is contingent on the security of GNSS satellites and the effectiveness of controls against jamming and spoofing, this is the remit of satellite operators and does not directly establish resilience within the UK airline industry, on which this study focusses. Until platform security can be guaranteed, international and national cyber security policy, industry regulations and internal airline guidance must be concentrated on preserving business and operational continuity in the event of a cyber-attack.

Many popular, globally accepted and agnostic cyber security frameworks exist, however, attempts to prescribe controls that are relevant and appropriate to the specialised systems used within airline operations have largely been overlooked. The vast majority of research into the area of GNSS resilience has focussed on satellite-based security, RFI detection and mitigation, whilst regulator guidance has been generic, non-prescriptive and rarely offers more than promotion of cyber awareness within the airline community, effectively putting the onus solely on the service-user to elect and implement protections.

Nevertheless, commonly used frameworks include those from the National Institute of Standards and Technology (NIST) and the International Organization for Standardization (ISO) that have many complimentary security controls.

ISO27001 and its associated controls suite outlined in ISO27002 guide an organisation through the process of establishing a robust cyber security program through the creation of an Information Security Management System (ISMS), with an emphasis on continuous improvement. It is internationally recognised and is applicable to any industry, regardless of size, however, due to its high cost and audit requirements it is mostly implemented by larger organisations.

NIST 800-53 is a comprehensive, US-centric, categorised controls catalogue mandated for implementation within federal agencies. It is more technical than ISO27001 and can be tailored to suit a particular organisation, industry, or risk profile [17].

These frameworks, whilst tailorable, are more suited to traditional ground-based IT infrastructure and do not adequately address the complexities and specificities of aerospace systems, nor have they been formally adopted or recommended by aviation regulators.

There exists a multitude of international and regional cyber security frameworks and guidance material that is relevant to aviation, some very basic and broad in scope such as the ICAO Aviation Cybersecurity Strategy that avoids anything prescriptive but instead emphasises the need for governance and international collaboration [49]. ICAO as the leading international body for aviation should be leading the way, so it is perhaps surprising that a keyword search using the words jamming, spoofing, GNSS, RFI, and satellite, returns nothing from this document or from their annex 17 of the Chicago Convention, which specifically looks at aviation security [50]. It seems ICAO are content to defer responsibility for the specifics to individual regions or airlines, whereas Nobles et al. [81] campaigns for a global aviation cyber security defence policy as the most sensible solution to combat a global threat [81]. Similarly, the International Air Transportation Association (IATA) 'Aviation Cyber Security Guidance Material Part 2: Aircraft', claims to be continuously updated yet it is already over 3 years old and very little has been written about spoofing, GNSS and RFI beyond basic definitions, providing only "minimum recommendations", with the responsibility for correct implementation being firmly placed on the airline operator [45, 46].

More specific frameworks such as ISA/IEC 62443 or NIST 800-82 look at the security of industrial controls systems (ICS), or ARINC Report 811 and EUROCAE ED-202A that specify certification and airworthiness standards for new avionics systems [46].

Arguably, the most appropriate and most comprehensive attempt at a framework that promotes resilience and offers mitigations to the increasing threat against GNSS is actually one that is specific to PNT users within CNI, yet it is not included in IATA's compilation of aviation cyber security documentation, unlike the aforementioned documents, nor has it been endorsed, referenced or recommended by any UK regulator or aviation organisation that this researcher could find. The NIST Internal Report (IR) 8270, NIST IR 8401 and NIST IR 8323 provide a space overlay for the

NIST Cybersecurity Framework (CSF), a similar framework to the NIST and ISO offerings, yet lightweight and voluntary with wider application. These reports cover the space segment, ground segment and user segment respectively. NIST IR 8323 provides the most value, being relevant to PNT users, including airlines. The controls and guidelines recommended, including actions against jamming and spoofing, are informed and aligned to NIST 800-53 amongst a plethora of other international cyber security frameworks and recommendations, and is designed to serve as a voluntary foundation upon which sector-specific strategy can be built [4].

The notable gap in cyber security discussion focussed on GNSS threats highlights the industry's inability to keep pace with the changing threat landscape. NIST IR 8323 provides a solid foundation to mitigation and one that could immediately be adopted by airlines globally, yet the lack of endorsement or regulatory requirement is troubling considering the increasing and widespread risk. UK guidance and regulation needs to follow suit, a responsibility that falls in large part to the Civil Aviation Authority (CAA) as the UK's regulator.

The Cyber Assessment Framework was developed by the UK National Cyber Security Centre (NCSC) and first published in 2018, with the current version v3.2 released in 2024 [77]. It was a response to the Network and Information Systems Directive (NIS-D), issued by the European Parliament in 2016 and adopted as the UK NIS Regulations in 2018, addressing vulnerabilities to CNI and placing enforceable responsibilities on Operators of Essential Services (OES) for nine subsectors, of which air transport forms one [40]. An update to the EU NIS directive will be enforced later in 2024, with UK NIS regulations changes anticipated to follow shortly after, that will likely include additional subsectors, amended reporting requirements and higher penalties [90].

The CAF is a voluntary assessment of NIS compliance and assesses an organisation against four broad cyber risk management objectives, fourteen targeted technical and administrative objectives and a set of indicators of good practice (IGPs) for each, which can be assessed as achieved, partially achieved or not achieved [78]. The CAA has adapted the CAF to create the CAF for Aviation, used in associated ASSURE audits [11]. It assesses an aviation organisation's cyber maturity, forming part of the CAA's Cyber Security Oversight Process for Aviation outlined in CAP 1753 and supported by a new oversight strategy, CAP 2976, released in 2024. Other supporting documentation includes CAP 1849, which helps an organisation with critical systems scoping and selection, yet none of these documents mention GNSS-dependent systems or highlight their threats, despite aviation resilience being core to the vision and mission of the CAA's cyber strategy [10] (Fig. 7.6).

The World Economic Forum have identified the CAF as a positive move towards establishing a common cyber resilience baseline for the aviation industry [114], however, it is apparent that more work must be done to improve, develop and expand upon current regulatory guidance in order to reinforce a suitable framework that encompasses complex aviation systems within the modern risk environment and promotes resilience within the industry; the outcome this study aims to achieve.

Fig. 7.6 Process diagram showing the relationship between regulations

Having discussed the infrastructure, its dependencies and threats, the importance of a comprehensive, appropriate and enforceable cyber framework for the UK aviation industry is obvious, the development and implementation of which will be beneficial to security, flight safety and regulatory compliance. The focus can now shift to achieving this.

7.3 Research Methodology

An appropriate resolution to the identified problem can only be truly achieved through a comprehensive understanding of the problem itself. Problem framing is a concept used extensively within project management and design thinking to aid understanding of complex business problems, identify scope, assess impact and determine root causes [111].

The Stanford school embraces problem framing within design project scoping, focussing on problem comprehension, user impact and recognition, an awareness of the business context, and the refinement of specific goals, ultimately uncovering the unknowns or the problem beyond a problem to formulate a strategy to navigate a challenge [8].

These approaches compliment decades of problem-solving discourse and have been collated by product discovery solutions company, New Haircut, to produce a simplified and usable model for problem framing, as shown in Fig. 7.7 [79].

A similar approach to this research problem would allow for a deeper understanding of the GNSS system security problem within the context of aviation operations whilst steering the research towards an appropriate and considered solution. This research has already discovered the initial problem of limited resilience within the aviation industry with respect to its GNSS-dependent systems and services. The

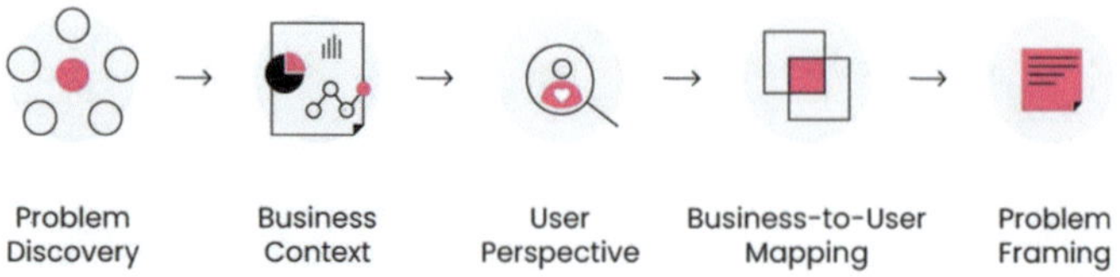

Fig. 7.7 The problem framing process. *Source* New Haircut

problem must now be reframed by assessing the business context and user perspective beyond that previously discussed in Chap. 2. The frame of reference in this case would be satellite-based CNS technology and the global aviation dependency.

A qualitative research methodology will be pursued for the most part, with an element of secondary analysis of quantitative data. This mixed method of research, advocated by Alan Bryman, will allow for the necessary observation, description and document analysis that accompanies a qualitative approach, whilst permitting the publication of supporting data that offers completeness, greater validity and an enhanced illustration of the problem [9].

Understanding the extent of the threat will be accomplished through the analysis of GNSS jamming data that is regularly captured and displayed on a world map to highlight areas of high concentration. This will then provide the basis for statistical analysis and discussion about the geographical distribution of the data, its patterns, trends and root causes. Spoofing events will be reviewed and analysed utilising a similar heatmap, readily available case studies and flight incident reports; secondary data collated once again by previous researchers and made available for public consumption. The differences and similarities between these attacks will be compared and the data associated with their system effects and wider impacts will be used for trend analysis, assessment of pilot and industry response, and guidance towards a policy-based mitigation.

This secondary data analysis will allow for the evaluation and appreciation of the wider threat to aviation operations today and the trend in incidents to justify the need for further aviation-related cyber security discourse and revised governance.

There is very little benefit in attempting to create a new dataset when experts, aeronautical organisations and regulators have already collected and collated data, and many researchers, analysts and commentators have in most cases discussed initial findings and presented them through open-source media. It is, however, imperative that due diligence is performed to ensure that the data is genuine, trustworthy, comprehensive and unbiased, and that the data providers are reputable and thorough in their research. This insurance check will be time-consuming but will save vast amounts of time when compared with the inevitably lengthy process of data collection, analysis and visualisation that will be necessary to produce a separate yet similar dataset, not to mention the requirement for specialised and often expensive equipment and the technical expertise to operate it.

Shortcuts can safely be taken with efficiency in mind without any perceivable detrimental impact on the quality of the data analysis, however, allowance should always be made for the possibility of the discovery of new and uncleaned, unprocessed data that could be incorporated into the existing dataset and analysis and presented for clarity and reinforcement using one or more data visualisation tools such as Tableau or Nvivo. There will be discussion focussed on the complex web of regulatory frameworks and technical controls that are currently enforced and those which will be developed for later implementation so, where possible, their relationship, dependencies and ownership has been presented diagrammatically, as seen in the previous chapter.

The aforementioned problem beyond a problem from the design thinking process will have been identified through root cause analysis, looking initially at the increase in GNSS attacks, leading to questions surrounding the increasing attack surface, acknowledgement of the lack of awareness and resilience, and would uncover omissions in regulatory compliance documentation that need to be addressed.

Business-to-User mapping will involve the evaluation of existing cyber security frameworks and associated controls for relevance, specificity, and comprehensiveness, conducted against the backdrop of increasing GNSS jamming and spoofing events within the airline industry. A concentrated SWOT Analysis will be conducted to assess the Strengths, Weaknesses, Opportunities and Threats, enabling the identification of the mitigating controls and guidance.

Lastly, there is a wealth of relevant knowledge available within aviation organisations, airlines and consultancy firms. Publications, guidance and subject matter experts within these institutions will be approached for advice and research direction, where possible.

7.3.1 Limitations and Ethical Considerations

In the interest of avoiding ethical issues, informed consent from these individuals and groups will be sought in a policy of openness and honesty.

Another significant consideration and a limitation of the research is that of the quality, reliability and validity of the secondary data used to convey the prevailing threat to GNSS, as previously alluded to. Military data will not be available due to its sensitive and confidential nature, whilst the accuracy of data reported by certain countries cannot be wholly relied upon. Hostile or politically unstable nations may attempt to skew or disguise their figures to feign compliance and collaboration whilst keeping potentially damaging figures from public view.

Subconscious bias must be avoided in the selection of data sources so as not to present an incorrect threat picture. Every attempt will be made to utilise impartial and professional sources that incorporate an inclusive data set. Nevertheless, these datasets and visualisations are limited in number, so there is an element of trust, however, the sources referenced have been scrutinised for quality and integrity, and have been awarded respect within the industry, enough to be confident of quality data output for the purpose of this research.

Finally, malicious actors could potentially make use of known vulnerabilities and susceptibility to attack if attributed to a particular person or organisation. Therefore, care must be taken to maintain anonymity and not divulge case study information that identifies those involved. Damage limitation is important, despite open-source information being used, as unnecessary harm brought about by a possible politically sensitive issue is totally avoidable.

7.4 Data Analysis and Critical Discussions

Following on from literature review, an opportunity is provided here to evaluate the extent of the current risk to GNSS by analysing GNSS jamming and spoofing heatmaps and assessing the distribution of events, any significant trends and potential offenders. Two case studies will be briefly reviewed before looking at gaps in UK aviation cyber security strategy and who owns the responsibility for improvements and the implementation of a new strategic direction. The study then acknowledges a successful approach to regulatory guidance and requirements and uses this as the benchmark against which to measure a new strategy proposal for the UK aviation industry. This new strategy is then validated by reviewing the previous case studies to show how it would have minimised the negative effects of the interference events. The chapter closes with speculation about future integration of emerging technologies into aviation security, whilst discussing the next steps in assuring resilience within the airline industry.

7.4.1 Scope of the GNSS Threat

The analysis in this section aims to act as justification for the singling out of the jamming and spoofing threat to GNSS and for the inclusion of its mitigations, controls and recommendations into revised UK aviation cyber security regulatory guidance material.

According to InsideGNSS (2024), 38.5% of European overflight air traffic operates in RFI affected regions, which provides significant opportunity for operational disruption with the potential to cost a considerable amount of money, damage or even lives. Understanding where these areas are is of the utmost importance to flight operations.

John Wiseman has collected GPS jamming data since 2022 and presented it in a heatmap showing areas of high, medium or low concentration. The popular flight tracking application, Flightradar24, has also very recently launched an equivalent heatmap. Despite using GPS in their titles, they both map jamming data for all approved GNSS systems and it is worth noting, that GPS and GNSS are often used interchangeably within the industry, although their differences are important and notable. This perhaps highlights a weak understanding of the satellite infrastructure utilised within aviation, a problem that will be revisited later, or reinforces the historic preference for the tried and tested GPS infrastructure, prior to relatively recent attempts to blend multi-constellation signals. Nevertheless, the heatmaps convey a good appreciation of the jamming threat, offering a data visualisation that can be easily interpreted and analysed by relevant airline flight planning departments and security teams.

The data comes from the Navigation Integrity Category, transmitted within ADS-B messages. It assesses the validity and reliability of an aircraft's position and will note instances of low navigation accuracy and heuristics, such that a jamming event can be assumed [32, 51].

Figure 7.8 is a snapshot of Wiseman's jamming map taken from 24th July 2024. A brief analysis shows swaths of green, representing a low concentration across most of Europe, with very concentrated areas of red, indicating a higher percentage of jamming events in these areas.

It shows that the highest percentage of jamming events took place around the Black Sea, including northern Turkey and coastal areas of Romania and Bulgaria. Substantial hotspots also exist on Russia's borders with Ukraine, Latvia, Estonia and Finland. The Middle Eastern states of Iraq, Iran, Syria, Lebanon and Israel are also shown in red along with a smaller, isolated pocket on the border between India and Pakistan.

Unsurprisingly, perhaps, these areas are all within or near to active conflict zones. The ongoing conflict within the disputed Kashmir region explains the concentration

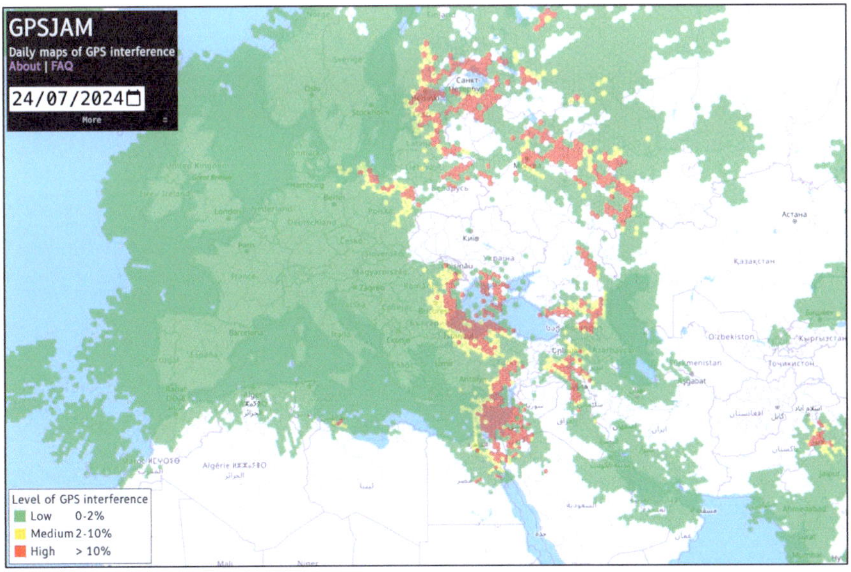

Fig. 7.8 Wiseman's GPS jamming heatmap from 24 July 2024 [116]

of jamming events within India and Pakistan, whilst ongoing political instability and conflict continues within Iraq, Iran and Syria. Undoubtedly the most militarily active regions are those embroiled in recent conflict brought about by the Russia/Ukraine war and the aggressive confrontations in Israel and Palestine, that have since crept into Lebanon.

Interestingly, there is almost nothing showing for Ukraine, despite being the epicentre of a major conflict. As the heatmap is based on ADS-B data it is most likely due to there being no civil air traffic due to the closure of the entirety of Ukrainian airspace by the State Aviation Administration of Ukraine on 24th February 2022 [93]. Those military or state-approved aircraft that continue to fly within Ukraine almost certainly avoid the use of ADS-B to prevent position data transmissions being intercepted by hostile forces. However, null areas could also show that there were no active receivers to collect the interference data.

A similar map has been produced by SkAI Data Services in conjunction with the Centre for Aviation at Zurich University of Applied Sciences showing areas of assumed spoofing and plots of position disagreement [98]. An OpsGroup report tells of a 400% increase in spoofing events since the first quarter of 2024, with nearly 900 events per day [85].

The geographical distribution of events is very similar to that of the jamming heatmap. Again, there is no data for Ukraine, for the same reasons as noted above, but this is not to say that the few aircraft that do operate in the region do not experience jamming or spoofing, it is simply that there is no data to confirm or deny this (Figs. 7.9 and 7.10).

The highest concentrations of assumed spoofing events are centred on Simferopol airport in Russian occupied Crimea and Ben Gurion airport in Tel Aviv, Israel.

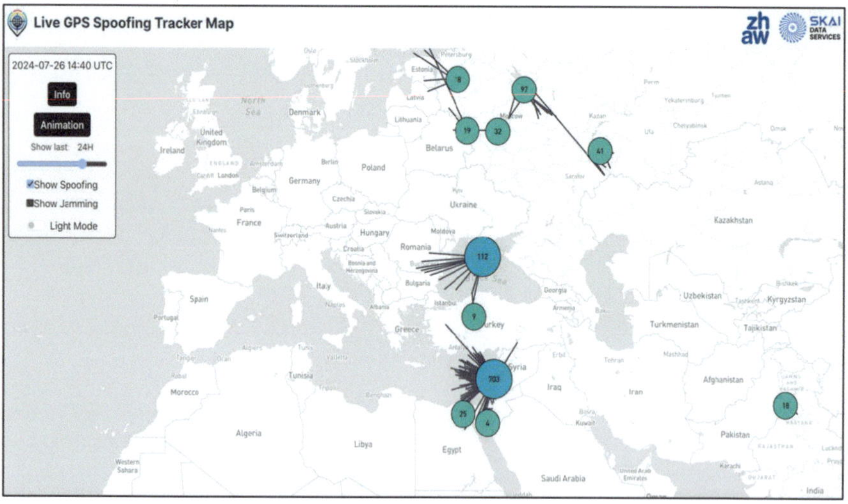

Fig. 7.9 SkAI data services map of assumed spoofing events

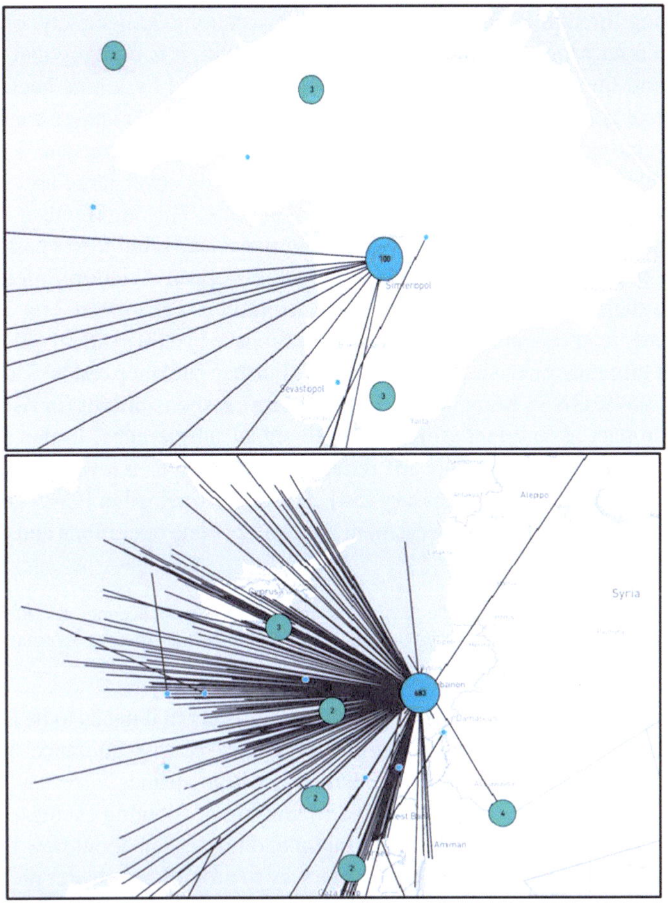

Fig. 7.10 Spoofing hotspots at Simferopol and Tel Aviv airports

Simferopol has been under Russian control since the annexation of Crimea in 2014 having once been a port of strategic importance to Ukraine [102]. It is possible that the large number of spoofing attacks could be the result of Ukrainian military efforts to disrupt Russian air and maritime operations, but equally as likely is a Russian attempt to mask its own politically sensitive operations, protect their critical infrastructure and disrupt navigation signals to hinder Ukrainian offensive efforts. It is easy to assume that an aggressor is the perpetrator of destructive cyber warfare, yet there are tactical reasons that a nation state or organisation may wish to self-inflict these interference events for countersurveillance purposes or to gain a strategic military or political advantage.

Deliberate attacks on CNI are becoming more prevalent and being used to disrupt or deny domestic operations and utilities, putting pressure on a population or its government, so the deliberate denial of GNSS-based services is likely to become an increasing threat that has the potential to devastate modern society, blurring the line between hacking and cyber-terrorism. To that end, it is unlikely that attacks as technical and impactful as this will have been conducted by a lone hacker, except in the case of accidental jamming like the personal jammer situation mentioned in Sect. 7.2.3, rather a well-funded criminal organisation or a nation state.

Whilst the maps' event zones are concentrated, they cover large areas that spill over into regions not involved in the military conflicts. This highlights a significant risk to civil aircraft operations not just within these areas, but also nearby. This is something that has been acknowledged by the European Aviation Safety Agency (EASA) in their Conflict Zone Information Bulletin (CZIB) dated 31st July 2024, advising that all air operators avoid Ukrainian airspace by 200 nautical miles to avoid intentional targeting or misidentification [23]. Further guidance can be found within ICAO Doc 9849: GNSS Manual, clarifying a State's responsibilities for risk management and impact assessment with respect to signal interference. It also calls for a multi-constellation infrastructure for redundancy as well as a reversion to conventional navigation aids when necessary [54]. This is also echoed in ICAO Doc 10084, which discusses essential risk assessment for conflict zone operations and details the requirement for:

> [...]alternate communication and navigation methodologies to account for the potential of conflict-associated jamming or electric interference that disrupts normal aircraft communications and navigation [55].

The need for analogue fallback is clearly recognised but it needs to be acted upon and emphasised within a national regulator's cyber security guidance material in order to promote the required level of resilience within airlines.

Airlines are encouraged to report GNSS jamming or spoofing events to the CAA to enable evaluation of the extent of the threat and discussion about how to mitigate the risk. This information sharing goes some way to providing a clearer picture of the GNSS risk, however the information needs to be brought into everyday flight planning and distributed to flight crews in a comprehensive and effective way, something akin to a military intelligence brief. NOTAMs (Notices to Air Missions) currently exist to disseminate important airport, flight service, hazard or CNS information relevant to the intended route [12]. Some GNSS event information is provided within NOTAMs, however a more significant route brief would afford the risk the attention it probably deserves.

7.4.2 Case Studies and Event Analysis

Detailed within this section are two anonymous examples of recent jamming and spoofing events that illustrate the impact of an attack, following on from the previous section that explored the risk. These cases will be revisited later for comparison and validation of the proposed new cyber security strategy.

The first is an account of an Embraer Phenom private jet departing Lahore, Pakistan in 2022 that experienced multiple and simultaneous system failures early in the climb. GNSS data was lost, and the electronic flight instruments became unreadable. The situation worsened with the loss of the autopilot function and a spurious terrain avoidance warning that produces aural and visual cockpit effects. This dangerous event was further compounded by the startle factor and a critical phase of flight, in which numerous radio calls will have provided a background distraction, and numerous departing and arriving aircraft would have been considerable hazards that would need to have been avoided, altogether creating an extremely confusing and stressful situation.

Lahore is located very close to the Indian border and the contested Kashmir region, where a propensity for jamming and spoofing events has already been discussed. Knowledge of these RFI regions and a rehearsed process to follow during the event may have mitigated the effects and minimised the fight-or-flight response. These involuntary neurological and physiological responses to a stressor have historically resulted in inappropriate pilot responses that have left aircraft in an undesirable state. Process, procedure and muscle memory would go some way to minimising the severity, which is something that will be addressed in the new strategy proposal.

The second event, shown in Fig. 7.11, involved a Boeing 737 flying southbound within Iraqi airspace that encountered a GPS clock jump and a 75 nm position discrepancy, such that an encroachment on Iranian territory was imminent. ATC advised the flight crew of the proximity of this restricted area and provided them with an accurate position fix and time. Whilst a Quick Reaction Handbook (QRH) procedure existed, it failed to instruct the crew to deselect GPS updating of the Inertial Reference System (IRS), used for position determination. GNSS signals are used to update and increase position accuracy of the analogue IRS system. A reconfiguration to a standalone system upon recognition of a GNSS signal error should take place in order to mitigate the effect of a spoofing attack.

Iraq and Iran were identified in Sect. 7.4.1 as regions with a high concentration of jamming and spoofing events. Tensions between these nations has escalated in recent years due to numerous historical, religious and geopolitical factors including recent border disputes and cross-border counterinsurgency operations against localised separatist groups [30]. An unannounced encroachment on Iranian airspace, therefore, could rapidly have turned into a hostile political situation with the end result being a fatal misidentification and interception.

A similar threat exists on the border of Israel and Lebanon with many aircraft experiencing RFI, inducing navigation errors that have forced aircraft to track towards Beirut, a hostile nation.

Awareness, communication and procedures must be improved to enable and empower crews and agencies to better deal with these incidents.

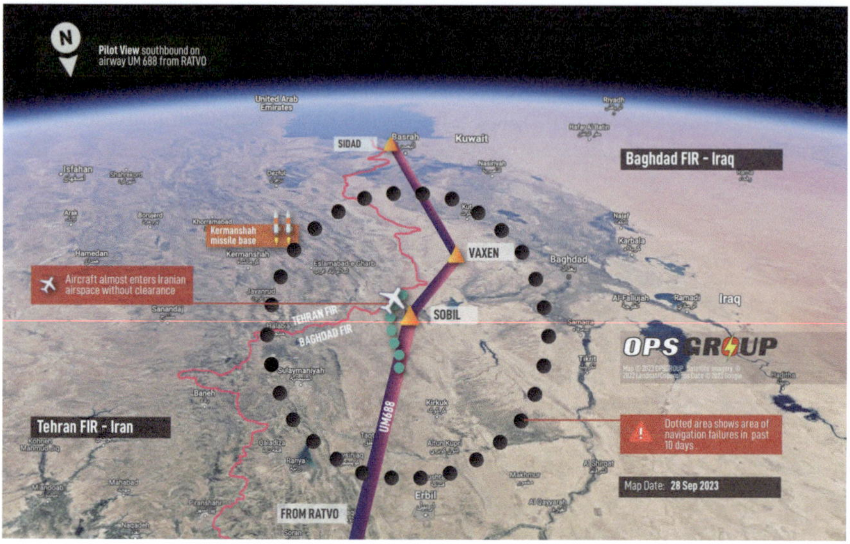

Fig. 7.11 An OpsGroup pictorial representation of the spoofing event that nearly resulted in Iranian airspace entry without clearance

7.4.3 Failings of the UK Aviation Cyber Strategy

There is extensive aviation, space and cyber security documentation within UK legislation, which on face value seems comprehensive and inclusive, yet on the whole, UK aviation cyber security policy is weak, non-prescriptive and non-committal. The National Space Strategy 2021 identifies the jamming and spoofing risk whilst acknowledging the significant economic impact of a GNSS outage [43], yet the UK Space Agency Cyber Security Toolkit directs asset owners to implement effective controls and fails to reference GNSS, PNT or the wider risk to industry anywhere in the document let alone emphasise the prevailing RFI threat [105]. Published in 2020, it is likely that policy hasn't caught up with the developing cyber risk, but regular review, reporting and collaboration are promoted within the policy, and so it too, should be updated in earnest to reflect this.

The Aviation Cyber Security Strategy expired in 2022 and was not specifically replaced, despite there being no mention of the jamming and spoofing threat to GNSS [20]. 'Aviation 2050: The future of UK aviation' establishes a commitment to cyber security within aviation as part of a wider review by the Department for Transport (DfT) and external consultation on the future of the sector up to 2050. 'Hardware interference' is mentioned, yet does not prescribe controls, and there is still no specific mention of the threats discussed in this paper [19].

CAA cyber security oversight strategy is outlined in CAP 2976 and is the document that should provide the most specific direction to address the jamming and

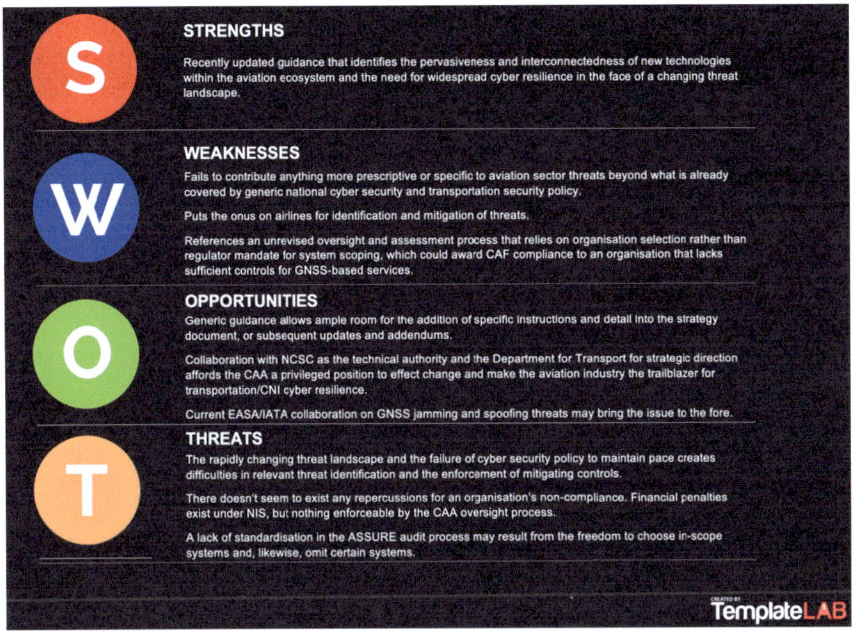

Fig. 7.12 SWOT analysis of CAP2976—CAA cyber security oversight strategy 2024–26

spoofing risk to CNS services. Figure 7.12 is a SWOT analysis of this document, identifying areas that can and should be improved to ensure industry resilience.

UK policy is not particularly well-supported by regional or international policy either. ICAO should lead the way, however, their 'Annex 17: Aviation Security' only mentions the need for asset identification, cyber risk management and protections without offering specific guidance.

International conventions that cover the suppression of unlawful seizure or criminal acts onboard aircraft, for example, were written decades ago and are not wholly adequate for the evolving cyber threat. It would be prudent, in order to keep pace and remain relevant, to develop a new concentrated convention for the suppression of cyber security threats to aircraft.

Yet positive steps are being made with EASA and IATA beginning to collaborate on combatting the specific issue of GNSS jamming and spoofing, which might hopefully galvanise the UK regulators and policymakers [24].

7.4.4 *Responsibility for Direction and Implementation*

A global threat requires a global response and therefore, in the absence of a single, legally binding instrument that promotes and ensures cyber security resilience within the aviation industry, ICAO should be zealously spearheading a strict, prescribed and collaborative strategy to fortify the industry against an escalating threat in an extremely safety conscious environment. Whilst ICAO cannot directly enforce aviation law, national regulators and governing bodies should be compelled to implement cyber strategy as a priority or face significant penalties and sanctions, especially when considering the global and interconnected nature of air transport and the risk of indirect exposure and exploitation brought about by the globalisation of technology and the ubiquitous aviation system of systems.

Successful implementation of international policy, however, is influenced by an individual nation's politics, laws, finances, infrastructure and exposure, so arguably the first step should be the establishment and strengthening of domestic policy before international alignment can occur.

In the interim the CAA should focus on enhancing and consolidating its oversight strategy to steer cyber security reform whilst engendering an ethos of self-regulation and accountability within airlines, who may otherwise be reluctant to engage due to focussing on the maximisation of revenue. National regulation will inevitably encounter bureaucratic delays that hinder expeditious implementation, so airlines should be proactively engaged in bolstering their own resilience, despite potentially high costs. Internal action would have the most immediate effect but would still require direction through policy, effective cyber security frameworks and an appropriate mechanism for self-assessment.

Cyber security within the aviation and space environments requires an understanding of operational technology (OT), the realm of cyber-physical systems, industrial control systems and the internet of things, yet the management of traditional IT infrastructure versus OT seems to remain siloed. Airline engineering teams must liaise with IT departments to understand their roles and responsibilities, whilst collaborating with hardware and software manufacturers to incorporate the latest component certification requirements and avionics communications protocols to harden onboard systems, contributing to continuing airworthiness as championed by EUROCAE and EASA [22]. Ongoing collaboration and discussion through internal cyber working groups also needs to take place to keep relevant stakeholders informed and involved.

Security is unquestionably easier to regulate for systems owned and managed by the airline, however, intricate supply chains and third-party contractors with their own integrated systems are commonplace within the aviation system, which complicates the quest for resilience. GNSS services are provided by satellites operated by one nation but offered as a global service, which brings responsibility, accountability and regulation into question. Defence in depth and comprehensive risk management can only be achieved through a thorough understanding of the assets and infrastructure. Perhaps again, additional powers of international regulation for a global critical asset should be granted to ICAO rather than left to the satellite operator, particularly with an increasing reliance on Commercial off-the-shelf (COTS) components [115]. This overreliance on COTS software, exaggerates the inherent coding vulnerabilities that have been identified as an attack vector within the SPARTA matrix. The danger is a widespread satellite network outage similar to that of the 2024 Crowdstrike worldwide system failure, caused by a logic error in an automated software update that affected Windows operating systems [35].

CNI is in everybody's best interest to defend and protect and cannot be allowed to collapse in such a manner because of technical overdependence and a poor appreciation of wider system dependency. Resilience must be inbuilt with strong IT leadership and direction as part of a collaborative and multi-disciplinary approach to risk management.

7.4.5 OFGEM—A Successful Implementation

Guidance exists for several industrial sectors in the form of a CAF overlay, detailing sector-specific interpretations of IGPs and the regulator's expectations from an auditee.

A CAF overlay for the National Health Service England was adopted in September 2024 replacing its existing Data Security and Protection Toolkit, supporting a culture of evaluation and improvement and allowing better security alignment for effective compliance and information governance [80]. There is also work currently being done by several UK government departments to develop a CAF overlay for local authorities [96, 104].

Of particular note, however, the UK energy regulator, Ofgem, has produced a CAF overlay for the downstream gas and electricity (DGE) sectors that assists OES to attain the necessary security outcomes required by the CAF by providing clarification and detailing actions and indicative behaviours that exemplify IGP achievement. The 300-page document provides substantial detail for each of the assessment criteria, explaining what exactly is required without prescribing specific controls or frameworks, leaving the organisation free to elect the most suitable defence and risk management strategy. It does, however, provide a useful mapping of CAF IGPs against several relevant control frameworks for reference. Furthermore, the Ofgem guidance makes reference to specific threats that should be addressed, such as phishing and the use of COTS, unlike the CAF itself, which is subjective by comparison [84].

Guidance of this kind would assist an aviation organisation by removing the audit guesswork and streamlining the self-assessment process required by the cyber oversight process outlined in CAP 1753. Making the requirements easier to digest and understand may even improve engagement, whilst providing additional benefit to any consultancy that is employed to help with the audit and controls implementation.

Similar prescriptive guidance from aviation governing bodies is lacking, so the industry needs to learn from this successful CAF overlay incorporation into governance and assessment strategy, develop a comprehensive and appropriate aviation-focussed narrative, and use it to promote industry alignment and compliance with the CAF and ultimately the NIS Regulations as a crucial step in establishing resilience within the aerospace sector.

7.4.6 A New Strategy Proposal

An updated strategy should undo existing policy failings and mitigate current threats to the industry. Whilst there is much that could be added to UK aviation cyber security strategy, the focus of this research has been on GNSS jamming and spoofing.

The CAA should produce a CAF overlay, setting out their expectations and assisting organisations with compliance with GNSS resilience IGPs that should augment and improve the CAF.

In Fig. 7.13, several IGPs have been identified from the CAF and suggested guidance has been added that could further enhance the suitability and comprehensiveness of an audit, ensuring that GNSS resilience is considered. This guidance should be included in a separate but complementary sector-specific overlay. The list is not exhaustive nor are all CAF principles and objectives covered.

Additionally, amendments need to be made to the CAA publications mentioned in Sect. 7.2.4 to improve the quality and suitability of cyber security oversight, compliance and ultimately, resilience, particularly as this document series is the foundation upon which the audit process and proposed CAF overlay is built (Figs. 7.14, 7.15 and 7.16).

These collective revisions are important for understanding and managing the GNSS cyber risk. Nonetheless, care must be taken that proposed amendments do not introduce over-governance into cyber assessment processes. Frameworks are intentionally broad and agnostic to enable tailoring and wider applicability and these attributes must be preserved, whilst highlighting critical threats supporting resilience within a global industry. Backing for this more direct, prescribed approach to assessment has come from the Kings Speech 2024 in which he announced a new Cyber Security and Resilience Bill that bestows new powers by "putting regulators on a strong footing to ensure essential cyber safety measures are being implemented" [89].

7.4.7 Framework Validation

Section 7.4.2 introduced two case studies. The revised guidance presented in Sect. 7.4.6 will be applied to these studies to demonstrate its positive impact and potential success in managing the GNSS events and reducing negative outcomes.

Training formed an important element of the changes to policy so it can be assumed for the purpose of re-evaluation and validation that the aviation organisations in question have implemented the guidance and, as such, have been successfully audited and trained, whilst knowledge of the GNSS threats and awareness of high-risk areas have been integrated into normal flight operations.

In both cases, flight operations and security teams will have disseminated pertinent information about the risk to GNSS services and to operations within conflict zones, meaning flight crews would be primed, somewhat mitigating the startle effect. It

CAF OBJECTIVE	CAF PRINCIPLE	CAF IGP	PROPOSED CAA GUIDANCE
MANAGING SECURITY RISK	A2: Risk Management	Use of established methods or frameworks	Evidence of use of sector-specific exploitation frameworks for risk identification e.g. SPARTA/MITRE.
		Use of current threat and vulnerability information in risk assessment process	Use of collaborative threat intelligence to inform flight operations within conflict zones and within regions of known RFI, to be published within NOTAMs, route briefs and other relevant security documentation.
	A3: Asset Management	Asset register and sample critical asset check through the lifecycle	GNSS-dependent systems to be identified.
	A4: Supply Chain	Detail of cyber security requirements imposed on suppliers	Relationships with third-party vendors and/or service providers should be conditional upon the existence of complementary aviation cyber security strategies.
PROTECTING AGAINST CYBER ATTACK	B4: System Security	Policy setting out design requirements for network architecture, segregation, and access	Up-to-date secure avionics standards and protocols should become part of the system procurement, maintenance, and upgrade processes and policies.
	B5: Resilient Networks and Systems	Documented Business Continuity and Disaster Recovery Strategy with evidence of practices / tests being carried out	To include tabletop exercises and live GNSS jamming/spoofing drills in conjunction with ATC.
			Trial of safe system reconfiguration and analogue failover, tested in a controlled environment
	B6: Staff Awareness and Training	Cyber security awareness training and/or education programmes	To include extensive simulator testing for flight crew in the GNSS jamming/spoofing case.
			Operational crew education on the threat and impact of a cyber-attack on GNSS.
	C1: Security Monitoring	Records of periodic monitoring	GNSS jamming/spoofing events logged and shared. Collaboration with relevant authorities and other aviation organisations to combat threat.

Fig. 7.13 Suggested GNSS guidance for a proposed CAF overlay for aviation

Proposed Amendments to CAP 2976

CAA Cyber Security Oversight Strategy 2024-26

The jamming and spoofing threat to GNSS must be acknowledged.

Management of the GNSS threats must be prioritised and policy must call for international collaboration to combat this flight safety risk.

Promote collaborative threat intelligence and information sharing.

Airlines should be called upon to engage in discourse with relevant aviation organisations, such as ICAO, OpsGroup, and the Aviation ISAC, in a collective effort to understand the vulnerabilities of GNSS and their mitigations and share operational data.

Formally recognise and endorse frameworks.

The CAA should endorse the SPARTA matrix for identification of avenues of exploitation relevant to GNSS assets and services, and NIST IR 8323 for appropriate guidance and mitigations. These documents will be invaluable in educating and equipping GNSS users whilst establishing a level of resilience and a structure upon which satellite operators, engineers and end users can align their security strategy.

Fig. 7.14 Proposed amendments to CAP 2976

would buy enough time to discuss initial pilot actions should RFI be experienced, a valuable asset in a safety critical environment.

Flight operations may even conclude that operations should not take place within regions with a high percentage of RFI and instead be rerouted around the regions, or perhaps cancel or delay the flight, keeping passengers, crews and aircraft safe from harm.

As soon as GNSS data started to degrade in the case of the Phenom departing Lahore, new system reconfigurations or manual crew actions will have led to the selection of an analogue backup as advised by the 'resilient networks and systems'

Proposed Amendments to CAP 1753 ***The Cyber Security Oversight Process for Aviation***
Audit mandate Airlines will be legally bound to undergo an ASSURE audit as soon as practicable, and then every 2 years thereafter.
Extend the process to include General Aviation NIS threshold requirements to identify an aviation organisation as OES is based on passenger numbers (Gov.uk, 2018), meaning that private pilots and business aviation is not obliged to comply, however, GNSS-dependent systems are still in use. The CAA should encourage audit completion by offering a lower cost and lower baseline for compliance.
Updated Process Update policy wording to reflect changes made to the CAF, CAP 1753 and CAP 1849.

Fig. 7.15 Proposed amendments to CAP 1753

principle within the revised CAF guidance. Exposure to the disorienting cockpit effects of an RFI event would have been acquired through simulator training, which would somewhat negate negative pilot inputs due to a honed muscle memory and pilot skills consolidation. Efficient and knowledgeable liaison with ATC will have prompted priority handling and a relatively simple in-flight return to the departure airfield or suitable enroute alternate.

Similar training and awareness would have prevented the risk of Iranian airspace encroachment and cockpit confusion in the second case study. Collaboration with ATC on detecting and mitigating RFI events would have led to an increasingly monitored flight path and a collective and expeditious problem-solving effort. Updated

Proposed Amendments to CAP 1849

Cyber Security Critical Systems Scoping Guidance

Compulsory GNSS System Inclusion

The current audit scope puts the responsibility for selection of in-scope systems on the aviation organisation in collaboration with any hired consultancy. In light of the prevailing threat to GNSS, the CAA should now mandate that GNSS-dependent systems are to be included in all cases.

Fig. 7.16 Proposed amendments to CAP 1849

company documents as part of the CAF 'staff training and education' principle would have ensured that correct procedures were followed, directing the deselection of GPS updating and a continuation of the flight, or a diversion to a suitable airport utilising conventional navigation aids where necessary.

These two events have provided a snapshot of the threats faced by the aviation industry, with a simple, yet significant proposal focussed on training and awareness being key to improved resilience, ultimately benefitting flight safety.

7.4.8 Preparing for the Future

Cyber resilience is ultimately the futureproofing of an organisation. It is almost impossible to predict when or how an attack might occur, or who may be responsible, so all avenues of exploitation should be considered during risk management discussions and the setting of corporate strategy.

It is reasonable to assume that with an increase in space activity and more political players in this arena, that more vulnerabilities will be exposed in line with the broadening threat landscape and attack surface.

Future threats could include direct kinetic attacks against space platforms utilising the kind of space-launched, nuclear, anti-satellite weapons being developed by Russia [18], or a ground-launched ballistic missile of the like tested by China back in 2007 [113]. This particular attack caused wider concern as it resulted in the destruction of a Chinese satellite, leaving over 3000 pieces of dangerous space debris in LEO, posing an indirect threat to other platforms.

An increased frequency and intensity of attacks against GNSS and other satellite infrastructure and services should also be planned for, most likely conducted as part of disruptive military or political operations such as the Viasat satellite communication network hack that took out strategic and tactical communications and data links the day before Russia invaded Ukraine [109].

International collaboration is key to the success of global satellite service provision and security, yet there is potential for political tensions to lead to the denial of access to a particular nation's segment of the GNSS or deliberate, selective interference that compromises the integrity of a multi-constellation infrastructure, putting strain on the remaining systems whilst disrupting aerospace operations. This is where inbuilt resilience and business continuity planning is important and will prove its worth.

The proposals within this paper, alongside the previously referenced secure engineering practices and traditional cyber hygiene will go a long way to protecting systems and increasing resilience yet, as Albert Einstein once said, "We cannot solve our problems with the same thinking we used to create them", so to overcome the threats inherent within GNSS and wider critical infrastructure, new models of detection and mitigation need to be embraced, harnessing emerging future technologies for a future threat. These must have better encryption and access controls and, where possible, the ability to autonomously identify irregularities, both mechanical and digital. In this era of modern technology and automation, this could be achieved through Artificial Intelligence, the groundbreaking technology that can simulate the humanlike abilities of learning and decision making [100]. AI is advocated for within the UK Government Cyber Security Strategy 2022–2030 [44], is already utilised within space-based systems and is being researched to achieve predictive threat intelligence, autonomous incident response and continuous anomaly monitoring [58].

Successful AI implementation into aviation systems could provide an autonomous program that can interrogate a conflict zone database to compare against an aircraft's present position and enter a defensive system configuration in anticipation of an RFI event, which could include system self-testing and validation in preparation for analogue failover. Any problems identified at this stage could be announced to the flight crew whilst offering and actioning an alternative strategy, such as a reroute through highlighted low-interference areas, with its associated flight time and fuel burn adjustments pre-calculated.

Fast and accurate detection and response to anomalous GNSS data would trigger not only a cockpit system alert but also an automated notification to Air Traffic Management services to promote priority handling and support for an affected aircraft, increasing pilot capacity and mitigating the startle effect.

Achieving this requires trust and reliability as placing the responsibility of flight safety and emergency decision making in the hands of unregulated AI structures comes with its own inherent dangers that need to be considered and addressed. Public opinion on the use of AI is already mixed [95] and its integration within aviation could spark more alarm and division. The security of AI systems must be ensured and managed through policies of governance, regulation and containment, yet carefully balanced so as not to stifle technological innovation.

This chapter has provided a detailed view of the distribution of GNSS jamming and spoofing events whilst analysing responsibility for implementation, management and direction. Proposals have been made to existing guidance that were proven to positively contribute to the referenced case studies before looking at what the future of GNSS threats may look like.

7.5 Conclusions and Future Work

UK aviation cyber security policy and guidance is weak. This study has shown that whilst there is much discussion surrounding the subject and, more recently, a general acknowledgement of the threat to GNSS and attempts to galvanise aviation organisations to implement robust structures and processes, there currently exists very little in the way of prescriptive guidance, technical controls and enforceable policy.

A new strategy must include a revised CAA cyber security oversight process that encompasses the compulsory inclusion of GNSS systems within scope of an ASSURE audit against the CAF for aviation, as well as a more comprehensive education surrounding the jamming and spoofing risk. The CAF framework itself should also be updated to reflect the GNSS components within the risk management, controls implementation and threat detection objectives, enforcing compliance with acceptable methods to boost resilience within flight operations.

Extensive and targeted training of flight crew and of flight planning personnel must also be mandated as a crucial and highly effective element of security, being operationally positioned on the frontline of mitigation, detection and response.

Airlines should take the initiative and adopt this study's proposed strategy, as with successful implementation comes a solid example of policy-based resilience that others can follow, and regulators and governments can endorse, if not mandate. However, it is unlikely they would do this of their own accord due to the financial, time and resource burden on the company, plus a lack of urgency with there being no regulatory requirement to do so, and no fatal incidents pinned on GNSS interference. Regulators need to make this compulsory before a hull-loss through a successful spoofing attack is tragically achieved, only then galvanising industry to react. Risk has generally been managed with the standard approach of impact versus likelihood. Now the likelihood is increasing, so too is the risk. The UK aviation industry must keep pace with the changing threat landscape and would benefit from the incorporation of NIST IR 8323 and the SPARTA exploitation framework into its risk management processes, whilst also publishing an aviation overlay for the CAF, following the successful example of Ofgem. Hopefully, this is something the new UK government might also encourage and address.

Overall direction, however, should come from ICAO as the international governing body, despite its inability to mandate policy, and the differences between nations when it comes to their ability to implement strategy. The arguably easier first step, though, is localised resilience, followed by international alignment.

Whilst the rise in RFI events evidenced within this paper may not subside, the negative impact hopefully will, whether it is the result of an attack from a nation state at war, or an accidental event caused by a personal jammer. With the future blending of signals and increased redundancy through the prospective multi-constellation infrastructure on the horizon, satellites, including current and future contributors to the GNSS will need to be robustly secured, technically, digitally and through policy, quite possibly with the help of AI, in order to defend and sustain critical infrastructure and fully promote resilience within the aviation industry, and beyond.

The ultimate goal of this research is an increase in resilience within airlines in the face of GNSS interference, which leads to increased flight safety by mitigating the risk of CFIT, pilot induced error or misidentification within a conflict zone. The avoidance of flight disruption obviously leads to operational efficiencies and cost-saving for an airline, which whilst important, comes second to safety. The added cost to implement this strategy will be noticeable, however the cost of disaster recovery would no doubt be far higher.

This study has presented the prevailing GNSS threat and assessed its impact through case studies, yet it must be recognised that the sample size is small. It is firmly believed that the assessment would be the same with a larger dataset and the validation, albeit subjective and hypothetical, is accurate.

Finally, it has been acknowledged here that bureaucratic processes are lengthy and whilst care has been taken to deliver an original piece, there may well be a policy being worked on within government or aviation organisations that is similar or perhaps even contradictory.

The threat landscape is ever evolving, and cyber security controls and frameworks must keep pace. The CAA strategy and process as well as future frameworks should be regularly updated in line with changing threat intelligence and industry observations. Further research could expand to include space operations and supporting ground infrastructure beyond the realms of traditional IT, as the aerospace industry continues to digitalise.

Additionally, further study into the practicality and effectiveness of quantum encryption within satellite communications and GNSS services would enable a review of its impact on future airline operations.

Another recommendation would be the evaluation of the benefit of granting ICAO further powers and formally recognising it as the international aviation regulator with the ability to demand international collaboration whilst approving, endorsing and mandating specific controls and frameworks, rather than simply acting as a figurehead offering guidance and unenforceable strategy. In unison, the goal of a wider governance structure and a global aviation cyber security policy, under ICAO, could be explored.

Research into the incorporation of AI within aviation could assess whether it can be regulated and can ensure the highest standard of flight safety demanded by the industry, whilst acknowledging the irony of a system implemented for added security, needing to be adequately secured against cyber-attack. The question of whether it would be accepted by passengers or even aviation insurers should be asked, whilst hypothesising as to whether extra buffers would need to be included within air traffic separation until it could be proven effective, reliable and accurate. Analysis could incorporate a review of the EU AI Act.

References

1. Ackerman SA, Platnick S, Bhartia PK, Duncan B, L'ecuyer T, Heidinger A, Skofronick-Jackson G, Loeb N, Schmit T, Smith N (2019) Satellites see the world's atmosphere. Meteorol Monogr 59:4.1–4.53
2. Air Sat One (2012) What is CPDLC—Air Sat One [Online]. https://www.airsatone.com/faqs/index/category_id:29#:~:text=CPDLC%20has%20been%20historically%20been,(control%20display%20unit)%20keypad. Accessed 15 July 2024
3. Airbus (2023) SBAS landing system takes off in Europe with A320 family operators, supported by the European Union [Online]. https://aircraft.airbus.com/en/newsroom/news/2023-07-sbas-landing-system-takes-off-in-europe-with-a320-family-operators-supported. Accessed 4 April 2024
4. Bartock M, Lightman S, Li-Baboud Y, Mccarthy J, Reczek K, Brule J, Meldorf K, Northrip D, Scholz A, Suloway T (2023) Foundational PNT profile: applying the cybersecurity framework for the responsible use of positioning, navigation, and timing (PNT) services NIST internal report 8323r1
5. Baselt G, Strohmeier M, Pavur J, Lenders V, Martinovic I (2022) Security and privacy issues of satellite communication in the aviation domain. In: 14th international conference on cyber conflict: keep moving! (CyCon). Tallinn, Estonia
6. BBC (2020) BeiDou: China launches final satellite in challenge to GPS [Online]. https://www.bbc.co.uk/news/business-53132957
7. Borio D, Dovis F, Kuusniemi H, Lo Presti L (2016) Impact and detection of GNSS jammers on consumer grade satellite navigation receivers. Proc IEEE 4:1233–1245
8. Both T (2019) Design project scoping guide. Hasso Plattner Institute of Design at Stanford
9. Bryman A (2012) Social research methods, 4th edn. New York, Oxford University Press
10. CAA (2024) CAP 2976: cyber security oversight strategy 2024–26. Civil Aviation Authority
11. CAA (n.d.) ASSURE cyber audit [Online]. https://www.caa.co.uk/commercial-industry/cyber-security/assure-cyber-audit/. Accessed 17 Aug 2024
12. CAA (n.d.) Notices to aviation (NOTAM) [Online]. Civil Aviation Authority. https://infringements.caa.co.uk/tutorial/notices-to-aviation-notam/. Accessed 27 Aug 2024
13. Cabinet Office (n.d) Secure by design principles [Online]. Central Digital and Data Office (CDDO), Cabinet Office. https://www.security.gov.uk/policy-and-guidance/secure-by-design/principles/# Accessed 10 Aug 2024
14. CBS News (2013) N.J. man in a jam, after illegal GPS device interferes with newark liberty operations [Online]. https://www.cbsnews.com/newyork/news/n-j-man-in-a-jam-after-illegal-gps-device-interferes-with-newark-liberty-operations/. Accessed 01 Aug 2024
15. Chai-Keong T, Li VOK (1998) Satellite ATM network architectures: an overview. IEEE Network 12:61–71
16. Cheng Y, Wei C, Sun S, You B, Zhao Y (2023) An LEO constellation early warning system decision-making method based on hierarchical reinforcement learning. Sensors 23:2225
17. Compliance Forge (n.d) NIST CSF vs ISO 27001/2 vs NIST 800-53 vs SCF [Online]. https://complianceforge.com/grc/nist-800-53-vs-iso-27002-vs-nist-csf-vs-scf. Accessed 12 Aug 2024
18. Debusmann Jr B (2024) Russia developing 'troubling' new anti-satellite weapon, US says [Online]. BBC. https://www.bbc.com/news/world-us-canada-68309496. Accessed 18 Sept 2024
19. Department for Transport (2018) Aviation 2050: the future of UK aviation. HM Government, London
20. Department for Transport (2018) Aviation cyber security strategy. Crown Copyright
21. Dovis F (2015) GNSS interference threats and countermeasures. Artech House

22. EASA (2023) Part-IS regulation published, completing regulatory framework for cyber-resilient aviation [Online]. https://www.easa.europa.eu/en/newsroom-and-events/news/part-regulation-published-completing-regulatory-framework-cyber-resilient. Accessed 05 Sept 2024
23. EASA (2024) CZIB-2022-01R10: operations in flight information regions: FIR LVIV (UKLV), FIR KYIV (UKBV), UIR KYIV (UKBU), FIR DNIPROPETROVSK (UKDV), FIR SIMFEROPOL (UKFV), FIR ODESA (UKOV), FIR MOSCOW (UUWV), ROSTOV-NA-DONU (URRV) and FIR MINSK (UMMV) [Online]. European Aviation Safety Agency. https://www.easa.europa.eu/en/domains/air-operations/czibs/czib-2022-01r10. Accessed 22 Aug 2024
24. EASA (2024) EASA partners with IATA to counter aviation safety threat from GNSS spoofing and jamming [Online]. https://www.easa.europa.eu/en/newsroom-and-events/press-releases/easa-partners-iata-counter-aviation-safety-threat-gnss-spoofing. Accessed 04 Sept 2024
25. Elmarady AA, Rahouma K (2021) Studying cybersecurity in civil aviation, including developing and applying aviation cybersecurity risk assessment. IEEE Access 9:143997–144016
26. ESA (2020) The current state of space debris [Online]. https://www.esa.int/Space_Safety/Space_Debris/The_current_state_of_space_debris. Accessed 12 June 2024
27. ESA (2024) Galileo satellites [Online]. https://www.esa.int/Applications/Satellite_navigation/Galileo/Galileo_satellites. Accessed 04 June 2024
28. ESA (2024) Galileo, now fit for aviation [Online]. https://www.esa.int/Applications/Satellite_navigation/Galileo/Galileo_now_fit_for_aviation#:~:text=Galileo%2C%20already%20the%20world's%20most,for%20the%20most%20critical%20operations. Accessed 27 June 2024
29. FAA (2023) ADS-B air traffic control (ATC) applications [Online]. United States Department of Transportation. https://www.faa.gov/air_traffic/technology/adsb/atc#:~:text=ADS%2DB%20is%20also%20transmitted,intervals%20in%20time%20and%20space. Accessed 23 July 2024
30. Fazil S, Tartir A (2023) Iraq in 2023: challenges and prospects for peace and human security [Online]. Stockholm International Peace Research Institute. https://www.sipri.org/commentary/topical-backgrounder/2023/iraq-2023-challenges-and-prospects-peace-and-human-security. Accessed 30 Aug 2024
31. Ferre RM, Richter P, Falletti E, Fuente ADL, Lohan ES (2020) A survey on coping with intentional interference in satellite navigation for manned and unmanned aircraft. IEEE Commun Surv Tutor 22:249–291
32. Flightradar24 (2024) GPS jamming map [Online]. https://www.flightradar24.com/data/gps-jamming. Accessed 27 Aug 2024
33. Forshaw JL, Aglietti GS, Navarathinam N, Kadhem H, Salmon T, Pisseloup A, Joffre E, Chabot T, Retat I, Axthelm R, Barraclough S, Ratcliffe A, Bernal C, Chaumette F, Pollini A, Steyn WH (2016) RemoveDEBRIS: an in-orbit active debris removal demonstration mission. Acta Astronaut 127:448–463
34. Frard J, Martin Sacristan L, Rambach D, Tranchet F, Warns T (2019) GNSS interference [Online]. Airbus. https://safetyfirst.airbus.com/gnss-interference/
35. Fung B (2024) We finally know what caused the global tech outage—and how much it cost [Online]. CNN. https://edition.cnn.com/2024/07/24/tech/crowdstrike-outage-cost-cause/index.html. Accessed 06 Sept 2024
36. Gallon E, Joerger M, Pervan B (2022) Robust modeling of GNSS orbit and clock error dynamics. Navig: J Inst Nav 69:navi.539
37. Gao Y, Li G (2022) A slowly varying spoofing algorithm avoiding tightly-coupled GNSS/IMU with multiple anti-spoofing techniques. IEEE Trans Veh Technol 71:8864–8876
38. Garg VK (2007) Chapter 23—Fourth generation systems and new wireless technologies. In: Garg VK (ed) Wireless communications and networking. Morgan Kaufmann, Burlington
39. Ge H, Li B, Ge M, Zang N, Nie L, Shen Y, Schuh H (2018) Initial assessment of precise point positioning with LEO Enhanced Global Navigation Satellite Systems (LeGNSS). Remote Sens 10:984

40. GOV.UK (2018) The network and information systems regulations 2018 [Online]. https://www.legislation.gov.uk/uksi/2018/506/regulation/1/made. Accessed 16 Aug 2024
41. Government Office for Science (2018) Satellite-derived time and position: a study of critical dependencies. Crown Copywright, London
42. Han K, Xu B, Shao F, Gong W, Ren Q (2021) An adaptive topology optimization strategy for GNSS inter-satellite network. Institute of Electrical and Electronics Engineers (IEEE)
43. HM Government (2021) National space strategy. Crown Copyright
44. HM Government (2022) Government cyber security strategy: building a cyber resilient public sector
45. IATA (2021) Aviation cyber security guidance material. Part 2: Aircraft [Online]
46. IATA (2022) Compilation of cyber security regulations, standards, and guidance applicable to civil aviation
47. ICAO (2007) Annex 10 to the convention on international civil aviation: aeronautical telecommunications—Vol. III: Communication systems, 2nd edn
48. ICAO (2017) Doc 9849: Global Navigation Satellite System (GNSS) manual, 3rd edn. Montreal, Quebec
49. ICAO (2019) Aviation cybersecurity strategy. Montréal, Quebec
50. ICAO (2022) Annex 17 to the convention on international civil aviation: aviation security, 12th edn. Montreal, Quebec
51. ICAO (2022) Parameters to monitor the performance of ADS-B systems, 1st edn. Mexico
52. ICAO (2023) Annex 10 to the convention on international civil aviation: aeronautical telecommunications—Vol. I: Radio navigation aids, 8th edn. Montreal, Quebec
53. ICAO (2023) Doc 9613: Performance-Based Navigation (PBN) manual, 5th edn. Montreal, Quebec
54. ICAO (2023) Document 9849: Global Navigation Satellite System (GNSS) manual, 4th edn. Montreal, QC
55. ICAO (2023) ICAO Doc 10084: risk assessment manual for civil aircraft operations over or near conflict zones, 3rd edn. International Civil Aviation Organization, Montreal, Quebec
56. INMARSAT (2017) The $3bn boost—how SATCOM has helped the aviation industry [Online]. https://www.inmarsat.com/en/insights/aviation/2017/the-3bn-boost-how-satcom-has-helped-the-aviation-industry.html. Accessed 4 April 2024
57. INSIDE GNSS (2022) Mitigating the threat of jamming and spoofing to aeronautics [Online]. https://insidegnss.com/mitigating-the-threat-of-jamming-and-spoofing-to-aeronautics/. Accessed 01 Aug 2024
58. Islam R (2023) Enhancing satellite security: leveraging GenAI for advanced cyber defense [Online]. Independently published
59. Jafarnia-Jahromi A, Broumandan A, Nielsen J, Lachapelle G (2012) GPS vulnerability to spoofing threats and a review of antispoofing techniques. Int J Navig Observ 2012:1–16
60. Kaplan ED, Hegarty CJ (2006) Understanding GPS: principles and applications [Online]. Artech House
61. Lang S, Füreder P, Riedler B, Wendt L, Braun A, Tiede D, Schoepfer E, Zeil P, Spröhnle K, Kulessa K, Rogenhofer E, Bäuerl M, Öze A, Schwendemann G, Hochschild V (2020) Earth observation tools and services to increase the effectiveness of humanitarian assistance. Eur J Remote Sens 53:67–85
62. Levigne N, Puskar C, Akos D (2019) Enabling high-integrity vehicular satellite navigation operations via automatic gain control. In: 2019 IEEE 90th vehicular technology conference (VTC2019-Fall), Honolulu, HI, USA, pp 1–5
63. Li C, Sun X, Zhang Z (2021) Effective methods and performance analysis of a satellite network security mechanism based on blockchain technology. IEEE Access 9:113558–113565
64. Lo S, Chen YH (2021) Automatic Dependent Surveillance-Broadcast (ADS-B) Universal Access Transceiver (UAT) transmissions for Alternative Positioning, Navigation, and Timing (APNT): concept and practice. Navigation 68:293–313
65. Maaz A (2023) What is ACARS and how do pilots and airlines use it? [Online]. Simple Flying. https://simpleflying.com/acars-guide/. Accessed 15 July 2024

66. Maral G, Bousquet M (2009) Satellite communications systems: systems, techniques and technology. Wiley, Chichester
67. Marques HA, Marques HAS, Aquino M, Veettil SV (2018) Accuracy assessment of precise point positioning with multi-constellation GNSS data under ionospheric scintillation effects. J Space Weather Space Clim 8:A15
68. Mclean D (2003) Automatic flight control systems. Meas Control 36:172–175
69. Meng L, Yang L, Yang W, Zhang L (2022) A survey of GNSS spoofing and anti-spoofing technology. Remote Sens 14:4826
70. Mi X, Zhang B, Yuan Y, Luo X (2019) Characteristics of GPS, BDS2, BDS3 and Galileo inter-system biases and their influence on RTK positioning. Meas Sci Technol 31:015009
71. Nardin A, Dovis F, Fraire A (2021) Empowering the tracking performance of LEO-based positioning by means of meta-signals. IEEE J Radio Freq Identif 5:244–253
72. NASA (2006) Landsat data base map for Google Earth [Online]. https://landsat.gsfc.nasa.gov/article/landsat-data-base-map-for-google-earth/. Accessed 11 June 2024
73. NASA (2022) ViaSat 1 [Online]. https://nssdc.gsfc.nasa.gov/nmc/spacecraft/display.action?id=2011-059A#:~:text=The%20satellite%20was%20launched%20by,for%20North%20America%20and%20Canada. Accessed 07 June 2024
74. NASA (2024) What are SmallSats and CubeSats? [Online]. https://www.nasa.gov/what-are-smallsats-and-cubesats/#:~:text=What%20are%20CubeSats%3F,%2C%206%2C%20and%20even%2012U. Accessed 07 June 2024
75. National Coordination Office for Space-Based Positioning, Navigation, and Timing (2022) The global positioning system [Online]. https://www.gps.gov/systems/gps/. Accessed 28 May 2024
76. National Research Council (2000) the role of small satellites in NASA and NOAA earth observation programs [Online]. The National Academies Press, Washington, DC
77. NCSC (2024) Cyber assessment framework [Online]. National Cyber Security Centre. https://www.ncsc.gov.uk/collection/cyber-assessment-framework. Accessed 16 Aug 2024
78. NCSC (2024) Cyber assessment framework V3.2
79. New Haircut (2024) Problem framing v2: Part 1 of 4 [Online]. https://www.newhaircut.com/blog/problem-framing-v2-part-1-of-4. Accessed 22 July 2024
80. NHS (2024) Changes to the DSPT for large NHS organisations in 2024–25 (2 September 2024) [Online]. National Health Service England. https://www.dsptoolkit.nhs.uk/News/DSPT-Changes-in-24-25. Accessed 06 Sept 2024
81. Nobles C, Burrell D, Waller T (2022) The need for a global aviation cybersecurity defense policy. Land Forces Acad Rev 27:19–26
82. Nowak A, Zajdel R, Sośnica K (2023) Optimization of orbit prediction strategies for GNSS satellites. Acta Astronaut 209:132–145
83. Oakley JG (2020) Cybersecurity for space. Apress
84. Ofgem (2023) NIS supplementary guidance and CAF overlay for DGE sector
85. OPSGROUP (2024) 400% increase in GPS spoofing; workgroup established [Online]. https://ops.group/blog/400-increase-in-gps-spoofing-workgroup-established/. Accessed 30 Aug 2024
86. OPSGROUP (2024) GPS spoofing: final report of the GPS spoofing workgroup
87. Orbiting Now (2024) Active satellite orbit data [Online]. https://orbit.ing-now.com/. Accessed 19 May 2024
88. PenTestPartners (n.d.) Aviation cyber security testing [Online]. https://www.pentestpartners.com/penetration-testing-services/aviation-cyber-security-testing/. Accessed 23 July 2024
89. Prime Minister's Office (2024) The king's speech 2024. Crown Copyright
90. PWC (2024) Is your organisation ready for NIS 2? [Online]. https://www.pwc.co.uk/issues/cyber-security-services/research/is-your-organisation-ready-for-nis-2.html#:~:text=The%20UK%20will%20not%20be,of%20social%20networking%20services%20platforms. Accessed 16 Aug 2024
91. Rathi V, Bhatia V, Suhag V (2014) Evolution of the global positioning system. Int J Res 1
92. ROSCOSMOS, About GLONASS [Online]. https://glonass-iac.ru/en/about_glonass/

93. Safe Airspace (2024) Conflict zone and risk database: Ukraine [Online]. OpsGroup. https://safeairspace.net/ukraine/#:~:text=Ukraine%20closed%20its%20entire%20airspace,between%20Russian%20and%20Ukrainian%20forces. Accessed 22 Aug 2024
94. Safran (n.d.) Aircraft Condition Monitoring Systems (ACMS) & Wireless Extension For ACMS (WEFA) [Online]. https://www.safran-group.com/products-services/aircraft-condition-monitoring-systems-acms-wireless-extension-acms-wefa. Accessed 15 July 2024
95. Satariano A, Kang C (2023) How nations are losing a global race to tackle A.I.'s harms [Online]. The New York Times. https://www.nytimes.com/2023/12/06/technology/ai-regulation-policies.html. Accessed 18 Sept 2024
96. Say M (2023) DLUHC teams works on local government 'overlay' for cyber assessment framework [Online]. https://www.ukauthority.com/articles/dluhc-teams-works-on-local-government-overlay-for-cyber-assessment-framework/. Accessed 06 Sept 2024
97. Shaikh F, Rahouti M, Ghani N, Xiong K, Bou-Harb E, Haque J (2019) A review of recent advances and security challenges in emerging E-enabled aircraft systems. IEEE Access 7:63164–63180
98. SkAI Data Services (2024) Live GPS spoofing tracker map [Online]. https://spoofing.skai-data-services.com/. Accessed 26 July 2024
99. Space Operations Command (2023) Factsheet: global positioning system. United States Space Force
100. Stryker C, Kavlakoglu E (2024) What is artificial intelligence (AI)? [Online]. IBM. https://www.ibm.com/topics/artificial-intelligence. Accessed 18 Sept 2024
101. The Aerospace Corporation (2024) Space Attack Research & Tactic Analysis (SPARTA) [Online]. https://sparta.aerospace.org/. Accessed 03 Aug 2024
102. Timsit A, Westfall S, Taylor A, Francis E (2022) Why Crimea is so important in the Russia-Ukraine war [Online]. The Washington Post. https://www.washingtonpost.com/world/2022/08/18/crimea-russia-ukraine-war/. Accessed 27 Aug 2024
103. Udani SK (1999) VENUS: a virtual environment network using satellites. University of Pennsylvania
104. UK Ministry of Housing, Communities and Local Government (2024) Cyber assessment framework for local government [Online]. UK Ministry of Housing, Communities and Local Government. https://www.localdigital.gov.uk/cyber/cyber-assessment-framework-for-local-government/. Accessed 06 Sept 2024
105. UK Space Agency (2020) Cyber security toolkit
106. Union of Concerned Scientists (2023) UCS satellite database [Online]. https://www.ucsusa.org/resources/satellite-database. Accessed 31 May 2024
107. Van Camp C, Peeters W (2022) A world without satellite data as a result of a global cyber-attack. Space Policy 59:101458
108. Varadharajan V, Suri N (2024) Security challenges when space merges with cyberspace. Space Policy 67:101600
109. Vasquez C, Groll E (2023) Satellite hack on eve of Ukraine war was a coordinated, multi-pronged assault [Online]. Cyberscoop. https://cyberscoop.com/viasat-ka-sat-hack-black-hat/. Accessed 18 Sept 2024
110. Vera-Soto P, Villegas J, Fortes S, Pulido J, Escaño V, Ortiz R, Barco R (2024) An event-driven link-level simulator for the validation of AFDX and Ethernet avionics networks. Aerospace 11:247
111. Vetan D (2023) What is problem framing? [Online]. Design Sprint Academy. https://www.designsprint.academy/blog/what-is-problem-framing. Accessed 22 July 2024
112. Viasat (2021) How big is that satellite? A primer on satellite categories [Online]. https://news.viasat.com/blog/scn/how-big-is-that-satellite-a-primer-on-satellite-categories. Accessed 07 June 2024
113. Weeden B (2010) 2007 Chinese anti-satellite test fact sheet [Online]. Secure World Foundation. https://swfound.org/media/9550/chinese_asat_fact_sheet_updated_2012.pdf. Accessed 18 Sept 2024

114. WEF (2021) Pathways towards a cyber resilient aviation industry [Online]. World Economic Forum. https://www3.weforum.org/docs/WEF_Pathways_Cyber_Resilient_Aviation_2021.pdf
115. Willbold J, Schloegel M, Vögele M, Gerhardt M, Holz T, Abbasi A (2023) Space Odyssey: an experimental software security analysis of satellites. In: IEEE symposium on security and privacy (SP), San Francisco, CA, USA, pp 1–19
116. Wiseman J (2024) GPSJAM: daily maps of GPS interference [Online]. https://gpsjam.org/?lat=49.93199&lon=23.42071&z=2.4&date=2024-07-24. Accessed 24 July 2024
117. Wu DL (2024) GNSS signal jamming as observed from radio occultation. IEEE J Sel Top Appl Earth Observ Remote Sens 17:8642–8645
118. Wu Z, Zhang Y, Yang Y, Liang C, Liu R (2020) Spoofing and anti-spoofing technologies of global navigation satellite system: a survey. IEEE Access 8:165444–165496
119. Wulder MA, Coops NC (2014) Satellites: make earth observations open access. Nature 513:30–31
120. Zhao H, Yan Y, Shi X (2019) A dynamic localization network for regional navigation under global navigation satellite system denial environments. Int J Distrib Sens Netw 15:155014771983442
121. Zidan J, Adegoke EI, Kampert E, Birrell SA, Ford CR, Higgins MD (2021) GNSS vulnerabilities and existing solutions: a review of the literature. IEEE Access 9:153960–153976

Chapter 8
Compatibility of Blockchains in the Internet of Vehicles

Clement Ayodeji Sanni and Umair B. Chaudhry

Abstract The aim of the study is to explore the use of blockchain technology in the Internet of Vehicles (IoVs), especially as regards matters of security and privacy in IoV ecosystems. The literature review from a number of studies which have been carefully examined and juxtaposed shows that decentralized nature, data immutability, and cryptographic protection make Blockchain well fit for meeting challenges in peer-to-peer energy trading, secure data management, and decentralized vehicle interactions. Among other advantages, the review avails tamper-proof data transfer, enhanced confidentiality, and reduction of reliance on middlemen. In the IoV environment, however, scalability issues, regulatory constraints, and resource limitations present significant barriers. This review has highlighted those tailored solutions, including IoV-specific consensus mechanisms, and updated legal frameworks, are necessary to address these challenges. This calls for future research on blockchain technology, which has the transformative potential to construct secure and transparent smart transportation systems.

Keywords Drones · Blockchain technology · Internet of Vehicles · UAV

8.1 Introduction

8.1.1 Background of Study

One of the most persistent challenges and issues for humanity in particular has been the matter of security and privacy. This challenge has been present and valid since time immemorial and is unlikely to vanish as long as people, regions, and territories exist [15]. While in the time of antiquity security was majorly a contest between sticks and stones and who had the bigger stick, in the modern era a region's

C. A. Sanni · U. B. Chaudhry (✉)
Northumbria University London, London, UK
e-mail: umair.chaudhry@northumbria.ac.uk

U. B. Chaudhry and H. Jahankhani, *Autonomous Revolution: Strategies, Threats and Challenges*, Advanced Sciences and Technologies for Security Applications,
https://doi.org/10.1007/978-3-031-96782-5_8

security and spy networks often determine the outcome of civilizations and conflicts. Taking this into account, a significant portion of modern warfare hinges on gathering information concerning military formations, the economics of the enemy region, and secret knowledge crucial to establishing domination [84].

With the significant amount of destruction in World War II and the advancement of weapons of mass destruction, particularly nuclear weapons, most of the major powers realized that large-scale conflict was no longer a viable means of resolving disputes [66]. Such conflicts would leave devastation for all parties, with any "winner" presiding over nothing but ashes and ruins. This understanding led to a shift toward information wars—conflicts fought subtly and quietly behind closed doors, leveraging superior access to and control over information [26]. One of the most significant examples of this was the Cold War between the USSR and the USA, which went on for 45 years from 1947 to 1991 [85]. This prolonged warfare of intelligence culminated in the USSR's eventual dissolution without the need for a full-scale war [69].

In historical times, gathering and securing knowledge was usually a method of showing strength and defeating an adversary [68]. However, with the arrival of the internet, things began to change in how information is distributed. The Internet has brought into existence rapid facilitation regarding knowledge sharing as it forms a part of our lives in the digital age [70, 83]. At the same time, however, its openness has imposed various kinds of security issues. Today, data is called "new gold," pointing out its huge value [35]. That is why the security of data and privacy became crucial in order to prevent any unauthorized access and misusing of information [15].

8.1.2 The Emergence of IoV and Security Challenges

Cybersecurity has grown into a wide range of applications from ordinary online data protection to protection of information emitted by devices like cars, smartwatches, and so on, including IoT technologies [49, 58]. Discussion of potential misuse of this data remains scant, and vulnerability to hacking has been one of the longest-standing concerns for IoT devices [33]. Despite the noted vulnerabilities, IoT devices have become commonly used to improve convenience and efficiency, thus data breaches happen quite often and signal a dire need for reliable means of security [10].

One of the most important trends brought about by the Internet of Things in the world is autonomous cars, also referred to as unmanned or self-driving vehicles [17]. With rapid technological advancements it has become clear that, within the current contemporary technological era, society has fallen in love with new technology, and autonomous vehicles are slowly taking the place of conventional cars [19]. The attractiveness of autonomous vehicles is due to their innovative capabilities, which provide convenience and at the same time improve safety and efficiency on the road [73].

In such a context, authentication stands out as a top priority, as it verifies the identities of vehicles, users, and service providers to ensure that only legitimate entities have access and can control the vehicle's systems. One of the main driving forces behind widespread application involves the increasing number of intelligent sensors, IoT devices, and 5G network support [74].

These technologies have the high-bandwidth, low-latency communications necessary for real-time data transmission. For example, 5G makes communication between autonomous vehicles faster, so as a result they can compute data from different sources in a very short amount of time and make immediate decisions based on that information [48]. That is important functionality if there is to be automated driving, lane-keeping, and collision avoidance systems. Apart from these, IoT devices installed within a vehicle gather information about performance, fuel consumption, or wear and tear, thus suggesting maintenance to avoid unexpected stoppages that may take long to get repaired; that decreases operational costs [12].

What the Internet of Vehicles (IoVs) represents is the integration of vehicles with advanced communication networks. These networks make it possible for there to be real-time data exchange among vehicles, infrastructure, pedestrians, and cloud-based platforms [72]. As one of the cornerstones of ITS, IOVs seek to enhance traffic efficiency, improve road safety, and optimize vehicle performance through seamless connectivity and data-driven decision-making. Vehicle-to-everything communication for IOVs involves V2V, V2I, V2P, and V2C capabilities. This, in a holistic way, is an eco-communication system that allows active cooperation among vehicles by constantly exchanging information about traffic situations, weather conditions, any other potential dangers, etc., to proactively reduce the chances of accidents or traffic congestion [93].

Regardless, looking at it from the point of functional and safety-related security risks, IOVs have a wide range of vulnerabilities against various types of cyber-attacks due to increasing connectivity and dependence on wireless communication protocols [90]. These might involve attacks on vehicle control, such as those handling acceleration, braking, and steering, thus creating possibilities for accidents or unauthorized vehicle control. Besides, the incorporation of many communication networks, for instance, cellular, Wi-Fi, V2V, and V2I, raises the bar for potential hackers to have additional attack surfaces [32].

8.1.2.1 Common Security Challenges of VANETs and IoVs

Fast-growing developments within Vehicular Ad Hoc Networks (VANETs) and the Internet of Vehicles have transformed transportation into one of the most intelligent, effective, and wireless information platforms [37]. VANETs represent a subclass of Mobile Ad Hoc Networks (MANETs), which are meant for providing wireless communication between vehicles-termed Vehicle-to-Vehicle (V2V)-and between vehicles and infrastructure-termed Vehicle-to-Infrastructure (V2I). These networks will be nothing short of pivotal in developing an intelligent transport infrastructure, which will over time improve road safety, manage and reduce traffic, and make driving much easier and better for everyone [27]. Vehicles act as nodes in a VANET-a dynamic wireless network. They are continuously sending and receiving data in real time. Some of this data include things like location, speed, traffic conditions, and safety alerts [44].

However, this increased connectivity also means that there are significant security challenges. The dynamic nature of these networks, along with the amount of sensitive data being transferred, opens the probability of vulnerabilities that malicious people can use. In sum, addressing these issues is impossible without a profound grasp of the underlying threats to be implemented through robust security measures [27].

Here are a few details about some of the high-priority security concerns and vulnerabilities that are often encountered in regard to VANETs and IoVs:

1. Network Instability: IoV communication can suffer catastrophic real-time delays due to unstable networks and bandwidth constraints. There should be a suitable security system in place to deal with these circumstances in order to prevent similar mishaps [24].
2. Insecure Communication Channels: Due to how much data is exchanged between IoV nodes, it is susceptible to cyberattacks. Delay in data processing and communication also creates opportunities for cyber attackers to intercept, manipulate, or destroy data packets en route. This poses a threat to both vehicle functionality and users' safety [71, 86].
3. Security versus Privacy Trade-off: IoV networks need to balance the need for secure communication with the protection of user privacy. Vehicles continuously broadcast sensitive information like real-time location, travel history, and authentication credentials. Unauthorized third parties may use this data as a window to breach the privacy of individuals [18].
4. Emerging Vehicle Theft Threat: In-vehicle IoV system integration opens up more attack vectors to vehicle security. Attackers can utilize security vulnerabilities in authentication mechanisms for hijacking control or falsifying unique identifiers of any vehicle [28]. This facilitates the act of theft, concealing the identity of a vehicle, reselling, and using it for any form of crime. Since car theft is a major issue for consumers these days, installing a suitable anti-theft system ought to be mandatory. The anti-theft system ought to be properly authenticated so that hackers might take use of its authentication mechanism. Only authorized users should be able to access authentication. Only users should have access to information on vehicle movement, condition, real-time locations, prior trip history, etc. [75].
5. Scalability Challenges: The increasing ratio of IoV-enabled vehicles leads to scalability issues in VANETs. This makes management of communication, data exchange, and resource allocation far more complicated with millions of interconnected vehicles. This scenario leads to performance bottlenecks and increased security vulnerabilities due to network expansion.

8.1.3 *Blockchain's Role in Securing IoV Systems*

8.1.3.1 How Blockchain Works

A digitally distributed database or tally that is shared by all of the nodes in a peer-to-peer network is called a blockchain [64]. The distributed ledger stores events or information as a collection of blocks that are secured and connected to one another using a hashing algorithm. Only the connected users could maintain these blocks, which are shared by all linked nodes [7]. As noted in [64] the server-client paradigm, in which the server holds the primary power to manage all of the network's linked nodes, has been superseded by blockchain technology. Blockchain operates on a client-to-client model, allowing all users or linked nodes to create and manage shared data inside the network, significantly lowering time and money-related expenses. The ever-expanding chains of blocks that make up a blockchain employ two different kinds of mechanisms: consensus and cryptography [89]. The Blockchain's cryptographic system encrypts data using a hashing technique, in which each block has a distinct hash value in addition to the hash value of the block that came before it on the chain. Digital signatures, a form of asymmetric cryptography used by blockchain, rely on public and private key cryptography to authenticate identities and supply data. Elliptical Curve Digital Signatures (ECSDA) are typically used in blockchain [53].

Blockchain verifies the privacy and integrity of data through a consensus process. The consensus mechanisms used in the blockchain to add new transactions, change the information already included in the blocks, and confirm identity identification include Proof of Work (PoW), Proof of Stack (PoS), and Practical Byzantine Fault Tolerance (PBFT) [55]. We can comprehend the blockchain's operation from Fig. 8.1. Public Blockchains are really decentralized distributed ledgers where everyone may take part in validating the transactions or data using its continuous consensus method, whereas private Blockchains can be based on permission [6]. It is regarded as unchangeable and offers participating users or nodes complete anonymity because of its consensus method. Islam [6] notes that a private blockchain is a centralized or permissioned blockchain in which the network is managed by a single company or network administrator, and only authorized users or nodes are permitted to participate in transaction validation.

8.1.3.2 Blockchain as a Security Mechanism

As previously stated, the appeal of autonomous vehicles lies in their innovative capabilities, which not only offer convenience but also enhance safety and efficiency on the road. However, one cannot ignore the security issues that come along with such technology, and while there have been several solutions offered, one of the most feasible responses has been the introduction of blockchain for advanced security [77].

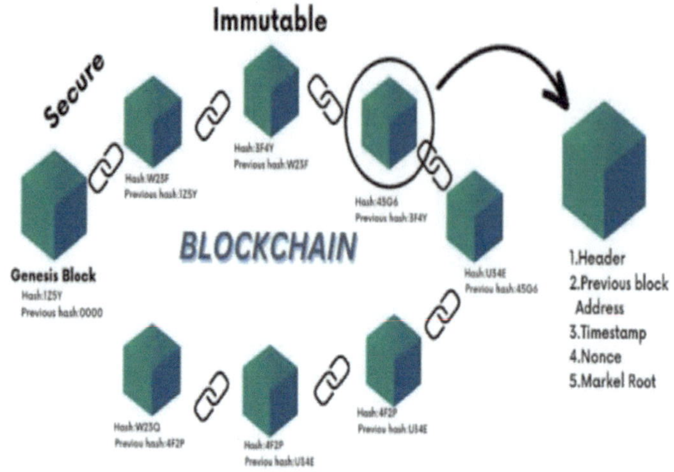

Fig. 8.1 Working mechanism of blockchain

The prominence of blockchain technology provides a promising solution to these challenges. As covered in Singh et al. [77], Blockchain operates on a client-to-client (C2C) architecture, where data is distributed across multiple nodes in a decentralized ledger. Each autonomous vehicle can be equipped with a blockchain-based identity that allows it to communicate securely with other vehicles, infrastructure, and service providers. By taking this approach, not only is there now a heightened level of security brought about by the elimination of the concept of a single point of failure, but there is also the provision of full transparency and traceability for all transactions and data.

Take the matter of data security, for instance; blockchain can provide a cryptographic mechanism for location, speed, and destination in general [89]. It makes the interception or manipulation of data by an unauthorized party very difficult since blockchain breaks down data into encrypted packets and distributes them across various nodes. Again, since the nature of blockchain is to distribute data in encrypted packets, the common fault of most traditional systems, where a single point of failure is possible, is invalid here, thus securing privacy and protecting against any unauthorized access.

Besides data security, blockchain technology can be used to avoid car theft-a vital factor, considering how much value an autonomous car has [77]. Vehicle theft alone is a multibillion-dollar problem each year, the main fate of stolen cars is either being dismantled for parts or sold in illegal markets [39]. Blockchain will ensure the problem is tackled with the creation of a safe, digital ledger tracking each vehicle's ownership and transaction history. Each time there is a sale or change of ownership in a motor vehicle, the information recorded on the blockchain makes it virtually unchangeable without the relevant authorization [14]. This would assist in both theft situations and loss vehicles recovery.

In conclusion, this research embarks on a timely and significant exploration into the intersection of blockchain technology and the Internet of Vehicles (IoV), with a focus on advancing security, safeguarding data privacy, and reducing vulnerability to cyber threats and vehicle theft. The increasing interconnectivity of vehicles within IoV systems, while offering transformative benefits in transportation safety and efficiency, also exposes these networks to escalating cybersecurity risks. This study is motivated by the urgent need to address these vulnerabilities through innovative and sustainable solutions. Blockchain's inherent properties of decentralization, immutability, and transparency position it as a promising candidate to strengthen trust and resilience within IoV ecosystems. Guided by a clearly defined research aim, objectives, and hypothesis, this work aims to propose and validate a comprehensive blockchain-based security framework. The outcomes of this research are expected to contribute meaningfully to both academic discourse and practical advancements in smart transportation, paving the way for a more secure and reliable digital infrastructure for modern mobility systems.

8.2 Literature Review

8.2.1 Introduction

The current rate of technological advancement has introduced IoV technologies that provide several new security challenges due to their inherent nature, which is interdependent and decentralized [77]. Unlike traditional IT systems, IoT devices often have limited security capabilities due to resource constraints and may be physically accessible in open environments [31]. Furthermore, the IoT system introduces new attack vectors that could compromise user privacy and security [16], because a vehicle is talking in real time with other vehicles, infrastructure, and cloud services, the attack surface grows drastically, thereby allowing room for malicious activities [20]. These are not only technical issues but also operational challenges: data integrity, privacy, and system resilience against threats that continuously evolve. For instance, hostile actors may use vulnerabilities in the communication protocols to intercept data or disrupt vehicular functions, leading to chaos in the traffic flow or worse, causing death [38].

Also, IoV is aiming to integrate both autonomous and semi-autonomous vehicles; thus, the stakes are huge. An unmanned vehicle will be dependent on advanced sensors and AI-driven decision-making capabilities, and it is highly susceptible in cases of hijacking and cyberattacks. Incidents in real life-like, the infamous Jeep Cherokee hack in 2015 [88], point out the potential consequences of a compromised vehicle security landscape. Such incidents highlight that it is high time there are more competent solutions to IoV's security issues, where traditional solutions mostly fall short. This section critically analyses such challenges by examining the findings of research and exploring conflicting perspectives to gain a comprehensive understanding of security in IoV.

8.2.2 *Concerns on the Security of IoVs; a Systematic Review*

8.2.2.1 Vulnerability in Communication and Man-in-the-Middle Attacks

IoV heavily relies on VANETs for communication between V2V, V2I, and other network entities. However, their decentralized and open architectures make such networks very susceptible to various types of attacks [8]. One of the key issues is the susceptibility to man-in-the-middle attacks, where malicious actors intercept or alter communications. The extent to which attacks of such nature might be feasible has been extensively documented by Petit and Shladover [67], in their work they identified spoofing and jamming as major risks to V2V communication. These vulnerabilities can lead to severe safety risks, including false traffic warnings or manipulated vehicle behaviours. However, research by Parno et al. [65] challenged the extent of these risks. The argument made was that real-time cryptographic protocols was more than capable of effectively mitigate MITM attacks if properly implemented. Yet, this optimistic view often underestimates the challenges of deploying cryptographic systems in resource-constrained vehicular environments, especially under real-world latency constraints.

8.2.2.2 Denial of Service (DOS)

Another critical threat to IoV communication networks is a Denial of service (DoS) attacks. These are attacks that involve the smashing of a system with so much traffic that it causes delays in communications or complete failures thereof. Comert et al. [25] showed how trials of autonomous vehicles recorded increased latency and immobilization during simulated DoS attacks. This result points out the weakness of IoV systems when network congestion occurs. On the other hand, Ksouri et al. [50] proposed the use of resilient routing protocols that reroute communications during active DoS incidents with their integrity intact [56]. Though this sounds promising, there is one critical critique-that the very pathways to which communication can be rerouted in the presence of an attack could also be subject to sophisticated attacks themselves [79].

8.2.2.3 Hacking, Hijacking and Data Breaches

Undoubtedly, perhaps the most sensational and most alarming worry about IoV are those concerning unauthorized access and hijacking of vehicles [86, 91]. The infamous 2015 Jeep Cherokee hack by Miller and Valasek remains a pivotal case study and classic example of this [28, 88].

In their study, Al-Soufi et al. [6] demonstrated how GPS jamming and spoofing, executed using a HackRF One PortaPack Software Defined Radio (SDR), can severely compromise a drone's navigational system, redirecting it to unintended locations or causing operational failure. The researchers proposed and evaluated a novel recovery mechanism called "Return-to-Start," implemented on a Raspberry Pi-based custom drone. This mechanism successfully mitigated the effects of GPS attacks by autonomously returning the drone to its starting position during signal disruptions.

By exploiting vulnerabilities in the Uconnect infotainment system, researchers remotely accessed critical controls, including steering and braking [29]. Drozhzhin [29] highlighted the inadequate authentication mechanisms and outdated software systems pervasive in many IoV deployments. However, a counterargument emerges from manufacturers who have since implemented over-the-air (OTA) updates to address such vulnerabilities. For example, Tesla has employed OTA updates in patching security vulnerabilities in a timely manner [76]. But researchers like Li et al. [51] advise that overdependence on OTA systems may be unwarranted because such mechanisms could also serve as attack vectors if they are not well-protected. The threat of privacy and data breaches also increases within the IoV environment, given the massive volume of sensitive information generated and stored [3, 33]. Centralized cloud systems create a single point of failure, and it is very common to use them for data storage [77]. Examples include the 2019 Honda data breach, which included the leakage of personal information about more than 26,000 customers [36].

8.2.2.4 Lack of Standardized Security Protocols

Privacy advocates say the best solutions to avoid exposure will be decentralized: federated learning and edge computing. For instance, Liu et al. [54] have indicated the capability of edge computing in processing information locally and reducing the possibilities of breaches. However, the critics have also pointed to the costs related to this decentralization regarding computational overhead and the potential compromise of edge nodes themselves. Besides this, the lack of standardized security protocols within IoV manufacturers and varied platforms has increased its vulnerability further [42, 78]. Several studies have been conducted on the security features of IoT modules in autonomous vehicles [41, 62, 81, 91], and these have demonstrated large differences in the security practices between the various modules.

8.2.2.5 Cyber-Physical Threats and Sensor Attacks

Another big challenge is the cyber-physical threats on IoV sensor systems. While the integration of LiDAR, radar, and cameras helps the vehicle understand its environment, these sensors are still vulnerable to spoofing, adversarial attacks or plain old error [94].

8.2.3 Advanced Solutions for the Challenges of Security in Internet of Vehicles-IoV

The integrating of vehicles, infrastructure, and users in IoV makes transportation more efficient and safer. The unauthorized access, data breach, and system weaknesses are the various vulnerabilities introduced by the interconnection of the objects in IoV. Machine Learning, AI, robotics, and are some of the Cryptographic techniques that have been studied in detail to handle these challenges are Machine Learning, AI, and robotics [2, 11, 57]. The aim here is to review these methodologies, talk about their mechanisms, strengths, and limitations, and their practical applications.

8.2.3.1 Machine Learning and Deep Learning Approaches

Intrusion Detection Systems (IDS)

The ML-based IDS detects malicious activity within vehicular networks due to unauthorized access or denial-of-service attacks [47]. These systems operate based on either supervised model trained by labelled datasets or unsupervised models that define deviations from an established norm. They assure real-time threat detection and scalability but are not without challenges such as dependence on quality training data and vulnerability to zero-day attacks [47]. The IDS model presented by Cheng and Chai [23] is a typical example, as seen that it was able to reach somewhat significant level of detection accuracy but showed poor performance when the dataset was imbalanced.

Anomaly Detection

Deep learning (DL) models, in particular, neural networks inspect the pattern of vehicular communications to detect anomalies that can indicate a potential threat. Though adaptive and effective against evolving threats, these methods are computation-intensive; thus, deployment challenges persist in resource-constrained IoV environments [40]. Eykholt et al. [30] emphasized that though DL models are efficient in detecting adversarial inputs, they also showed their vulnerability to crafted adversarial attacks.

Federated Learning

One of the well-known advantages of Federated learning is that it enables decentralized learning across IoV devices, preserving data privacy while fostering collaborative security solutions. Despite its advantages, such as enhanced privacy and adaptability, FL is susceptible to poisoning attacks and requires reliable communication infrastructure [52]. Research by demonstrated FL's potential for IDS in IoV but noted performance degradation in high-mobility scenarios, plus even though FL mitigates direct data sharing, sharing model updates can still leak sensitive information [21, 22].

8.2.3.2 Artificial Intelligence Solutions

AI-Based Authentication

AI uses behavioural biometric and contextual information to develop enhanced authentication. This helps in reducing the dependence on static credentials and provides some resistance against spoofing attacks, but it also raises several privacy concerns and suffers from issues related to overlapping behavioural patterns [61]. For instance, a framework of AI-driven authentication might be highly resistant but suffer from false rejections in edge cases and might not cater to the user's needs [61].

Predictive Maintenance

The way this works is that the AI algorithms analyse vehicle performance data in order to predict possible system failures, sort of like a pre-emptive measure rather than a reactive one which helps to proactively fix vulnerabilities and enhance reliability [1]. The drawback however is significant, in that while this method enhances security, it requires a lot of historical data and does not directly counter cyber threats [60].

8.2.3.3 Robotics and Sensor Fusion

Autonomous Navigation

Robotics enable navigation through a complex environment autonomously, hence improving efficiency in operation and reducing human error. Their systems are prone to adversarial attacks and spoofing [9]. Advances by Nvidia in robotics address improvements in autonomous navigation but face challenges in environments whose inputs have been manipulated [82].

Sensor Fusion

According to Yeong et al. [92], the Integration of multi-sensors will improve the environment perception and security. Such redundancy can potentially avoid single-point failures but also introduces significant computational demands and possible integration errors. Sensor fusion has also been shown to enhance accuracy and robustness in detection against spoofing attacks with added complexity and higher costs.

8.2.3.4 Cryptographic Techniques

Public Key Infrastructure (PKI)

PKI authenticates the messages in IoV with digital certificates and encryption for secure communication. However, PKI, while providing strong security, faces challenges in the implementation of IoV dynamics because of certificate management and scaling [45]. A very good example is the PKI for vehicular networks, effective but with latency in the validation of certificates and they are prone to bottlenecks [13].

Lightweight Cryptography

Designed for resource-constrained IoV devices, lightweight cryptographic algorithms secure communications without overburdening system resources. While efficient, these methods usually have poor balancing between security and performance, and they are not strong enough to resist powerful adversaries [80].

8.2.4 Blockchain Technology for AVs

Blockchain technology has been an emerging strong solution for problems in the field of automated vehicle security, considering that it is decentralized, tamper-proof, and public in nature [4]. Basically, it was developed for cryptocurrency. In less than a decade, the technology grew to cover key sectors like health and supply chain management. Huge applications in IoV assure integrity, authentication, privacy, and communication security of data. From a theoretical approach to real-world applications, blockchain technology has gathered momentum over the last decade and found applications in almost every major industry-be it food supply chains, healthcare, or IoV [46]. As the numbers of AVs and other connected devices continue to grow on the road, so too will demand for secure, transparent, and efficient data sharing. Much of this can be realised with blockchain, thanks to its decentralised, immutable, and cryptographically secure framework-a good alternative to more traditional centralized security models.

8.2.5 *Advantages of Blockchain over Traditional Solutions in AVs*

Blockchain provides several formidable benefits over common centralized systems in IoV security. Traditional solutions are usually based on central servers, which are single points of failure and might be exposed to cyberattacks and data breaches. Blockchain's distributed architecture ensures resilience and redundancy by spreading synchronous data across multiple nodes, thereby reducing risks related to a central storage of data [4].

Another strong feature of blockchain and arguably the most important is immutability. As seen in Molina et al. [59], Once posted on blockchain, data can neither be removed nor changed without an agreement of the network. This thus provides evidence of the data integrity and an auditable line, thereby engendering confidence among the stakeholder community. These traditional solutions are amenable to tampering and unauthorized change since they are naturally centralized.

Transparency is one of the best features of blockchain. Nguyen et al. [63], cover how that all transactional information can be traced on the distributed ledger, and all stakeholders of the supply chain can check the same. This makes transactions more accountable. These are not always open in the context of centralized systems. Lack of transparency may result in inefficiency and disputes.

Finally, the inherent design of blockchain is fully capable of decentralizing most transactions between peers without the need for intermediaries. This nature of blockchain reduces costs and accelerates transaction settlement with minimum probabilities of manipulation. Traditional systems, therefore, are related to inefficiency in processes and increased operational costs due to the involvement of intermediaries [59].

8.2.6 *Limitations of Blockchain in IoV Applications*

Despite the transformational potentials of blockchain technology, there are several challenges ahead to integrate it into IoV systems, most of which are covered in Kapassa et al.'s work of 2021, such as below.

8.2.6.1 Legal and Regulatory Challenges

Regulatory and legal challenges are there that blockchain adoption in IoV faces. Thus, the legal framework so far does not provide for a decentralized interaction and P2P trade. Besides, the concept of smart contract requires major revision of the legal framework [34]. GDPR, on the other hand, has posed an added complication over consumer data protection, wherein privacy is required for sensitive data while blockchain was conceived with transparency in mind [43].

8.2.6.2 Performance-Based Constraints

IoV systems have an operational basis in the real-time processing of big volumes of data. Scalability and throughput, on the other hand, remain big problems for blockchain technologies. For example, according Kapassa et al.'s study in 2021, where VISA processes thousands of transactions per second, Bitcoin processes only seven. Better speed and low latency is expected by IoV applications at large scales, and therefore the blockchain performance gap limits applicability in this sense.

8.2.6.3 Resource Limitations

Most of the current IoV devices, sensors, and charging stations are usually resource-constrained in their computation and storage capabilities. High energy consumption of PoW-oriented consensus algorithms is unrealistic for resource-constrained IoV devices [43]. In addition, it is also not feasible to store the whole blockchain in every device. This calls for finding an alternative strategy to minimize the storage challenges.

8.2.6.4 Suitable Consensus Algorithm

Most of the generally accepted consensus algorithms, such as PoW and PoS, are inefficient and unsuitable for IoV due to their high computation or due to the centralization problem [87]. However, IoV-specific algorithms have just started developing and need further testing and optimization to show reliability and scalability [43].

8.3 Conclusions

This literature review focused on blockchain technology integrated with IoV and AV with the purpose of enhancing security and privacy. Blockchain technology has come up as an evolutionary solution to security, privacy, and transparency issues in the ecosystem of IoV. Due to the inherent decentralized nature of a blockchain, enhanced by advanced cryptographic techniques, this is the clear winner for resilience in single-point-of-failure attacks, data immutability, and advanced trust mechanisms over more conventional systems. Challenges to widespread adoption still persist at several critical barriers: scalability challenges, legal issues, constraints of resources, and optimization specific to IoV. Overcoming these limitations with regulatory innovation, consensus mechanisms tailored to fit purpose, and optimized storage strategies will be important in realizing full value from blockchain for IoV applications. As research and development continue to get better, the integration of blockchain into IoV can revolutionize how vehicles and their supporting infrastructure function and thus lay the bedrock for secure and efficient smart transportation systems.

References

1. Aboshosha A, Haggag A, George N, Hamad HA (2023) IoT-based data-driven predictive maintenance relying on fuzzy system and artificial neural networks. Sci Rep 13(1):12186
2. Al-Garadi MA, Mohamed A, Al-Ali AK, Du X, Ali I, Guizani M (2020) A survey of machine and deep learning methods for internet of things (IoT) security. IEEE Commun Surv Tutor 22(3):1646–1685
3. Al-Soufi L, Demirsoy T, Soyak EG (2022) An implementation-based study of the detection of and recovery from GPS spoofing attacks for unmanned aerial vehicles. Int J Adv Eng Pure Sci 36(3):211–223
4. Ali B, Awad AI (2018) Cyber and physical security vulnerability assessment for IoT-based smart homes. Sensors 18(3):817
5. Alladi T, Chamola V, Sahu N, Venkatesh V, Goyal A, Guizani M (2022) A comprehensive survey on the applications of blockchain for securing vehicular networks. IEEE Commun Surv Tutor 24(2):1212–1239
6. Aminul Islam M (2022) Blockchain technology: a tool to solve the challenges of education sector in developing countries. arXiv-2211
7. An M, Fan Q, Yu H, Zhao H (2023) Blockchain technology research and application: a systematic literature review and future trends. arXiv:2306.14802
8. Anzer A, Elhadef M (2018) A multilayer perceptron-based distributed intrusion detection system for internet of vehicles. In: 2018 IEEE 4th international conference on collaboration and internet computing (CIC). IEEE, pp 438–445
9. Arafin MT, Kornegay K (2021) Attack detection and countermeasures for autonomous navigation. In: 2021 55th annual conference on information sciences and systems (CISS). IEEE, pp 1–6
10. Arshad J, Azad MA, Amad R, Salah K, Alazab M, Iqbal R (2018) A review of performance, energy and privacy of intrusion detection systems for IoT. Electronics 9(4):629
11. Asharf J, Moustafa N, Khurshid H, Debie E, Haider W, Wahab A (2020) A review of intrusion detection systems using machine and deep learning in internet of things: challenges, solutions and future directions. Electronics 9(7):1177
12. Aydin O, Guldamlasioglu S (2017) Using LSTM networks to predict engine condition on large scale data processing framework. In: 2017 4th international conference on electrical and electronic engineering (ICEEE). IEEE, pp 281–285
13. Bellur B (2008) Certificate assignment strategies for a PKI-based security architecture in a vehicular network. In: IEEE GLOBECOM 2008—2008 IEEE global telecommunications conference, New Orleans, LA, USA, pp 1–6. https://doi.org/10.1109/GLOCOM.2008.ECP.355
14. Benarous L, Kadri B, Bouridane A, Benkhelifa E (2022) Blockchain-based forgery resilient vehicle registration system. Trans Emerg Telecommun Technol 33(10):e4237

15. Bertino E (2016) Data security and privacy: concepts, approaches, and research directions. In: 2016 IEEE 40th annual computer software and applications conference (COMPSAC), vol 1. IEEE, pp 400–407
16. Bharati S, Podder P (2022) Machine and deep learning for IoT security and privacy: applications, challenges, and future directions. Secur Commun Netw 2022(1):8951961
17. Blowers M, Iribarne J, Colbert E, Kott A (2016) The future internet of things and security of its control systems. arXiv:1610.01953
18. Boualouache A, Soua R, Tang Q, Engel T (2020) Software-defined location privacy protection for vehicular networks. In: Machine intelligence and data analytics for sustainable future smart cities. Springer International Publishing, Cham, pp 395–411
19. Cao H, Zou W, Wang Y, Song T, Liu M (2022) Emerging threats in deep learning-based autonomous driving: a comprehensive survey. arXiv:2210.11237
20. Cathey G, Benson J, Gupta M, Sandhu R (2021) Edge centric secure data sharing with digital twins in smart ecosystems. In: 2021 third IEEE international conference on trust, privacy and security in intelligent systems and applications (TPS-ISA). IEEE, pp 70–79
21. Chellapandi VP, Yuan L, Brinton CG, Żak SH, Wang Z (2023) Federated learning for connected and automated vehicles: a survey of existing approaches and challenges. IEEE Trans Intell Veh
22. Chellapandi VP, Yuan L, Żak SH, Wang Z (2023) A survey of federated learning for connected and automated vehicles. In: 2023 IEEE 26th international conference on intelligent transportation systems (ITSC). IEEE, pp 2485–2492
23. Cheng Z, Chai S (2020) A cyber intrusion detection method based on focal loss neural network. In: 2020 39th Chinese control conference (CCC). IEEE, pp 7379–7383
24. Chu J, Wu Q, Fan Q, Li Z (2023) Enhanced C-V2X mode 4 to optimize age of information and reliability for IoV. In: 2023 IEEE 23rd international conference on communication technology (ICCT). IEEE, pp 1082–1086
25. Comert G, Chowdhury M, Nicol DM (2020) Assessment of system-level cyber attack vulnerability for connected and autonomous vehicles using Bayesian networks. arXiv:2011.09436
26. Crilley K (2001) Information warfare: new battle fields terrorists, propaganda and the Internet. In: Aslib proceedings, vol 53, no 7. MCB UP Ltd., pp 250–264
27. Cunha F, Villas L, Boukerche A, Maia G, Viana A, Mini RA, Loureiro AA (2016) Data communication in VANETs: protocols, applications and challenges. Ad Hoc Netw 44:90–103
28. Dibaei M, Zheng X, Jiang K, Maric S, Abbas R, Liu S, Zhang Y, Deng Y, Wen S, Zhang J, Xiang Y (2019) An overview of attacks and defences on intelligent connected vehicles. arXiv:1907.07455
29. Drozhzhin A (2015) Black Hat: Jeep Cherokee hack explained. Kaspersky Daily, 6 August. https://www.kaspersky.com/blog/blackhat-jeep-cherokee-hack-explained/9493/. Accessed 30 Dec 2024
30. Eykholt K, Evtimov I, Fernandes E, Li B, Rahmati A, Xiao C, Prakash A, Kohno T, Song D (2018) Robust physical-world attacks on deep learning visual classification. In: Proceedings of the IEEE conference on computer vision and pattern recognition, pp. 1625–1634
31. Fan J, Yang W, Liu Z, Kang J, Niyato D, Lam KY, Du H (2023) Understanding security in smart city domains from the ANT-centric perspective. IEEE Internet Things J 10(13):11199–11223
32. Farivar F, Haghighi MS, Jolfaei A, Wen S (2020) On the security of networked control systems in smart vehicle and its adaptive cruise control. IEEE Trans Intell Transp Syst 22(6):3824–3831
33. Feng C, Celdran AH, Sanchez PMS, Kreischer J, von der Assen J, Bovet G, Perez GM, Stiller B (2023) CyberForce: a federated reinforcement learning framework for malware mitigation. arXiv:2308.05978
34. Finck M (2018) Blockchains and data protection in the European Union. Eur Data Protect Law Rev 4(1):17–35. https://doi.org/10.21552/edpl/2018/1/6
35. Forbes Council (2023) How to make use of the new gold: data. Forbes. https://www.forbes.com/councils/forbestechcouncil/2023/03/27/how-to-make-use-of-the-new-gold-data/#:~:text=Conclusion,valuable%20insights%20and%20fuel%20innovation. Accessed 20 Dec 2024

36. Gatlan S (2019) Honda exposes 26,000 records of North American customers. Bleeping computer, 24 July. https://www.bleepingcomputer.com/news/security/honda-exposes-26-000-records-of-north-american-customers/. Accessed 30 Dec 2024
37. Guerna A, Bitam S, Calafate CT (2022) Roadside unit deployment in internet of vehicles systems: a survey. Sensors 22(9):3190
38. Hasan M, Mohan S, Shimizu T, Lu H (2020) Securing vehicle-to-everything (V2X) communication platforms. IEEE Trans Intell Veh 5(4):693–713
39. IBC (2024) End auto theft. https://www.ibc.ca/stay-protected/theft-prevention/end-auto-theft. Accessed 21 Dec 2024
40. Ingham KL, Somayaji A (2007) A methodology for designing accurate anomaly detection systems. In: Proceedings of the 4th international IFIP/ACM Latin American conference on networking, pp 139–143
41. Jha S, Jha N, Prashar D, Ahmad S, Alouffi B, Alharbi A (2022) Integrated IoT-based secure and efficient key management framework using hashgraphs for autonomous vehicles to ensure road safety. Sensors 22(7):2529
42. Kang J, Xiong Z, Niyato D, Ye D, Kim DI, Zhao J (2019) Toward secure blockchain-enabled internet of vehicles: optimizing consensus management using reputation and contract theory. IEEE Trans Veh Technol 68(3):2906–2920
43. Kapassa E, Themistocleous M, Christodoulou K, Iosif E (2021) Blockchain application in Internet of Vehicles: challenges, contributions and current limitations. Future Internet 13(12):313. https://doi.org/10.3390/fi13120313
44. Khan FY (2023) Analysis and design of VANET protocols for Srinagar City. arXiv:2301.03227
45. Khan S, Luo F, Zhang Z, Rahim MA, Ahmad M, Wu K (2022) Survey on issues and recent advances in vehicular public-key infrastructure (VPKI). IEEE Commun Surv Tutor 24(3):1574–1601
46. Khanfar AA, Iranmanesh M, Ghobakhloo M, Senali MG, Fathi M (2021) Applications of blockchain technology in sustainable manufacturing and supply chain management: a systematic review. Sustainability 13(14):7870
47. Kimanzi R, Kimanga P, Cherori D, Gikunda PK (2024) Deep learning algorithms used in intrusion detection systems—a review. arXiv:2402.17020
48. Kountouris M, Pappas N (2020) Semantics-empowered communication for networked intelligent systems. IEEE Commun Mag 59(6):96–102
49. Koupaei ANA, Nazarov AN (2020) A hybrid security solution for mitigating cyber-attacks on info-communication systems. In: 2020 international conference engineering and telecommunication (En&T). IEEE, pp 1–4
50. Ksouri C, Jemili I, Mosbah M, Belghith A (2022) Towards general Internet of Vehicles networking: routing protocols survey. Concurr Comput Pract Exp 34(7):e5994
51. Li T, Sahu AK, Talwalkar A, Smith V (2020) Federated learning: challenges, methods, and future directions. IEEE Signal Process Mag 37(3):50–60
52. Li B, Hu W, Da L, Wu Y, Wang X, Li Y, Yuan C (2024) Over-the-air upgrading for enhancing security of intelligent connected vehicles: a survey. Artif Intell Rev 57(11):314
53. Linke B (2014) The fundamentals of an ECDSA authentication system. Tutorial article, maxim integrated
54. Liu S, Liu L, Tang J, Yu B, Wang Y, Shi W (2019) Edge computing for autonomous driving: opportunities and challenges. Proc IEEE 107(8):1697–1716
55. Lohar P, Rathi K (2023) Evaluation of Non-Fungible Token (NFT). arXiv:2308.14389
56. Long M, Wu CH, Hung JY, Irwin JD (2004) Mitigating performance degradation of network-based control systems under denial of service attacks. In: 30th annual conference of IEEE Industrial Electronics Society, 2004. IECON 2004, vol 3. IEEE, pp 2339–2342
57. Macedo EL, De Oliveira EA, Silva FH, Mello RR, França FM, Delicato FC, de Rezende JF, de Moraes LF (2019) On the security aspects of Internet of Things: a systematic literature review. J Commun Netw 21(5):444–457
58. Makhdoom I, Abolhasan M, Lipman J, Liu RP, Ni W (2019) Anatomy of threats to the Internet of Things. IEEE Commun Surv Tutor 21(2):1636–1675. https://doi.org/10.1109/COMST.2018.2874978

59. Molina F, Betarte G, Luna C (2020) A blockchain based and GDPR-compliant design of a system for digital education certificates. arXiv:2010.12980
60. Mourtzis D, Angelopoulos J, Panopoulos N (2020) Intelligent predictive maintenance and remote monitoring framework for industrial equipment based on mixed reality. Front Mech Eng 6:578379
61. Mustafa MA, Abidin A, Rúa EA (2018) Frictionless authentication system: security and privacy analysis and potential solutions. arXiv:1802.07231
62. Nanda A, Puthal D, Rodrigues JJ, Kozlov SA (2019) Internet of autonomous vehicles communications security: overview, issues, and directions. IEEE Wirel Commun 26(4):60–65
63. Nguyen LT, Nguyen LD, Hoang T, Bandara D, Wang Q, Lu Q, Xu X, Zhu L, Popovski P, Chen S (2023) Blockchain-empowered trustworthy data sharing: fundamentals, applications, and challenges. arXiv:2303.06546
64. Park A, Li H (2021) The effect of blockchain technology on supply chain sustainability performances. Sustainability 13(4):1726
65. Parno B, Perrig A, Gligor V (2005) Distributed detection of node replication attacks in sensor networks. In: 2005 IEEE symposium on security and privacy (S&P'05). IEEE, pp 49–63
66. Perry WJ (2004) Military technology: an historical perspective. Technol Soc 26(2–3):235–243
67. Petit J, Shladover SE (2015) Potential cyberattacks on automated vehicles. IEEE Trans Intell Transp Syst 16(2):546–556
68. Raven BH (1999) Kurt Lewin address: influence, power, religion, and the mechanisms of social control. J Soc Iss 55(1):161–186
69. Richterova D, Telepneva N (2021) An introduction: the secret struggle for the Global South–espionage, military assistance and state security in the cold war. Int Hist Rev 43(1):1–11
70. Rosmani AF, Mutalib AA, Sarif SM (2020) The evolution of information dissemination, communication media and technology in Malaysia. J Phys: Conf Ser 1529(2):022044. IOP Publishing
71. Salek MS, Khan SM, Rahman M, Deng HW, Islam M, Khan Z, Chowdhury M, Shue M (2022) A review on cybersecurity of cloud computing for supporting connected vehicle applications. IEEE Internet Things J 9(11):8250–8268
72. Sanguesa JA, Fogue M, Garrido P, Martinez FJ, Cano JC, Calafate CT, Manzoni P (2013) An infrastructureless approach to estimate vehicular density in urban environments. Sensors 13(2):2399–2418
73. Scurt FB, Vesselenyi T, Tarca RC, Beles H, Dragomir G (2021) Autonomous vehicles: classification, technology and evolution. IOP Conf Ser: Mater Sci Eng 1169(1):012032. IOP Publishing
74. Sharma N, Chauhan N, Chand N (2018) Security challenges in Internet of Vehicles (IoV) environment. In: 2018 first international conference on secure cyber computing and communication (ICSCCC). IEEE, pp 203–207
75. Sharmila P, Dev B, Dhanvarsha D, Pavai A (2021) Charter authenticated anti-theft secure motor system. In: ITM web of conferences, vol 37. EDP Sciences, p 01009
76. Singh K (2024) Tesla resolves 700K vehicle recall via OTA update; NHTSA now highlights OTA recalls. Not a Tesla App, 29 December. https://www.notateslaapp.com/news/2450/tesla-resolves-700k-vehicle-recall-via-ota-update-nhtsa-now-highlights-ota-recalls. Accessed 30 Dec 2024
77. Singh PK, Nandi S, Nandi SK, Ghosh U, Rawat DB (2021) Blockchain meets AI for resilient and intelligent internet of vehicles. arXiv:2112.14078
78. Sisodia D (2020) On the state of internet of things security: vulnerabilities. Attacks, and recent countermeasures. University of Oregon
79. Sun Y, Apostolaki M, Birge-Lee H, Vanbever L, Rexford J, Chiang M, Mittal P (2021) Securing internet applications from routing attacks. Commun ACM 64(6):86–96
80. Thakor VA, Razzaque MA, Khandaker MR (2020) Lightweight cryptography for IoT: a state-of-the-art. arXiv:2006.13813
81. Thing VL, Wu J (2016) Autonomous vehicle security: a taxonomy of attacks and defences. In 2016 IEEE international conference on internet of things (IThings) and IEEE green computing

and communications (GreenCom) and IEEE cyber, physical and social computing (CPSCom) and IEEE smart data (SmartData). IEEE, pp 164–170
82. Vardhan PH (2024) Development and integration of an autonomous robotic system for enhanced lane detection and following using Jetson NX. Master's thesis, North Carolina Agricultural and Technical State University
83. Vidyasagar D (2006) Digital divide and digital dividend in the age of information technology. J Perinatol 26(5):313–315
84. Wenger A, Mason SJA (2008) The civilization of armed conflict: trends and implications. Int Rev Red Cross 90(872):835–852. https://doi.org/10.1017/S1816383109000277
85. Whitfield SJ (1996) The culture of the Cold War. JHU Press
86. Xing L, Wang K, Wu H, Ma H, Zhang X (2023) Intrusion detection method for internet of vehicles based on parallel analysis of spatio-temporal features. Sensors 23(9):4399
87. Xu R, Lan Q, Pokhrel SR, Li G (2022) A survey of distributed ledger technology for IoT verticals. arXiv:2208.10120
88. Yadron D, Spector M (2015) Hackers show they can take control of moving Jeep Cherokee. Wall Street J. https://www.wsj.com/articles/hackers-show-they-can-take-control-of-moving-jeep-cherokee-1437522078. Accessed 30 Dec 2024
89. Yaga D, Mell P, Roby N, Scarfone K (2019) Blockchain technology overview. arXiv:1906.11078
90. Yang L, Moubayed A, Shami A (2021) MTH-IDS: a multitiered hybrid intrusion detection system for internet of vehicles. IEEE Internet Things J 9(1):616–632
91. Yang L, Shami A, Stevens G, De Rusett S (2022) LCCDE: a decision-based ensemble framework for intrusion detection in the internet of vehicles. In: GLOBECOM 2022—2022 IEEE global communications conference. IEEE, pp 3545–3550
92. Yeong DJ, Velasco-Hernandez G, Barry J, Walsh J (2021) Sensor and sensor fusion technology in autonomous vehicles: a review. Sensors 21(6):2140
93. Zadobrischi E, Cosovanu LM, Dimian M (2020) Traffic flow density model and dynamic traffic congestion model simulation based on practice case with vehicle network and system traffic intelligent communication. Symmetry 12(7):1172
94. Zhou H, Li W, Kong Z, Guo J, Zhang Y, Yu B, Zhang L, Liu C (2020) DeepBillboard: systematic physical-world testing of autonomous driving systems. In: Proceedings of the ACM/IEEE 42nd international conference on software engineering, pp 347–358

Chapter 9
Securing Systems from Aerial Threats: Cybersecurity in the Drone Era

Enejda Torba and Hamid Jahankhani

Abstract This research addresses the critical need for robust cybersecurity frameworks for drone operations as they begin to be essential components of sensitive applications including surveillance, logistics, and crisis response. The study proposes a multi layered defense framework motivated by the vulnerabilities drones introduce to cyber threats such as GPS spoofing, Denial of Service (DoS) attacks, and unauthorized access. This framework was developed using MATLAB Simulink and it integrates many subsystems such as Drone Dynamics and Control, Communication Processing, GPS Spoofing Detection, Anomaly Detection, Threat Response, and Data Logging. The individual subsystems are responsible for focusing on the different aspects of the drone security domain, offering real time threat detection, operational logging and response. The framework was evaluated through simulated attack scenarios, it successfully detected and mitigated various cyber threats with high anomaly detection accuracy and no false positive. When threats were identified, emergency protocols such as Hover, Return Home, and Land modes served to protect the drone. A modular structure would enable subsystems to operate independently, while still contributing towards an integrated, resilient defense. This work showcases that a layer defense approach can provide amplified drone security with an adaptable and functional response to threat mitigation in both military and civilian drone implementations. Efficiencies in processing will be optimized and coverage to more threats will be increased in future improvements to increase drone resilience.

Keywords Drones · Aerial threats · Cyber security · Matlab · Framework

E. Torba · H. Jahankhani (✉)
Northumbria University London, London, UK
e-mail: hamid.jahankhani@northumbria.ac.uk

U. B. Chaudhry and H. Jahankhani, *Autonomous Revolution: Strategies, Threats and Challenges*, Advanced Sciences and Technologies for Security Applications,
https://doi.org/10.1007/978-3-031-96782-5_9

9.1 Introduction

Drone technology has advanced rapidly, bringing with its big benefits in terms of military operations, commercial delivery systems, agriculture, and infrastructure inspection. Amazon is developing Prime Air, a drone delivery service, bringing in fully automatic drones that will substantially reduce the delivery time from days to a few hours. Something similar is undertaken by military forces around the world to deploy drones for surveillance, reconnaissance, and even combat operations because they are versatile, and cheap [26]. Though they help provide many of these benefits, drones' reliance on wireless communication systems for navigation, control, and communicating data make them vulnerable to a variety of cyberattacks such as GPS jamming, signal spoofing, and eavesdropping [47, 48].

A real-world example of a cyber-attack on drones is when the GPS signals for the drone used in the U.S. Navy drone operation in Iran were manipulated in 2011 forcing the drone to land in enemy territory [41]. Hobbyist drones have also disrupted air traffic, such as at Gatwick Airport in 2018 when thousands of passengers were stranded as hundreds of flights were cancelled due to the use of drones to cause disruption [42]. These are cases where malicious use of drones posed a risk to public safety and critical infrastructure.

This research also came about because there was a need to tackle such vulnerabilities. Since drones are moving into civilian airspace along with bespoken infrastructure, it is crucial to have robust cybersecurity frameworks to tackle these risks. The objective of this research is to establish a resilient cybersecurity system that can protect drones from Denial of Service (DoS), GPS jamming as well as unauthorized access using modern technologies. The research proposes a multi layered defense system that can ensure secure deployment of drones in different sectors and ensure operational reliability as well as security.

A multi-layered framework for drones cybersecurity was developed, for detection, response and mitigation against GPS spoofing, DoS and unauthorized access threats The implementation was done in MATLAB Simulink because it has features that facilitate simulating complex dynamics of drones and cybersecurity threats in an isolated yet highly controlled environment. Real time testing and integration of subsystems such as Drone Dynamics and Control, Communication Processing, GPS Spoofing Detection, Anomaly Detection, Threat Response, and Data Logging were enabled by this approach.

The drone security was designed by each subsystem to fill a particular aspect of drone security. For example, the Communication Processing subsystem encrypted data integrity using encryption while the GPS Spoofing Detection subsystem introduced interference to test navigational robustness. Unusual patterns were detected by the Anomaly Detection subsystem which triggered emergency protocols such as Hover, Return Home, or Land, depending on threat severity, based on the Threat Response subsystem.

The results demonstrated that the framework is able to successfully detect and counter threats under simulated attack conditions. The Threat Response subsystem was made capable of timely responses by the Anomaly Detection subsystem that yielded a high detection rate with no false positives. The drone's resilience was increased with the framework's modular design, which enabled each subsystem to perform independently but in an integrated manner. The overall result showed that a layered, real time defense strategy was successful for securing drones from multiple cyber threats and can be used as a means of enhancing drone security in places where there is likely to be a lot of interference from cyber threats.

9.2 Literature Review

9.2.1 Introduction to Drones and Cybersecurity

Drones or Unmanned Aerial Vehicles (UAVs) have evolved from being a niche military tool to be mere general purpose assets used in multiple industries such as military, commercial and recreational activities. In the military they use drones for surveillance, reconnaissance and even combat operation where they can give a lot with terms of operational efficiency and safety. Besides, companies like Amazon are also looking at a drone based delivery system in the commercial logistics sector, that is, to simplify logistics and thereby decrease the delivery time. Furthermore, drones are increasingly being used to monitor crops in agriculture, and in infrastructure inspection, and in aerial photography in a recreational context. People are adopting Drones technology in all kinds of fields as it would be helpful for them in all the matter.

However, as drones become a more integrated part of the fabric of critical infrastructure and everyday work, cybersecurity becomes paramount. Since UAVs depend on wireless communication for navigation, control and the data transmission, they are exposed to various cyber threats; including GPS jamming, signal spoofing and unauthorized access. As the use of drones becomes more critical for tasks like military operations and infrastructure surveillance, the risk of cyberattack increases. Therefore, securing of UAV systems is a critical challenge to ensure UAV systems to be operated safely and reliably.

This literature review seeks to analyze the existing vulnerabilities of existing drone system and defense mechanisms. The review examines current cyber threats and emerging mitigation technologies to make an effort towards providing a better basis for building more robust cybersecurity frameworks for UAVs in various sectors.

9.2.2 Overview of Different Types of Drones

9.2.2.1 Fixed-Wing Drones

Drones with fixed wings, designed to look like airplanes, lift their way forward to move over long distances quickly. As is the case for the propeller-based rotary-wing drones, fixed wing UAVs require runways on which to take off and land. Greater flight endurance and range can be achieved by this design and renders suitable for long duration missions, particularly surveillance, mapping and military reconnaissance [11]. As fixed wing drones such as Parrot's ANAFI USA tend to be more efficient for fuel or battery when performing tasks that need broad area coverage, their aerodynamic wingspan supports extended flights time, which is very useful (Fig. 9.1).

In terms of use cases, fixed wing drones are used extensively for military use in ISR (intelligence gathering, surveillance and reconnaissance) missions. Additionally, they are used in environmental monitoring and large scale mapping operations for the image quality necessary to capture high resolution imagery over large swaths of the planet [47, 48]. For example, in such scenarios as disaster response, fixed wing UAVs can perform an important aerial imagery role (for damage assessment and rescue) that is critical.

Despite their operational benefits, however, fixed wing drones are highly vulnerable to cybersecurity risk. GPS and wireless communication systems for command and control depend on GPS for navigation, leaving them susceptible to cyberattacks such as GPS jamming or signal spoofing, resulting in loss of control or misdirection during mission critical operations [11]. Attackers can exploit these vulnerabilities to disrupt military operations or sabotage long range surveillance activities. As an example, Parrot [28, 29] makes fixed wing drones susceptible to breaches of communication channels that could leak information resulting from reconnaissance

Fig. 9.1 Show Parrot ANAFI USA test [28, 29]

missions. As these risks emphasize, fixed wing UAVs on long range deployments have strong need for good cybersecurity.

9.2.2.2 Rotary-Wing Drones (Quadcopters, Helicopters)

Rotary wing drones such as quadcopters and helicopters have multiple rotors to utilize in order to generate lift and maneuverability, and to maintain themselves hovering with high precision and high precision off set motions. Unlike fixed wing drones, rotary wing UAVs can take off and land vertically and so they are ideal for use in confined spaces and complex terrain [35]. Their versatility and agility in turn have made rotary wing drones very popular for a whole range of civilian and commercial applications (Fig. 9.2).

In terms of use cases, rotary wing drones, such as Amazon Prime Air are used in commercial services to deliver goods to customers in the last mile. They're good at hovering, which makes them good for jobs like inspecting infrastructure, where an operator can get very close to bridges, power lines and buildings. They are also commonly employed in recreational applications, in particular for aerial photography and videography, being stable and controlled [24].

However, wireless communication as a premise for rotary-wing drones, holds significant cybersecurity problems. Denial of Service (DoS) attacks are possible, where attackers can invade the communication channels and flood the drone, thus preventing the drone from communicating with the control station. It can eventually cause operational failures or crashes. In addition, rotary wing drones are prone to signal jamming and hijacking, particularly in use with sensitive data collection or high-value delivery [35]. These vulnerabilities are not just risks for commercial use as they are also for recreational use cases and call for adequately secured interfaces.

Fig. 9.2 Showing electric helicopter drone [25]

9.2.2.3 Hybrid Drones

Hybrid drones are UAVs that share features of both fixed-wing UAVs and rotary-wing UAVs. This aspect that allows them to hover and maneuver vertically while simultaneously sustaining their flight as long as the fixed wing designs [13]. As such, hybrid drones are superior since they can perform more dissimilar tasks, and they can be invaluable for tasks with prolonged duration and high accuracy (Fig. 9.3).

Hybrid drones are among the more widely used types of activity for a military reconnaissance. By allowing hovering over targets for detailed observation without covering large distances quickly, they offer [23]. Hybrid drones are the perfect middle ground between range and agility and are ideal for commercial delivery because they can make dead accurate drop offs into tight urban environments while delivering much further. Additionally, hybrid drones can also quickly reach the disaster places and likewise convey services as important as search and rescue or supply conveyance, utilising their capacity to work in open and congested territories.

However, hybrid drones have complex cybersecurity issues. In particular, these UAVs are particularly vulnerable to GPS spoofing and communication jamming to be able to work over long ranges. By using sophisticated software systems such as iPad, they are exposed to flight control anomalies with an attacker that could exploit software vulnerabilities to manipulate the flight behavior causing crashes [13]. This means that these systems must be secure against multiple threats.

Fig. 9.3 Showing quantum systems' vector [38]

Fig. 9.4 VOXL® 2 AI and 5G development drone—Sentinel [36]

9.2.2.4 Autonomous Drones

Advanced artificial intelligence (AI) and machine learning (ML) systems are the autonomous drones' tool to work without human supervision. In particular, these drones can sense and make real time decisions, autonomously navigate complex environments, avoid obstacles, and perform missions [17]. With onboard AI, autonomous drones are highly efficient for precision, rapid response tasks in the field, as they are able to adapt to dynamic conditions as needed (Fig. 9.4).

Autonomous drones are applied in a vast range of applications. Used by the military in combat and reconnaissance missions, they alleviate the need for human operators in combat zones. Delivery systems are also being developed for autonomous drones that can ship goods long distances without human assistance. These drones are also crucial to drone swarm technology and their autonomy when coordinating with other UAVs to work on more complex tasks, including search and rescue missions, field surveillance [14].

While the autonomous drones are sophisticated, they are still vulnerable to a lot of cybersecurity threats. The primary concern of AI tampering or data poisoning is that adversaries will corrupt the drone's training data to make the wrong or dangerous decisions. Additionally, these drones are vulnerable to ML model spoofing, an attack where attackers can trick the onboard AI into mistaking sensor inputs or making bad navigational choices putting the mission at risk or crashing [17]. For autonomous drone to become reliable in real world applications, their AI systems and data integrity must be safeguarded.

9.2.2.5 Swarm Drones

A swarm drone is a group of UAVs that works together either in a centralized control system or with decentralized communication protocols so that the drones are able to autonomously perform the packages coordinated tasks. These swarms work as a

Fig. 9.5 Icarus Swarms—swarm flight system for emergency response forces [28, 29]

single unit to have the capability to perform complex missions more efficiently than individual drones by sharing information in real time and adapting to the changes of the environment [28, 29]. Swarm technology recreates biological behaviors in nature like bird or insect movement to achieve improved group efficiency and fault tolerance (Fig. 9.5).

Military applications of swarm drones include tasks such as surveillance, reconnaissance, and combat operations, due to their ability to cover large areas and quickly respond to evolving threats, giving a strategic advantage. In search and rescue missions, where they need to coordinate their movements to cover an area effectively in order to identify survivors in disaster zones. More significantly, swarm drones are adopted in environmental monitoring where they contribute to the data acquisition from various sites at one time and hence, enhance the scope and accuracy of assessments [49].

However, swarm drones face significant cybersecurity challenges. Denial of Service (DoS) attacks are one of the most potent threats, as the communication network inside the swarm can be overwhelmed, resulting in loss of coordination and mission failure. Furthermore, due to the decentralized nature of many swarm systems, command spoofing is also possible which means an attacker could use the control of one or more drones to disrupt the entire swarm [2]. Key operational success factors for swarm drone systems are ensuring secure communication protocols and resilient swarm coordination.

9.2.2.6 Nano and Micro Drones

Nano and micro drones refer to very small UAVs which are usually as small as the palm or even a little smaller for specific operations. Due to their relatively small size, when it comes to surveillance in spaces that are either constrained or difficult to infiltrate, such as buildings, or forests, these UAVs become extremely useful. Nano drones are often used in surveillance and spying in the military since they are not easily detected. These miniature drones are characterized by their small form factor, and thus possess limited payloads, to perform surveillance rather than logistics [49] (Fig. 9.6).

There are many possible uses for these drones. They are used for intelligence gathering in the military and provide real-time surveillance without detection. In civilian sectors, they work as a building inspector and specifically looking for damage in hard to reach locations such as roof tops or ventilation systems. In biological and environmental monitoring, nano drones are used to collect data without disturbing sensitive ecosystems [2].

However, nano and micro drones are faced by serious cybersecurity issues. Often, their limited computational resources prevent them from employing strong encryption protocols, which make their signal interception and eavesdropping easily possible. Furthermore, battery exhaustion attacks, in which malicious actors repeatedly broadcast noisy signals at the drones to exploit their battery resources, lead to mission failure [40]. The concerns illustrate the need for lightweight but powerful security tools to secure nano drones in missions that are critical.

Fig. 9.6 Showing nano drone technology [40]

9.2.3 *Cybersecurity Threats to UAVs*

9.2.3.1 Unauthorized Access and Control

Wireless communication systems are crucial to drones' navigation, control and data transmission, making them very susceptible to unauthorized access and control. These UAVs operate over a large distance and they depend on radio frequency (RF), Wi-Fi or satellite communications to stay in contact with ground stations. However, it is a reliance that brings significant cybersecurity risks, as adversaries can lever their knowledge of weaknesses in the communication channels to capture the drones and take control of them, disrupt an operation, and keep control of the flight path [41]. Unauthorised access to data can result in sensitive data being stolen, takeover of a drone, which could end in crashes, spying, or malicious use.

For example, the 2011 U.S. RQ-170 Sentinel drone hijacking incident in Iran was a notable example of such a vulnerability. Iranian forces during this event have successfully captured a U.S. military drone through the use of its GPS system. They used GPS spoofing to carry out the trick, the drone's navigation system was manipulated so that it thought it was following legitimate orders, and ultimately ended up landing in Iranian territory [9]. The incident clearly showed that the UAV communication and navigation systems needed to be secured against sophisticated attackers.

Military drones are not the only things exposing you to the risk of unauthorized access. All civilian drones used in commercial and recreational settings are vulnerable. For instance, adversaries can utilize the lack of good authentication capabilities in consumer grade drones and gain unauthorized control of them as well as employ them for use in illegal acts or industrial espionage [5]. Robust encryption and secure communication protocols are necessary to ensure that the drones in both military and civilian applications are not able to be accessed or controlled without authorization. As drones are used more and more for sensitive purposes and for public safety, they are getting more critical than ever to keep drones secure.

9.2.3.2 GPS Spoofing and Jamming

One of the most common cyberattacks on drones is actually GPS spoofing and jamming, since GPS is such a highly used tool for navigation and position for the drones. In GPS jamming, an attacker sends out signals on the same frequency as the legitimate GPS satellites, but the drone receiver is unable to use all of the navigation data. The drones can lose their position, and may therefore crash or have to land. In contrast, GPS spoofing, which is when fake GPS signals are transmitted in an attempt to make the drone miss its course, is used to hijack or destroy the mission [32].

GPS jamming was also used during 2008 Russia–Georgia war, when the Russian forces jammed Georgian UAVs by using GPS jamming. Georgian surveillance capabilities were effectively rendered ineffectual during key military operations as the drones either malfunctioned or crashed rendering them unable to maintain accurate position [43]. The serious nature of this incident highlights the strategic importance of GPS jamming as a means to degrade the capabilities of drones both in the military and civilian arenas.

GPS jamming can be highly destructive to drones. In military, applications of jamming eliminates ease of surveillance, reconnaissance, or combat mission of UAVs causing loss of key intelligence and could jeopardize operational buzzards eyes. Jammed in civilian airspace, drones that deliver packages, inspect infrastructure, or assist in search and rescue could malfunction causing accidents or mission failure. Additionally, jamming drones could illegally enter no fly zones, which could have an impact on manned aircraft, public safety [10]. However, anti-jamming technologies like adaptive antennas, or even signal encryption, are of critical importance to mitigate the risks allowing drone operation in crowded environments.

9.2.3.3 Denial of Service (DoS) Attacks

DoS attacks on drones overload their communication channels thus disrupting the operations and causing the UAVs to lose their ground control station connection. Such attacks can expose drone communications to DoS (denial of service) attacks in which the communication system is flooded with excessive data or requests that a drone is incapable of processing legitimate instructions. As a result, during critical missions, this can result in the drone malfunctioning, returning to its base, or crashing [24].

Drone swarms consisting of multiple UAVs working together to accomplish a task are amongst the most threatening. A DoS attack on a swarm's communication network can pull the plug on the entire operation by triggering drones to go out of sync. As an example, coordinated DoS attacks against a swarm of drones are able to result in losing several drones to crash or scattering in an uncontrolled manner, all the way from compromising the ability of the swarm to perform surveillance and reconnaissance missions [26]. These attacks represent a serious threat for military or civilian swarms for search and rescue or disaster relief applications. As drone network complexity grows so does their vulnerability to DoS attacks in hostile environments and robust network security and redundancy are needed to protect drones from such disruptions.

9.2.3.4 Eavesdropping and Data Breaches

Drones are particularly exposed to eavesdropping because their wireless communication, but this can also allow the data of a drone to be stolen for blackmail and other financial losses later on. Data coming from such information, which includes telemetry, video feeds or control signals, is sent by UAVs in the form of radio frequencies, Wi-Fi or satellite links. The communications of these systems to real time video streams or critical sensor data are not encrypted, and attackers can intercept and access these communications without permission [50]. In high stakes applications like surveilling military or law enforcement targets, or overhangs on power plants or bridges, such a large risk is especially acute.

One of such, is pertaining to the use of drones for surveillance of sensitive infrastructure. Drones shoot and relay high res imagery and data on the structural health of critical assets for these applications. Data intercepted by an attacker could reveal vulnerabilities in the infrastructure that could be used for sabotage or espionage. This risk is demonstrated in a study that shows unencrypted drone communication can be exploited to steal sensitive data from infrastructure monitoring operations [47, 48]. Strong encryption protocols and protected communication channels are needed to prevent access by unauthorized persons and protect private information collected by drones.

9.2.4 Technologies to Mitigate Cybersecurity Threats

9.2.4.1 Cryptographic Mechanisms

Cryptographic security in communication between the drones and ground control stations is essential to thwarting unauthorized access to and data tampering as well as eavesdropping. That makes them very attractive for cyberattacks, but UAVs need to be able to transmit sensitive information like telemetry, video feeds and navigation commands. AES (Advanced Encryption Standard) is undoubtedly one of the ways used to assure data transmitted between drones and ground stations is surely secured as symmetric encryption techniques. Symmetric encryption for real time transmission data is efficient and easy to use and is compatible with a UAV system because the speed and resource optimization of the communication [3].

However, asymmetric encryption like RSA (Rivest–Shamir–Adleman) is often used for operations as secure, but slower, as initial key exchanges or command authentication. This method utilizes a public and private key pair whereby the public key is used to encrypt the data, while the private key will permit reading the encrypted data only to the intended recipient. It protects critical command and control operations in a man-in-the-middle attack and unauthorized instructions [4].

Cryptography is key within the UAV system since it is the sole mechanism for safeguarding the integrity, confidentiality, and authentication of communications. Encrypted command signals protect adversaries from hijacking drones or injecting malicious commands. In addition, cryptography helps keep the data collected by UAVs, such as surveillance footage or sensor readings, intact by using digital signatures via hashing to verify that the data has not been compromised [24].

9.2.4.2 Blockchain for Data Integrity

Blockchain technology provides a strong solution to ensure the integrity of drone data by creating tamper proof records of everything that happens and is communicated. The ledger, which contains each transaction (such as flight commands or sensor data), is secure, decentralized, and extremely resistant to a modification or tampering. Decentralized structure of Blockchain provides no single entity to eliminate or manipulate the recorded data and is very useful in critical applications like military operations, supply chain logistics or carry log of infrastructure [18]. The blockchain ensures accountability and transparency by offering an immutable log of drone activities while maintaining the integrity of anything that concerns operation data even in adversarial environment.

Blockchain-enabled authentication is one promising application of blockchain in UAV systems; blockchain-enabled authentication is a secure command and control for UAV systems. In this scenario, blockchain is used to verify that commands sent to the drones have been authorized, that is, only authorized drone operators can send instructions. Each command is logged in the blockchain ledger where it can be verified by multiple nodes, thus avoiding a centralized control point or command spoofing [8]. Take for example, where this system is going to be applied using it to teach only, such people in all military and reconnaissance missions, while a tail of all done order orders will be linked. As the ability of blockchain to secure the communication between nodes, it can also maintain data integrity which makes it an amazing utilities in securing modern UAV systems, particularly in mission critical or sensitive operations.

9.2.4.3 Machine Learning for Anomaly Detection

The use of artificial intelligence (AI) and machine learning (ML) that can integrate drone networks has changed the way UAVs can detect anomalies and results in dramatically increased security and operational reliability of UAVs. Anomaly detection is the finding of behaviors, or patterns, which do not fit the normal courses or operations of a drone system such as intrusions and flights beyond permissions, or strange sensor readings. Drones can do so autonomously, using real time large amounts of data in the case of AI and ML to detect and although to possible threats. For instance, these methods are especially suitable for security by frequency of UAV systems from cyber attacks, operational failure and other security breaches [46].

The deep learning techniques provide an application of machine learning based anomaly detection to process the radar, optical and acoustic inputs. Through the process of training of deep learning algorithms such as CNN's the automatic detection of abnormal drone activity can be achieved through the detection of drone activity based on radar signatures, visual patterns or audio footprints. For example, radar data can be used to track drones in real time, and deep learning models can quickly identify drones flying differently than expected [39]. For instance, video feeds could then be passed to image recognition algorithms checking whether a drone is behaving suspiciously, unless permitted to fly in certain areas or deviating from the pre-programmed path.

Another useful source of input are acoustic inputs, and in addition to radar and optical data. Drones draw unique sound profiles and there is nothing wrong in using machine learning to train models to identify such profiles and which patterns should be flagged as abnormal. This results in finding acoustic anomaly detection to be useful where visual or radar inputs are unavailable, such as in dense urban environments and poor weather [24]. By fusing data from multiple sources, machine learning models can assist improving anomaly detection accuracy and reducing number of false positives.

One powerful approach to improving reliability of anomaly detection is through the use of sensor fusion algorithms. Through data fusion of multiple sensors, the radar, optical and GPS, and acoustic sensors, sensor fusion provides greater insight into the operation environment of the drone. Teixeira et al. [39] explain that with machine learning algorithm capturing multiple data streams, anomalous behavior is more accurately detected while keeping the probability of false alarms low. An example would be that the drone runs into a GPS intermference and instead of going by its planned route it deviates. A standalone GPS based anomaly detection system might fire the red flag to look out for possible threats, but a sensor fusion approach from radar and optical data proves the drone is not going off track and is not a false positive.

A case study based on sensor fusion and machine learning towards the development of advanced UAV systems for military reconnaissance was presented. In the case of radar data and optical data, deep learning models were trained, as well as sensor fusion algorithms, to detect anomalous flight behavior in real combat scenarios. GPS Spoofing and Signal jamming instances were successfully identified by the system and operators would be alerted to correct this activity before the drones were compromised [46]. A significant reduction in the rate of false positives was achieved using this method, compared to traditional anomaly detection that largely relies on a single data source.

9.2.4.4 Edge Computing

Edge computing helps drones increase their operational resilience by enabling drones to process critical data locally without dependency on central cloud servers or remote systems. More specifically, this capability is very useful to mitigate Denial of Service (DoS) attacks and network failures, since drones can still function autonomously even when there is no communication of any kind (ground control or central servers). Since drones can operate in persistent semi-disconnected mode, data can be processed at the edge by drones to analyze sensor inputs, make decisions, and complete tasks in real time, thus ensuring mission continuity in such cases where connectivity may come under strain [31].

Edge computing in UAVs offers one of the key advantages of improving resilience against DoS attacks. A successful DoS attack on the communication channels used by drones to interact with remote cloud servers for data processing could knock out the entire system. Whereas, with edge computing, drones can handle flight control, navigation and sensor data themselves, independent of the network environment, even under degraded network conditions [34]. Especially for drones in monitoring critical infrastructure, disaster response, or military applications where mission failure would be catastrophic if the network drops out, this is important.

Edge computing research also demonstrated the support from edge computing for autonomous decision making in a drone mission. As an example, AI and machine learning algorithms deployed throughout a team of drones with edge computing can detect obstacles, avoid collisions and adapt to ever changing environments all in real time, without this centre constantly feeding them with data. By decentralizing decision making, the drone is better able to respond to dynamic environments and helps overall improve the efficiency and safety of autonomous UAV operations [16].

9.3 Real Time Detection Methodology

This research developed an integrated cybersecurity framework for drones by using the methodology of real time detection, response and log security threats. The framework consisted of several interconnected subsystems: Drone Dynamics and Control, Communication Processing, Navigation and GPS spoofing, Anomaly Detection, Threat Response, Data Logging. Simulating and securing data transmissions, identifying anomalies and triggering of emergency protocols was all very critical on each subsystem. The design enabled a modular approach to protect drone operations, robust performance in the presence of cybersecurity challenges, and detailed post flight analysis.

9.3.1 Drone Dynamics and Control

A drone dynamics and control subsystem was created to simulate drone movement and orientation in the three dimensional space. An objective of this subsystem was to develop a model of the drone's position, velocity, and orientation (pitch, roll, and yaw) as the control inputs, throttle, pitch, roll and yaw commands. It implemented the subsystem in MATLAB Simulink using a MATLAB function block to encapsulate the core dynamics and control logic. It gave the flexibility of handling the complex mathematical computations and also the adjusting dynamics parameters like gravity, mass, and drag.

Equations of motion and orientation control were represented by a MATLAB function on the subsystem. Within this piece of code, they computed acceleration values given the throttle input for vertical motion and scaled roll and pitch inputs for lateral motion. The drone's acceleration values were integrated to get the values of the drone's velocity and the velocity values were then integrated to find the position in each axis. To make the simulation continuous, without resetting values, persistent variables within the MATLAB function were used to keep record of position and velocity across time steps.

The function also had a control mechanism for the drone's orientation. Pitch, roll, and yaw inputs were also proportional controllers used to modify the orientation rate as a crude example of a control response for stability. These orientation rates were integrated to calculate pitch, roll and yaw cumulative angles within time. The code segment below illustrates the MATLAB function used:

```
function [Position, Velocity, Orientation] = drone_dynamics_control(Throttle,
Pitch, Roll, Yaw)

  % Constants

  g = 9.81; % Gravity

  mass = 1.5; % Drone mass

  drag_coeff = 0.1; % Drag coefficient


  % Accelerations from Throttle and control inputs

  accel_z = Throttle / mass - g;

  accel_x = drag_coeff * Roll;

  accel_y = drag_coeff * Pitch;


  % Persistent variables for integration

  persistent vel_x vel_y vel_z pos_x pos_y pos_z;

  if isempty(vel_x)

    vel_x = 0; vel_y = 0; vel_z = 0;

    pos_x = 0; pos_y = 0; pos_z = 0;
```

```
end

% Update velocities and positions
vel_x = vel_x + accel_x * 0.01;
vel_y = vel_y + accel_y * 0.01;
vel_z = vel_z + accel_z * 0.01;

pos_x = pos_x + vel_x * 0.01;
pos_y = pos_y + vel_y * 0.01;
pos_z = max(0, pos_z + vel_z * 0.01); % Ground check

% Outputs for Position and Velocity
Position = [pos_x; pos_y; pos_z];
Velocity = [vel_x; vel_y; vel_z];

% Orientation control
kp_pitch = 1.0; kp_roll = 1.0; kp_yaw = 1.0;

% Calculate orientation changes
pitch_rate = kp_pitch * Pitch;
roll_rate = kp_roll * Roll;
yaw_rate = kp_yaw * Yaw;

% Persistent variables for orientation
persistent orientation_pitch orientation_roll orientation_yaw;
if isempty(orientation_pitch)
   orientation_pitch = 0; orientation_roll = 0; orientation_yaw = 0;
end
```

```
        orientation_pitch = orientation_pitch + pitch_rate * 0.01;
        orientation_roll = orientation_roll + roll_rate * 0.01;
        orientation_yaw = orientation_yaw + yaw_rate * 0.01;

        Orientation = [orientation_pitch; orientation_roll; orientation_yaw];
    end
```

9.3.2 *Communication and Signal Processing*

A Communication and Signal Processing Subsystem was designed to test for communication integrity under potential cyber threat (e.g. Denial of Service, or DoS) by data exchange between drone and the control center. The Dynamics and Control subsystem of the drone provided data about the drone's position, velocity, and orientation to this subsystem, which was then encrypted, transmitted with simulated noise interference and decrypted. The objective was to guarantee signal integrity in communication process, evaluating how noisy channels may affect data reliability.

Encryption, decryption and noise simulation was implemented by a MATLAB Function block within Simulink for secure communication. It combined position, velocity, and orientation into a single data array and scaled it so that it is compatible with integer based encryption. The data was encrypted by using a bitwise XOR operation with a predefined key. It offered a basic and computationally efficient encryption mechanism that followed real world encryption processes, but within Simulink. The function was defined as follows:

```
function [EncryptedData, DecryptedData] = CommProcessing(Position, Velocity,
Orientation)
    % Constants for encryption
    encryption_key = uint16(5); % Simple key for XOR encryption (use uint16)
    noise_level = 0.1; % Noise level for DoS simulation
    do_dos = true; % Flag to simulate DoS attack

    % Combine inputs into a single data vector for encryption
    data = [Position; Velocity; Orientation];

    % Ensure data is converted to integer for XOR operation
    data_scaled = data * 100; % Scale the data
```

```
    data_int = uint16(max(0, min(data_scaled, 65535))); % Scale and saturate
within uint16 range

    % Step 1: Encrypt data
    EncryptedDataInt = encrypt(data_int, encryption_key);

    % Step 2: Simulate noisy channel (DoS)
    NoisyDataInt = add_noise(EncryptedDataInt, noise_level, do_dos);

    % Step 3: Decrypt data after passing through the noisy channel
    DecryptedDataInt = decrypt(NoisyDataInt, encryption_key);

    % Convert decrypted data back to double for output
    EncryptedData = double(EncryptedDataInt) / 100; % Scale back to original range
    DecryptedData = double(DecryptedDataInt) / 100;

    % Helper functions for encryption/decryption and noise addition
    function encrypted = encrypt(data, key)
      encrypted = bitxor(data, key); % XOR encryption
    end

      function decrypted = decrypt(encrypted_data, key)
        decrypted = bitxor(encrypted_data, key); % XOR decryption
      end

      function noisy_data = add_noise(data, noise_level, do_dos)
        if do_dos
          noise = noise_level * randn(size(data)); % Generate noise
          noisy_data = data + uint16(noise); % Convert noise to uint16
```

```
        else
          noisy_data = data;
        end
      end
    end
```

The DoS attack was simulated by encrypting data stream using Gaussian noise added by the add_noise function, implying interferences in the communication channel. Then, decrypted data was compared to the original inputs to verify that integrity remained intact in the presence of noise. Testing resilience under noisy conditions, this subsystem effectively modeled secure communication dynamics.

9.3.3 Navigation and GPS Spoofing

The Navigation and GPS Spoofing Subsystem was developed to provide realistic GPS data for the drone and a source of spoofing noise to test the system's robustness against GPS manipulation attacks. Decrypted data assumed to be representing real time position coordinates of the drone was passed to this subsystem and noise was applied to simulate natural GPS errors as well as possible spoofing attacks.

First, a base GPS signal with slight random noise to simulate natural GPS inaccuracies was generated, and the base GPS signal acted as a reference to the true GPS position. Minor deviations in position data and corresponding a low noise level were applied, which approximate real world GPS variability. Then an additional, more substantial noise injection mechanism was then added, to use when spoofing. The spoofing noise was meant to mimic deliberate interference or manipulation of GPS signals to confuse the drone's navigation system.

The MATLAB function implemented for this subsystem included two key parameters: It uses standard GPS noise (gps_noise_level) and spoofing interference (spoof_intensity). The function accepted two inputs: enable_spoofing, DecryptedData The baseline gps coordinates were the former and the latter controlled if spoofing noise was used. The GPS data were spooked with additional spoofing noise if enable_spoofing was true, otherwise, standard noise was applied.

The MATLAB function used in this subsystem was as follows:

```
function ManipulatedGPS = GPSSpoofing(DecryptedData, enable_spoofing)
  % Parameters
gps_noise_level = 0.001;   % Standard GPS noise level
spoof_intensity = 0.005;   % Spoofing noise level

% Decrypt data to retrieve GPS coordinates
GPS_Coordinates = DecryptedData; % Base GPS coordinates

% Step 1: Add natural GPS noise
gps_noise = gps_noise_level * randn(size(GPS_Coordinates));
GPS_With_Noise = GPS_Coordinates + gps_noise;

% Step 2: Apply spoofing if enabled
if enable_spoofing
  spoof_noise = spoof_intensity * randn(size(GPS_Coordinates));
  ManipulatedGPS = GPS_With_Noise + spoof_noise;
else
  ManipulatedGPS = GPS_With_Noise;
end
 end
```

9.3.4 Anomaly Detection

The Anomaly Detection Subsystem was designed to detect potential cybersecurity threats to drone operational integrity by detecting anomalies in sensor data and navigation signals. This subsystem used data inputs from various sensors including manipulated GPS coordinates to find deviances or unusual patterns in the data inputs in order to look for clues to spoofing, or some other type of attack on the drone's navigational system. The drone reported real time security threats to the subsystem, which quickly reacted accordingly, as needed, by activating its safety protocols.

The process of anomaly detection was implemented as a MATLAB function block in the Simulink environment. The detection logic was based on a set of predefined thresholds and limits on rate of change. In particular, the subsystem which monitored the positional coordinates of the drone, as expressed by the first three elements of the ManipulatedGPS data vector, verified these coordinates were acceptable. For each input sample, the system looked to see if the GPS values were either above or below upper or lower thresholds that represent the normal operation range. It further calculated the rate of change for each of the coordinates so that it could detect sudden shifts—that may be due to spoofing or data manipulation. The run to achieve the occurrence rate exceeds the defined rate of change limit was flagged to trigger the AnomalyAlert condition where it exceeded any coordinate.

The MATLAB function used in this subsystem was as follows:

```
function AnomalyAlert = AnomalyDetector(ManipulatedGPS)
  % Extremely sensitive parameters for anomaly detection
  threshold_upper = [0.1; 0.1; 0.1];   % Very narrow threshold range for XYZ
  threshold_lower = [-0.1; -0.1; -0.1]; % Very narrow threshold range for XYZ
  max_rate_of_change = 0.005;          % More sensitive to small changes

  % Persistent variable to hold previous GPS data
  persistent last_GPS;
  if isempty(last_GPS)
    last_GPS = ManipulatedGPS;
  end

  % Initialize alert flag to 0 (normal)
  AnomalyAlert = 0;

  % Check for threshold violations in the first 3 elements (assumed XYZ
coordinates)
  if any(ManipulatedGPS(1:3) > threshold_upper) || any(ManipulatedGPS(1:3)
< threshold_lower)
    AnomalyAlert = 1; % Trigger anomaly alert due to threshold violation
  end

  % Check for abnormal rate of change in the first 3 elements
  rate_of_change = abs(ManipulatedGPS(1:3) - last_GPS(1:3));
  if any(rate_of_change > max_rate_of_change)
```

```
        AnomalyAlert = 1; % Trigger anomaly alert due to rapid change
    end

    % Update last GPS for next comparison
    last_GPS = ManipulatedGPS;

    % Diagnostic output for debugging (optional, can be removed in final
version)
    disp("ManipulatedGPS: "), disp(ManipulatedGPS(1:3));
    disp("Rate of change: "), disp(rate_of_change);
    disp("AnomalyAlert: "), disp(AnomalyAlert);
end
```

9.3.5 Threat Response

The Threat Response Subsystem has been designed for drone operational response to detected anomalies, thereby providing a safe cybersecurity threat response. The Anomaly Detection Subsystem generated an input signal AnomalyAlert, which was fed into this subsystem as input. When the AnomalyAlert was triggered, which could indicate a security breach or irregularity in navigation data, the Threat Response Subsystem then triggered a given set of predefined safety actions designed to stabilize the drone. They consisted in putting the drone in emergency modes, for example, in hovering, go back to a pre defined home location, or safe landing. This incremental buildup of response gave structure to the threat mitigation process, maintaining the safety of the drone in each case.

In Simulink, implementation of the Threat Response logic was carried out as a MATLAB Function Block that allowed real time assessment and response to the AnomalyAlert signal. Each level of response was defined within a set of emergency modes within the MATLAB function. The set of modes included Normal Mode (0), Hover Mode (1), Return Home Mode (2), and Land Mode (3). Upon AnomalyAlert signal of a high value, the subsystem would escalate through these modes based on the current state. For instance, if the drone was in Normal mode and was detecting an anomaly, it would come to Hover mode in order to remain its current position. Successive detections would cause Return Home mode to take place and eventually Land mode to be allowed with returning configurations becoming progressively more secure. The function was defined as follows:

```
function EmergencyMode = SafeModeController(AnomalyAlert)
  % Define the emergency modes
  NORMAL_MODE = 0;      % No anomaly, normal operation
  HOVER_MODE = 1;       % Hover in place
  RETURN_HOME_MODE = 2; % Return to predefined home position
  LAND_MODE = 3;        % Land safely

  % Persistent variable to hold the current emergency mode
  persistent currentMode;
  if isempty(currentMode)
    currentMode = NORMAL_MODE;
  end

  % Check for AnomalyAlert and set EmergencyMode based on logic
  if AnomalyAlert == 1
    % Set the emergency mode in response to detected anomaly
    % Example logic: go to HOVER_MODE first, then RETURN_HOME_MODE, and
finally LAND_MODE
    switch currentMode
      case NORMAL_MODE
        % If an anomaly is detected from normal mode, switch to hover
        currentMode = HOVER_MODE;

      case HOVER_MODE
        % If in hover mode, escalate to return home mode
        currentMode = RETURN_HOME_MODE;

      case RETURN_HOME_MODE
        % If in return home mode, escalate to land mode
        currentMode = LAND_MODE;

      case LAND_MODE
```

```
            % Already in land mode, maintain it
            currentMode = LAND_MODE;
        end
    else
        % If no anomaly is detected, stay in normal mode or maintain the current
mode
        if currentMode ~= LAND_MODE
            currentMode = NORMAL_MODE; % Reset to normal if safe and not already
landing
        end
    end

    % Output the current mode
    EmergencyMode = currentMode;

    % Optional: Display current mode for debugging
    disp("Current Emergency Mode: "), disp(EmergencyMode);
end
```

9.3.6 *Data Integrity*

A Data Integrity Subsystem was created to secure and reliably record critical operations of drone, emphasizing especially the logging of changes in EmergencyMode. This subsystem was a simplified logging mechanism to record each transition in the drone's operational state in a structured format for post flight analysis and data integrity verification. The subsystem aimed to increase traceability and accountability of drone actions taken during threat events by maintaining a persistent log of emergency responses. This approach was a very efficient way to record data without the overhead of anything more complex than two pointers to most common data types.

The Data Integrity Subsystem was implemented in Simulink using a MATLAB Function Block, where each EmergencyMode change was stored in a pre-allocated matrix for fixed size storage consistent with a Simulink's real-time processing constraints. In the code, a persistent variable named log_data was set to a fixed size matrix with 100 entries capacity, which was used to log up to 100 changes in EmergencyMode in sequence. When logging, also logs the entry count and the EmergencyMode value. The log was a circular buffer, and after the log size reached its maximum, the oldest entries were overwritten. Therefore, this design achieved memory efficiency and long term data retention while continuous logging was not at risk of overflow or memory related issues. The code segment below illustrates the MATLAB function used:

```
function LogArray = SimpleLogger(EmergencyMode)
  % Parameters for the log
  max_entries = 100;       % Maximum number of entries to log
  persistent log_data;      % Persistent variable to store the log
  persistent entry_count;    % Count of entries

  % Initialize log_data and entry_count if they are empty
  if isempty(log_data)
    log_data = zeros(max_entries, 2); % Pre-allocate [EntryCount,
EmergencyMode]
    entry_count = 1;            % Start with the first entry
  end

  % Log the EmergencyMode with a simple entry counter
  log_data(entry_count, 1) = entry_count;  % Log the entry number
  log_data(entry_count, 2) = EmergencyMode; % Log the EmergencyMode

  % Increment the entry count (wrap around if it exceeds max_entries)
  entry_count = entry_count + 1;
  if entry_count > max_entries
    entry_count = 1; % Wrap around if we exceed max_entries
  end
  % Output the entire log
  LogArray = log_data;
end
```

9.3.7 System Integration

System integration of this cybersecurity framework involved the structured flow of such data through an interconnected set of subsystems to monitor, react to, and log cybersecurity threat to drone operations. In fact, the system's workflow begins with the Drone Dynamics and Control Subsystem, generating and managing basic flight

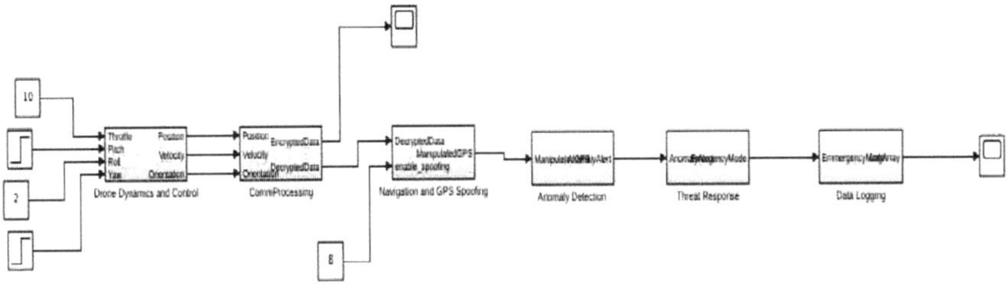

Fig. 9.7 Drone cybersecurity and threat response framework

parameters, such as throttle, position, velocity, and orientation. The Communication Processing Subsystem receives these parameters then encrypts them for secure transmission, simulating data exchange between a drone and a control center.

The data is encrypted, then sent to the Navigation and GPS Spoofing Subsystem. In order to simulate spoofing scenarios, manipulated GPS signals are introduced to test the system's robustness to navigation based attacks. This subsystem produces real and spoofed GPS data that is transmitted to the Anomaly Detection Subsystem. This subsystem monitors the position data and when data does not meet expected behavior, will raise an AnomalyAlert. When an AnomalyAlert is received the Threat Response Subsystem takes over driving the drone through emergency modes such as hovering, flying home or landing to make sure the drone is safe.

The Data Logging Subsystem finally stores the transition in EmergencyMode history to be used for post flight analysis of both for response behaviour and integrity over time. This integration of modules within the cybersecurity framework gives a complete panoramic view of the drone's operational integrity under adverse conditions, where they are secured from any potential cybersecurity threats with real time (Fig. 9.7).

9.4 Data Analysis and Critical Discussions

Data Analysis and Critical Discussions of the study are presented in this section, in which the effectiveness of each subsystem in the multi layered defense framework is evaluated. The integrated system is subjected to simulated attack scenarios (i.e. GPS spoofing, Denial of Service (DoS) attacks, and unauthorized control access) in order to assess how well the system performed in these situations. In addition, there is critical discussion of the strengths and weaknesses of the system and suggestions for improving areas of the system. Finally, it discusses potential real world application, and how this framework can aid in security in drone operations.

9.4.1 Data Collection and Preprocessing

Data collection and preprocessing was required in order to evaluate the performance of the multi layered defence system under simulated attack scenarios [33]. Real time data was collected from each subsystem of the defense framework as the framework responded to different cybersecurity threats such as GPS spoofing, Denial of Service (DoS) attacks and unauthorised attempts of accessing. EmergencyMode state, anomaly alert, position coordinates along with other telemetry data from Drone Dynamics and Control, Communication Processing, Navigation and GPS Spoofing subsystems were key variables recorded. It was used as the basis for assessing the accuracy and timeliness of threat detection, response initiation and drone stability during attacks.

The collected data went through a series of preprocessing steps in order to guarantee consistency and reliability. In the first place, any incomplete or inconsistent data entries were removed, and these included some that resulted from communication lags or interruptions that would artificially skew results [12]. Defined thresholds were set for each variable to determine outliers to identify any anomalies in data collection itself, only valid responses to simulated threats were included in the analysis. The accuracy of the evaluation depended on this cleaning process which removed noise and focused the evaluation on the real system responses to threats.

The data was organized chronologically to capture the sequential flow of each event in which specific transitions occurred between different EmergencyMode states. It provided an in depth look at the timing and effectiveness of each subsystem's response to detected anomalies. Further, logged telemetry data synchronized with the threat response was used to visualize overall system behavior under continuous or overlapping threats. Response times and subsystem performance were normalized across different metrics to facilitate the comparative analysis.

The processed dataset gave a clean, structured look at how the integrated defense system performed during simulated attacks. The structured dataset allowed for this accurate analysis, leading to insights regarding what drove each subsystem's contribution to the drone's overall resilience [1]. Based on these data, further analysis and critical discussions took place on the successes and aspect for improving in the multi layered defense framework.

9.4.2 Drone Dynamics and Control

The performance of the Drone Dynamics and Control Subsystem under normal conditions and under simulated attack conditions including GPS spoofing and unauthorized control access was evaluated [1]. The goal was to assess the subsystem's ability to maintain control parameters such as throttle, position, and orientation, as baseline to detect deviations during suspected cybersecurity threats (Fig. 9.8).

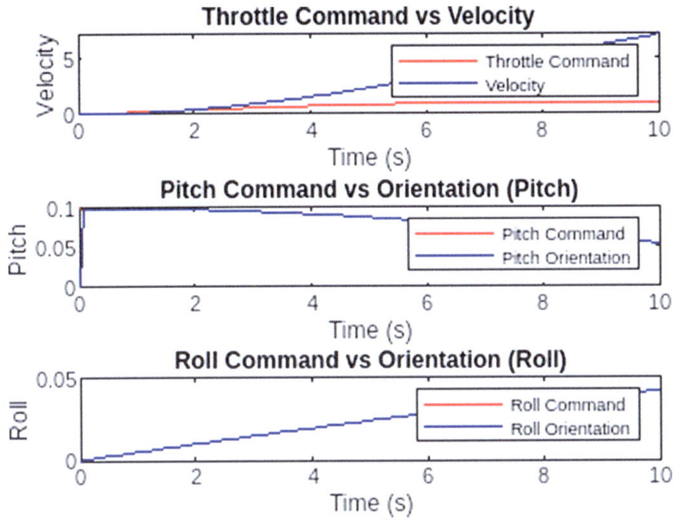

Fig. 9.8 Showing plots of control accuracy under normal conditions

The first set of plots demonstrate the subsystem's ability to control throttle, pitch, and roll under normal operating conditions. The Throttle Command versus Velocity plot demonstrates that drone velocity (blue line) closely tracked throttle (red line) with very little lag. The subsystem was able to effectively translate throttle commands into velocity changes necessary for smooth turning and control, as is shown by this alignment. The Pitch Command versus Orientation (Pitch) and Roll Command versus Orientation (Roll) plots also indicate that the subsystem was capable of stable orientation control. The commands in pitch and roll orientation were accurately tracked, and the drone was capable of controlling its attitude as commanded. These results indicate that under normal conditions the subsystem modeled and controlled dynamics of the drone in three dimensional space with high accuracy (Fig. 9.9).

When subjected to attack simulations such as GPS spoofing, unauthorized access varied drastically from the normal parameter set and response. The Normal Position versus Spoofed Position plot shows a divergence from the true (blue) position through Normal Position versus Spoofed Position (red line) because it is exposed to spoofed GPS data. The potential for navigational drift due to GPS spoofing is quantified in this discrepancy, which is measured by a Mean Absolute Error (MAE) of 1.2957. Without detection, spoofed data could make the drone stray from its intended path with a substantial position error, threatening mission objectives.

The Normal Pitch versus Unauthorized Pitch plot similar indicate the influence of allowing unauthorized control access on the pitch orientation. Compared to the normal pitch (blue line), the unauthorized pitch (red line) has erratic fluctuations with an MAE = 0.024484. Similar error (0.025372) was obtained for roll under unauthorized access (but not shown in the presented plots). While these errors are low, they imply that any unauthorized control inputs can destabilise the drone's

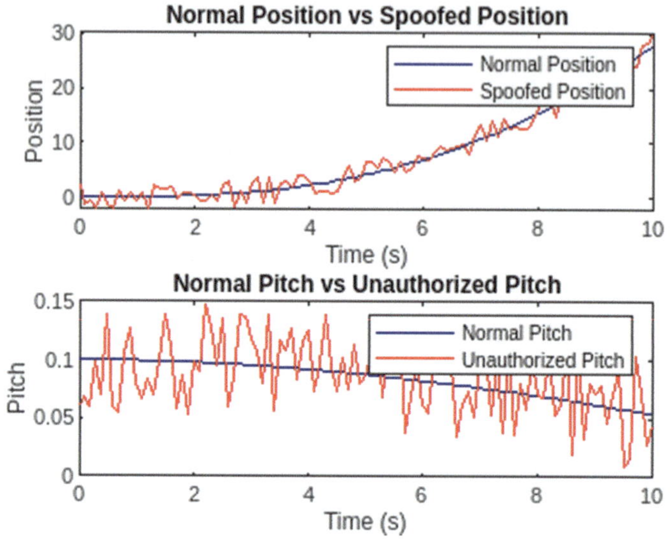

Fig. 9.9 Showing plots of performance variations under attack conditions

attitude to a potentially dangerous extent over relatively prolonged time or during any (precision) manoeuvre. The observed erratic behavior during unauthorized access further underscores the need for robust anomaly detection mechanisms to mitigate instability induced by unauthorized control commands.

The Drone Dynamics and Control Subsystem is the baseline for evaluating anomalies, a reference against which anomalies are measured during threat scenarios [1]. Under nominal condition, the subsystem performed reliably in control of position, velocity and orientation. However, the subsystem exhibited measurable deviations when spoofed and exposed to unauthorized access, demonstrating control vulnerabilities in position and orientation.

9.4.3 Communication Processing

The Communication Processing Subsystem was developed as a means of providing secure and reliable data exchange between the drone and the control center by restricting DoS attacks, encrypting the data and ensuring its integrity [22] (Fig. 9.10).

Original Position versus Decrypted Position plot shows the accuracy of encryption and decryption of the subsystem. The decryption (red dashed line) aligns nearly perfectly with the original Position data (blue line), this reflects the fact that the encryption algorithm preserved the data's integrity during transmission. A simple XOR based encryption was implemented, providing basic data confidentiality by hiding the original values. The encrypted data matched the position data from the

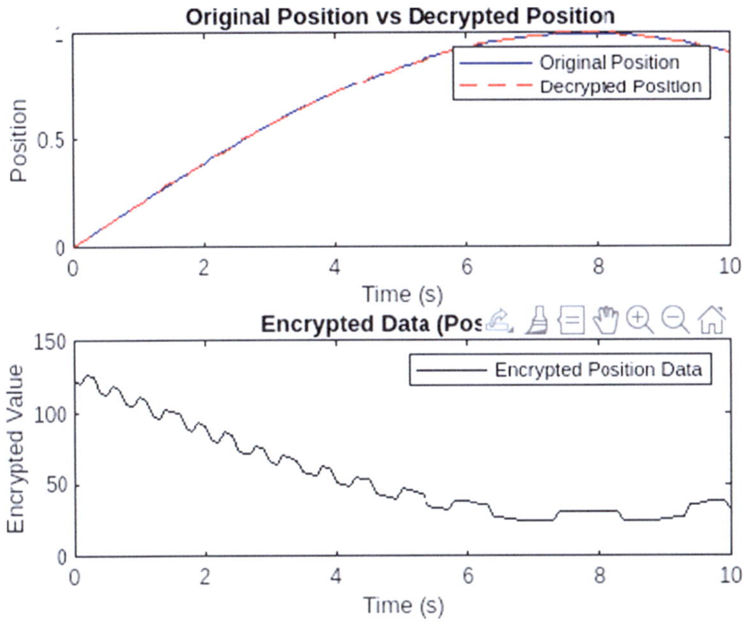

Fig. 9.10 Showing plot of original position versus decrypted position and plot of encrypted data (position)

original data, which proved that the encryption algorithm works well for secure data transmission. Encrypted Data (Position) plot shows encrypted position values. These values actually differ significantly from the original data which shows the encryption effectively obfuscated the location information and preservative their privacy (Fig. 9.11).

Packet Loss and Delay with Encrypted Data plot features an original encrypted data (black line) at the top, and the encrypted data under DoS attack conditions (red dashed line) at the bottom. During the data transmission introducing packet loss and random delays was simulated, as a means of DoS attack. These disruptions highlight the impact of the differences between the original and transmitted data. Gaps in the transmitted data were due to packet loss, and the misalignment of received data points with the original data points was caused by delay events. This plot emphasizes the precariousness of communication system to DoS attacks as it will interrupt continuity or timeliness of data for relayed data, subsequently affecting the reliability of the data transmitted into the control inputs of the drone.

Furthermore, Packet Transmission Success plot gives us a visual representation of the ways in which DoS attacks effect data reliability. 1 signifies successful packet reception, 0 a packet loss. The drops to 0 indicate packet loss during the simulation of the DoS attack. These interruptions' random nature show the variability of DoS impacts, which can be a significant threat to the mission critical data exchange, especially if lost or delayed packets hold critical control information (Fig. 9.12).

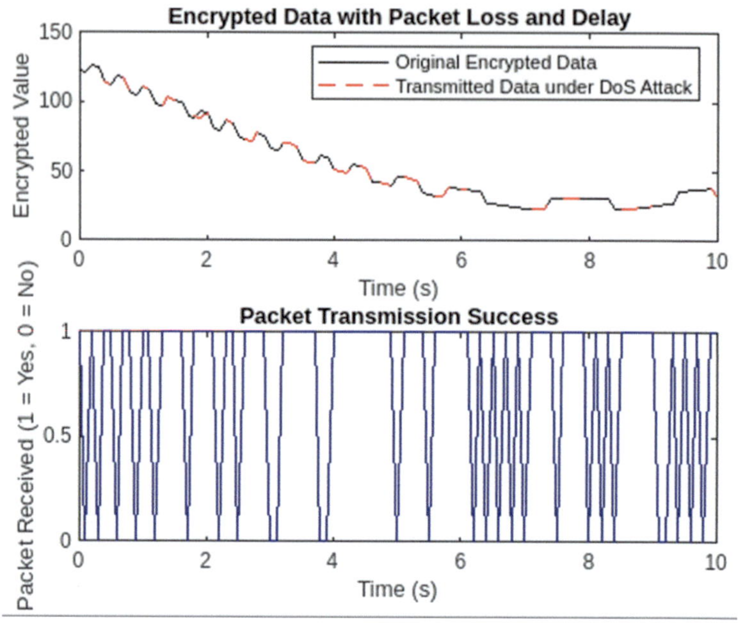

Fig. 9.11 Plot of encrypted data with packet loss and delay and plot of packet transmission success

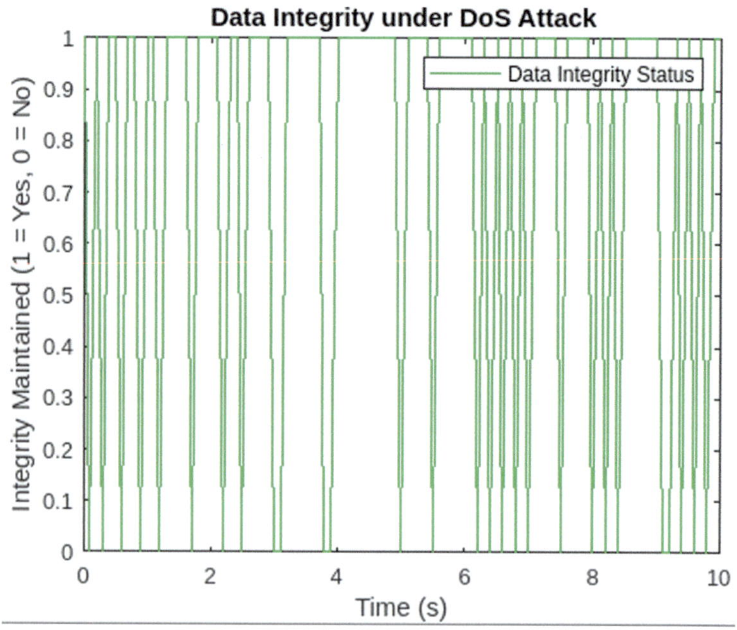

Fig. 9.12 Data integrity under DoS attack

Data Integrity under DoS Attack plot shows the insights on how successfully the subsystem maintains data integrity under the attack conditions. This plot then tracks Data that has been received, and is then decrypted and compared to the Original Position Data. A value of 1 means the data integrity has been maintained, a value of 0 means the data integrity has failed. Frequent back and forth from 1 to 0 indicates that the subsystem could not maintain data integrity during the DoS attack consistently. Received information was not guaranteed reliable as often, the decrypted data was out of position, had packet loss and was often delayed.

Under normal conditions, the Communication Processing Subsystem successfully encrypted and decrypted data, preventing the transmitted information from inadvertently being seen. While the DoS attack simulation revealed weaknesses in the ability of the subsystem to reliably exchange data, all subsystems performed well enough to allow the tower to proceed with operations as planned. Packet loss and delays disrupted data integrity causing occasional integrity failure. The results of these findings imply that protections provided by the subsystem are sufficient in normal conditions; however, improvements to resilience against DoS attacks are needed to insure data transmission and maintain reliability in the adversarial environment [33].

9.4.4 Navigation and GPS Spoofing

The Navigation and GPS Spoofing Subsystem was designed to assess the drone's ability to withstand GPS spoofing where manipulated GPS data could mislead the drone's navigation [6] (Fig. 9.13).

The X Position plot depicts the error between the red dashed line (spoofed data) and the blue line (original data) increasing steadily as time progresses. The fact that this drift does indeed occur in the X axis position suggests that spoofing noise can indeed pull the drone off its course in a slowly increasing way that, if not corrected, could result in huge navigation errors.

Similarly, the Y Position plot also displays significant deviation in the spoofed Y position from the real data. Surprisingly, the deviation profile is not the same as the X-axis, with the controlled Y position starting at a higher value but decreasing towards the initial path over time. This dynamic behavior may represent a situation where GPS spoofing interference in the Y direction changes with time, which is more realistic than a fixed noise pattern.

The Z Position plot, which is going to be the altitude, shows a small variation than both the X and Y positions. Despite some visible shift in the spoofed Z position, the offset is quite small and does not vary significantly. This means that the altitude spoofing noise was either intentionally lower or less effective in producing the drift. Altitude accuracy is usually crucial for avoiding obstacles or ensuring flight stability, and the minimal deviation observed shows that the subsystem can manage the spoofing intensity selectively [33] (Fig. 9.14).

Deviation in Position due to GPS Spoofing plot, shows the numerical results of the comparison between the real trajectory and the spoofed one in each direction

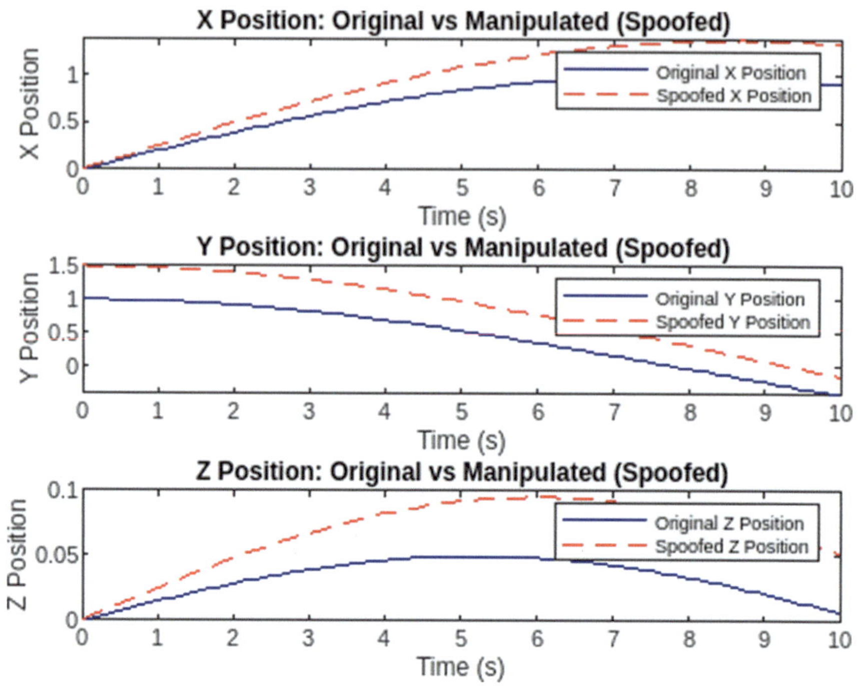

Fig. 9.13 Showing plot of X position, Y position and Z position

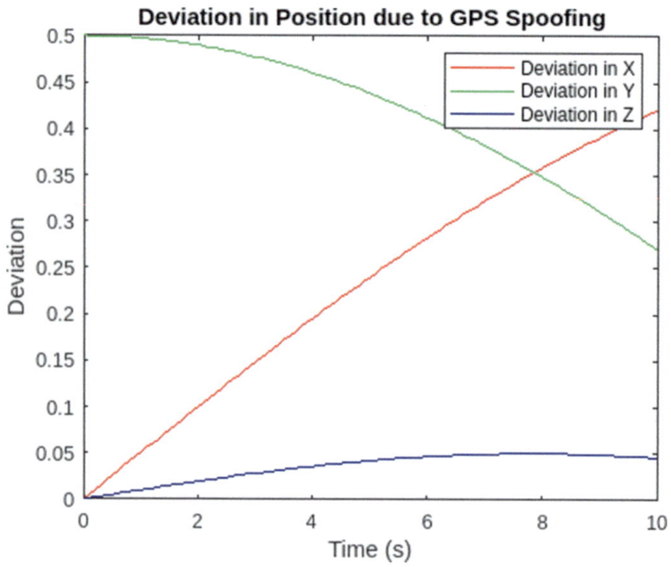

Fig. 9.14 Showing deviation in position due to GPS spoofing

[22]. The red line which denotes the X deviation rises progressively, showing the error accumulation as a result of spoofing. This is similar to the Y Position plot as the green line which shows the Y deviation starts high and then drops. The graph of this deviation also shows a decreasing trend, which could mean that the spoofing of Y reduces with time, even in simulated conditions of occasional interference. The blue line, which represents the Z deviation, is close to zero during the whole simulation, which proves that the altitude was changed only slightly.

The Navigation and GPS Spoofing Subsystem demonstrates an ability to emulate a large range of spoofing conditions with varied levels of error in each of the spatial dimensions. The perceived coordinates of X, Y, Z, as measured in the drone are obviously affected by GPS spoofing and this problem highlights the severity of the problem if left unattended. Therefore, this subsystem puts forward a necessary reliability of an anomaly detection system to avoid navigational drift and to meet the precise flight trajectory in complex situations.

9.4.5 Anomaly Detection

Due to the possibility of attack that may affect the system's ability to function (e.g. GPS spoofing, etc.), such an attack could be detected by the Anomaly Detection Subsystem. This subsystem has the primary goal of evaluating the changed GPS data and if certain discrepancies are observed relative to predetermined norms, to issue AnomalyAlert [6]. This function is to detect any abnormal behavior of the drone from normal movement and prevent the drone to go away from the path (Fig. 9.15).

X Position and Y Position are the first two plots which are grounded with GPS spoofing effect on the drone's position. The GPS data is plotted with the blue line where the actual position is depicted and the red dashed line depicts the spoofing. In both plots the GPS data are distorted from the real path and this is to depict a condition where the actual location of the drone is spoofed due to interference. This inconsistency results in a shift that is precisely the kind of variation that the Anomaly Detection Subsystem is meant to identify.

The third plot, Anomaly Detection Alert, shows the result of AnomalyAlert over time. This binary signal is used to show 1 in case there is an abnormality observed in GPS data. In this simulation, the AnomalyAlert correctly captures the deviations introduced by the spoofing noise to raise alarms during the entire GPS manipulation phase. The effectiveness of the anomaly detection is further supported by quantitative metrics: has a Detection Rate of 99.0099%, it has a False Positive Rate of 0% and a Missed Detection Rate of 0.9901%. The high detection rate shows that the subsystem was able to accurately detect almost all the spoofed data thereby alerting the anomalies in the GPS data. The low percentage of missed detection (only 0.9901%) indicates that not many of the anomalies have been missed by the subsystem, which proves that the subsystem is effective in identifying the problems with GPS.

Specifically, the False Positive Rate is 0%, which indicates that the subsystem raised no false alarms during periods of no anomalies. This lack of false positives

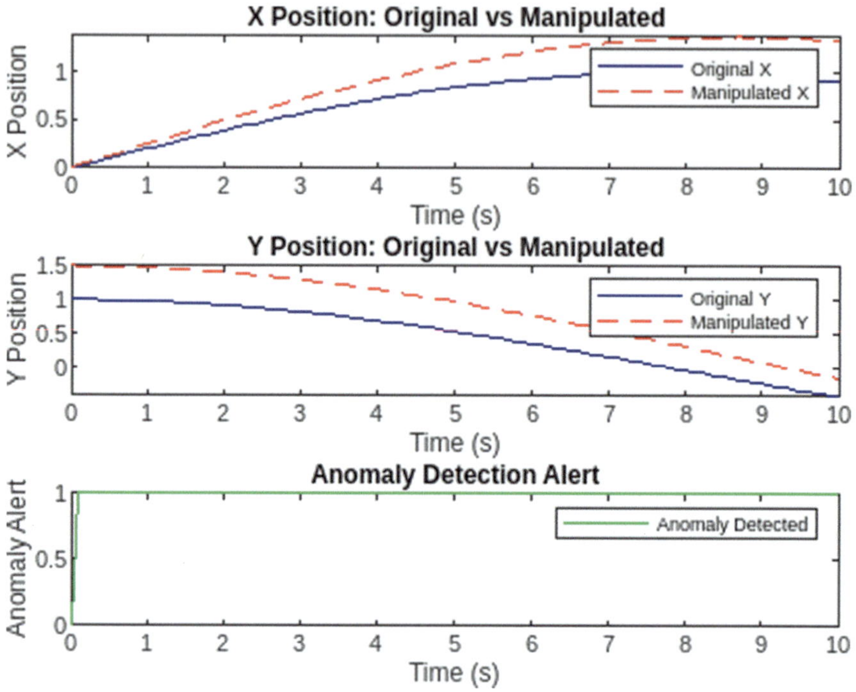

Fig. 9.15 Plot of X and Y position and plot of anomaly detection alert

means that the threshold for AnomalyAlert is well set as only real anomalies are detected.

The Anomaly Detection Subsystem delivers high efficiency in detecting anomalies in GPS data in attack scenarios. The high detection rate and zero false positive prove the efficiency of this sub system in differentiating between the normal and the jammed GPS signal. These results demonstrate the significance of the subsystem in preserving navigational integrity since it is able to identify GPS spoofing attacks while also assisting in shielding the drone's navigation system from threats.

9.4.6 Threat Response

The Threat Response Subsystem is one of the most important sub systems of the drone since it helps in coming up with the right measures to be taken when there is an anomaly detected while flying [4] (Fig. 9.16).

From the Anomaly Alert Signal plot, it is possible to notice that AnomalyAlert is activated (marked as 1 on the plot) at one or several points during the simulation. Every time the subsystem encounters an anomaly, the AnomalyAlert signal comes on, which shows that there is something wrong with the drone environment or system

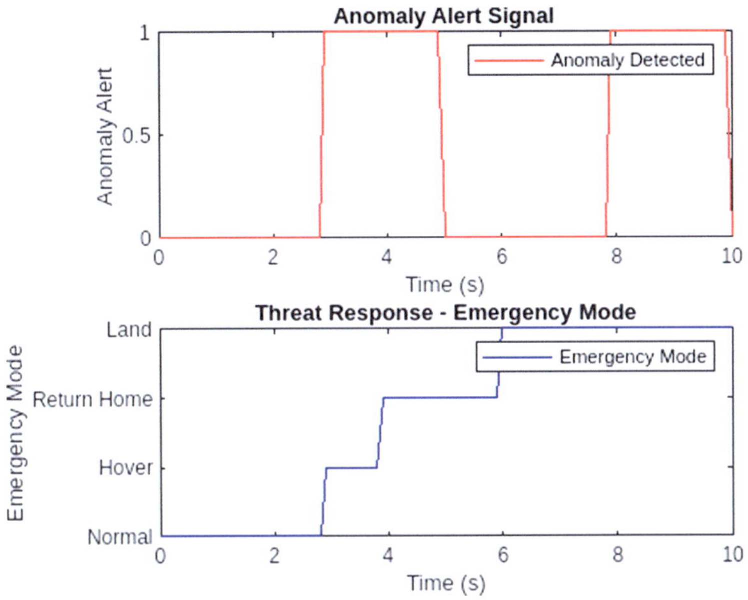

Fig. 9.16 Showing plot of anomaly alert signal and plot of threat response—emergency mode

data. This alert triggers the Threat Response Subsystem to initiate its sequence of emergency actions.

The Threat Response—Emergency Mode plot demonstrates how the Threat Response Subsystem escalates the different emergency modes in the face of the AnomalyAlert. At the start of the system, the system is in Normal Mode 0. When the first anomaly is detected, the subsystem changes to Hover Mode (1) and has the drone stop and hover. However, if the anomaly persists, the system will advance to the 2nd Return Home mode, telling the drone to start flying back to a home point. Finally, when there are long standing or sustained discrepancies, the subsystem drives the drone to land (3) to reduce risk.

Ensuring the safety of the drones is paramount in this sequential escalation and protecting their components was paramount in this sequential escalation because of the damage that could come from a collision with the other drone [43]. In this way the subsystem is equipped to cope with the anomalies at different severity levels initially approaching them from the modes in succession. Return Home is a guarantee to get the drone safely back in the event of problems, while Hover Mode is a short break to stabilize drone in case it has some problems. The last protection mechanism is Land Mode, that prevents the drone flying where it does not belong.

Detecting and managing the anomalies is reflected on the time intervals between successive levels of Threat Response Subsystem. The system is robust enough that if some disturbance knocks it off the intended paths, all the transitions from Normal to Hover, from Hover to Return Home and from Return Home to Land are accurate in time.

The Threat Response Subsystem is able to show its ability to address threats that has been detected through a systematic progression of emergency states. Such a response format is useful in preserving the drone integrity and enabling the system to cope with anomalies while having the potential for handling different levels of threat [1]. This subsystem's design is a beneficial layer of redundancy to ensure the safe operations of drones in possible challenging conditions.

9.4.7 Data Logging

A Data Logging Subsystem was designed to log all changes in the Emergency Mode in response to anomalies detected by the system. This capability is needed for post analysis, because detailed history of response of the drone is provided which in turn enables the engineers to evaluate how beneficial and timely each response was during threat conditions [27]. Each mode change is logged by the subsystem with a timestamp, stored as a structured array of log array, LogArray, for memory efficiency and fast retrieval (Fig. 9.17).

The Emergency Mode plot shows the data of drone mode transition time step by time step from Normal Mode, to Hover Mode, to Return home Mode, to Land Mode. The escalation sequence represents the system response to anomalies detected, from least sensitive and least severe mode to more sensitive and more severe mode in the order. This plot exhibited the smooth, stair step pattern and confirms that the

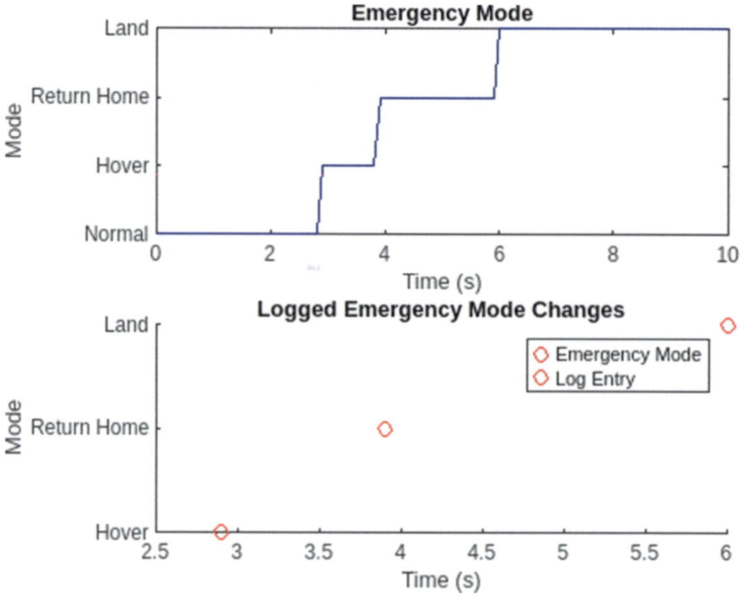

Fig. 9.17 Showing plot of emergency mode and plot of logged emergency mode changes

Threat Response Subsystem works as expected and escalates from mode to mode as a function of sustained anomalies.

The Logged Emergency Mode Changes plot shows red circles for each log entry of mode change. This plot reveals that all three mode transitions (Hover, Return Home, and Land) were recorded by LogArray. The log entries are seen to align with the times of mode changes, and thus confirm the subsystem's reliability of capturing each important event. A subsystem keeps a clear log structure, by recording only changes rather than continuously logging the mode, and eliminating redundant data, resulting in minimized memory use, while only focusing on critical events.

In addition, the Data Logging Analysis statistics shed more light on the subsystem's efficiency. All 3 mode changes were logged with the system, logging all transitions, creating a full record of emergency responses. LogArray, which allocates 0.79 KB memory usage, demonstrates the subsystem in terms of storage that allows being used in resource constrained environments such as onboard drone systems. It empowers fast retrieval of this compact logging format for capturing motivational use cases as real-time monitoring or post-mission analysis without overburdening system resources.

The Data Logging Subsystem has been consistently and efficiently effective. It reliably captures all transitions to Emergency Mode and the resulting structured log allows for step by step review of the drone's response. For embedded systems, the log structure has low memory footprint to avoid interference with the drone's main functionality with logging operations. As an integral part of the drone's safety framework, this subsystem records important data to facilitate post analysis while at the same time improving the traceability, and accountability to the responses taken in the event of a threat.

9.4.8 Overall System Performance

This multi-layered defense system features separate and coordinates subsystems, through a modular architecture, to ensure protection of multiple aspects of possible attacks [33]. Anomaly Detection, Threat Response, and Data Logging subsystems are combined into a single defense system to automatically detect, respond to, and log threats, including GPS spoofing and Denial of Service (DoS) attacks (Fig. 9.18).

The first plot shows the successful detection of irregularities by the Anomaly Detection subsystem. This subsystem monitors GPS data integrity and faulty packets, and raises alerts evidencing the deviation or packet loss that may indicate spoofing or DoS attacks. AnomalyAlert signal introduced early in simulation indicates system was quick to detect spoofed or compromised data. The critical part of this capability is that it can detect things very early, the earlier, the better, allowing all of the subsequent subsystems to respond in time to keep the drone safe.

The second plot shows the Threat Response subsystem, which shows how the system escalates through emergency modes—Hover, Return Home, Land—in response to detected anomalies. Based on duration of the detected threat, the plot

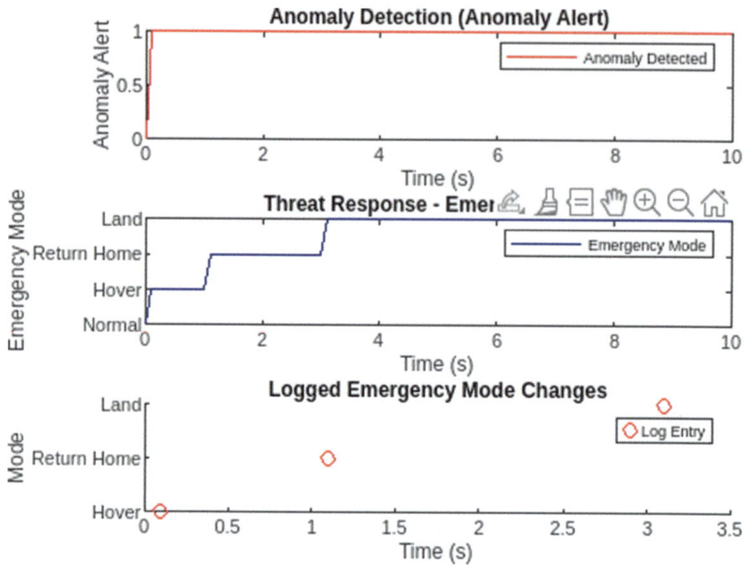

Fig. 9.18 Showing plots of anomaly detection, threat response and data logging function works coordinately in multi-layered defense system

shows a clear sequence of escalations, beginning with Hover mode, to Return Home mode, and finally Land mode. The subsystem depicts its threat management in a layered fashion, with this escalation sequence. The Hover mode gives us a first pause so that we can stabilize temporarily, and the Return Home mode makes sure our drone returns to a safe place if the anomaly continues. Finally, in Land mode, which is a last resort, it grounds the drone if it experiences prolonged anomalies. The structured approach enables the system to manage threats well, tying caution back in with operational continuity [27].

The third plot is to show the effectiveness of the Data Logging subsystem: Logged Emergency Mode Changes. Each logged entry corresponds to a change in emergency mode, each red marker. Each significant event is recorded by this subsystem along with a timestamp, enabling detailed post mission analysis of the drones response. The log shows that there were reliable records of every mode transition recorded by the system when it reacted to the detected anomalies. This makes the logging mechanism structure lightweight while still capturing all the key information, and use memory effectively.

The multi-layered defense system effectively provides robust protection against complex threats. Anomaly detection triggers threat responses while data logging provides accountability for a unified threat response. The system is resilient in this modular design, with each component able to function successfully independently with coordinated action. Its ability to find and respond in real time as well as an effective logging scheme make clear the system's usefulness in maintaining operational integrity in adverse conditions.

9.4.9 Comparative Analysis with Existing Solutions

In this Comparative Analysis with Existing Solutions section, the performance of the developed multi layered defense system is compared to existing drone security solutions with special attention to anomaly detection using automated methods, response timing, and data integrity. This is a case where this comparison shows the system's capability to improve resilience for complex attack scenarios like GPS spoofing and unauthorized access, DoS, and other similar attack scenarios which are crucial in autonomous drone security [6, 44].

Conventional systems for anomaly detection tend to use predefined thresholds or static rules, resulting in delayed response or missed detection in dynamic environments. However, the developed system is a flexible system that takes into account spoofing and DoS conditions by real time monitoring of deviations and packet loss. It significantly reduces detection latency, thus allowing the system to recognise anomalies almost instantly once it is detected. According to Yang et al. [45], rapid detection is necessary to prevent GPS spoofing and preserve navigational integrity. Real time response and sensitivity to irregularities of the developed system provide a substantial improvement over existing methods typically having limited adaptability to simultaneous threat scenarios.

The timing of response is another area of comparison that is critical. The way the developed system structures its escalation of emergency modes (Hover, Return home, and Land) is that each anomaly is addressed with appropriate level of caution according to the intensity and persistence of the threat. In contrast, many existing solutions do not provide such granular control, or overreact to minor anomalies, leading to abrupt landings or return to home commands that are unnecessary in the middle of a mission [15]. The developed system balances caution with operational continuity through staged responses ensuring mission success rates without sacrificing safety.

The developed system also has the feature of data integrity which is also a prominent feature. Every mode transition and response action is logged with a timestamp through its integrated data logging subsystem, providing a transparent, traceable record of the drone's responses. However, this level of logging detail is not traditionally available in conventional systems making it harder to analyze incidents after the mission. The developed system also has an efficient logging that helps in forensic analysis while also ensuring there is no significant event lost [37].

9.4.10 Implications of the Study

This study is of significant importance for both military and civilian sectors that rely on drone security. Since drones are now a critical aspect of operations from sensitive environments, a strong cybersecurity frame should be in place to avoid compromising security of data and ensure operational control as well as the public safety [2]. This study emphasizes how comprehensive protection against vulnerabilities such as those determined in this study for GPS, wireless communication and control systems are required. Real time anomaly detection, response and logging as part of a multi layered defense model that addresses critical security gaps and serves as a blueprint for future drone cybersecurity systems [17].

This framework could potentially increase mission reliability for military applications by protecting drones from sophisticated GPS spoofing and jamming attacks, and therefore save critical intelligence gathering and reconnaissance functions [26]. In the commercial sector, secure and robust drone operations can also support logistics and infrastructure inspection applications by minimizing the DoS and data interception risks and data confidentiality [5]. The established study approach not only strengthens drone security but also systematizes UAV cybersecurity, paving the way for scalable and adaptable security strategies to advance the integration of drone use within new industry sectors [47, 48].

9.4.11 Limitations of the System

In the limitations of the system section, some aspects of chief design and testing challenges facing the multi layered defense system are addressed. While it provides enhanced capabilities to detect and respond to GPS spoofing, Denial of Service (DoS) attacks, and unauthorized access, some shortcomings to real time performance, operational flexibility and threat coverage limit its application [19, 30].

A major drawback is the computational complexity incurred by also having real-time anomaly detection and threat response mechanisms for the system. While effective, the system's multi layered architecture brings a substantial processing load, particularly during simultaneous threats. For example, the anomaly detection subsystem continuously monitors packet integrity deviations, necessitating real-time data processing and comparing operations. However, this complexity with the need of continuous decision making for the threat response logic based on input from the detection subsystem introduces processing overhead. This potential overhead can degrade system response time in high stakes high speed drones or multiple concurrent threat scenarios [21]. These delays can be mitigated by implementing optimized algorithms or hardware accelerations but at a cost of extra resources and infrastructure.

Another limitation is the reliance of the system on emergency modes, Hover, Return Home and Land. These modes are essential to keeping the drone safe, but they sometimes lead to too careful responses that spoil the mission tasks. Testing revealed that even minor persistent anomalies could lead to critical, and often severe emergency modes, including operational restrictions or premature landings. This is especially problematic where intermittent signal disturbance may cause a false positive leading to disruption. By incorporating adaptive thresholds or contextual awareness into future iterations of the system, benign and critical anomalies can be better distinguished [20].

Finally, this study does not cover all the conceivable attack vectors. For example, this implementation did not account for sensor jamming, which interferes with sensor based measurements like altitude or obstacle detection. Consider sensor jamming, it incurs the unique problems of specialized detection mechanism which will add both computational demands and system complexity. Moreover, the system offers no specific measures to preventing cyber vulnerability such as firmware tampering and data exfiltration that enables drone operations in a non physical manner. These limitations make it clear that there is a demand for a more encompassing security framework that would address a wider range of threat scenarios for enhancing system's robustness against evolving security problems [7].

9.5 Conclusion and Future Work

In conclusion, this study demonstrates the necessity of a multi-layered cybersecurity framework for drone protection from ever more complex and pervasive cyber threats. Furthermore, the proposed framework, combining real time anomaly detect, threat response, and data logging subsystems, has resisted specific attacks of GPS spoofing, Denial of Service (DoS), and unauthorized access. Through the development of a modular and scalable system architecture, the work offers a foundational model for securing drones in different operational environments, ranging from military to commercial and civilian. Such a system allows the drones to operate safely, maintain data integrity, and remain in functional failure even under hostile conditions, filling a critical requirement for unmanned aerial vehicle (UAV) technology.

The future work will be to refine the real time processing capabilities of the system to better cope with high speed drones or simultaneous threats with little loss in response time. Processing overhead could be reduced through the optimization of algorithms and utilizing hardware accelerations for more efficient threat detection and response. Furthermore, the framework could be extended to improve resilience to additional attack vectors, including sensor jamming and firmware tampering. Additionally, integrating adaptive thresholds and machine learning models into anomaly detection systems may help improve the system's ability to discriminate between benign and critical anomalies while decreasing the unnecessary emergency responses.

The framework could be extended to address multi-UAV or swarm operations enabling insight into coordinated threat response, where drones act in a decentralized way to mitigate risk within a network. Data security and accountability would be provided in areas with poor connectivity through enhanced data encryption and distributed logging mechanisms. Further research should also involve collaboration with regulatory bodies to develop drone cybersecurity standards for UAVs, and help create safer integration of UAVs into national airspaces. Through these areas, the framework can adaptively and resiliently evolve to improve future developments to support the continued drone growth and safe deployment in various sectors.

References

1. Abdulrazak C (2024) Cybersecurity threat analysis and attack simulations for unmanned aerial vehicle networks. arXiv:2404.16842
2. Ahmad BI, Rogers C, Harman S, Dale H, Jahangir M, Antoniou M, Fioranelli F (2023) A review of automatic classification of drones using radar: key considerations, performance evaluation and prospects. IEEE Aerosp Electron Syst Mag. https://doi.org/10.1109/MAES.2023.3335003
3. Ahmadi M, Kaur J, Rani Nayak D, Nutan R, Taw S, Afaq Y (2023) A review of various symmetric encryption algorithms for multiple applications. Kilby 100:7. https://doi.org/10.2139/ssrn.4491217
4. Aissaoui R, Deneuville JC, Guerber C, Pirovano A (2023) A survey on cryptographic methods to secure communications for UAV traffic management. Veh Commun 100661. https://doi.org/10.1016/j.vehcom.2023.100661
5. Al Qathrady M, Almakdi S, Alshehri MS, Alqhtani SM (2023) Security challenges in multi-UAV systems communication network. In: Unmanned aerial vehicles applications: challenges and trends. Springer International Publishing, Cham, pp 289–321. https://doi.org/10.1007/978-3-031-32037-8_10
6. Alhoraibi L, Alghazzawi D, Alhebshi R (2024) Detection of GPS spoofing attacks in UAVs based on adversarial machine learning model. Sensors 24(18):6156. https://doi.org/10.3390/s24186156
7. Alturki N, Aljrees T, Umer M, Ishaq A, Alsubai S, Saidani O, Ashraf I (2023) An intelligent framework for cyber-physical satellite system and IoT-aided aerial vehicle security threat detection. Sensors 23(16):7154. https://doi.org/10.3390/s23167154
8. Alvares P, Silva L, Magaia N (2021) Blockchain-based solutions for UAV-assisted connected vehicle networks in smart cities: a review, open issues, and future perspectives. In: Telecom, vol 2, no 1. MDPI, pp 108–140. https://doi.org/10.3390/telecom2010008
9. BBC News (2013) Iran shows 'hacked US spy drone' video footage. https://www.bbc.com/news/world-middle-east-21373353
10. Burbank J, Greene T, Kaabouch N (2024) Detecting and mitigating attacks on GPS devices. Sensors 24(17):5529. https://doi.org/10.3390/s24175529
11. Chae MH, Park SO, Choi SH, Choi CT (2023) Commercial fixed-wing drone redirection system using GNSS deception. IEEE Trans Aerosp Electron Syst 59(5):5699–5713
12. Chen B (2022) Data collection and preprocessing. In: Wineinformatics: a new data science application. Springer Nature Singapore, Singapore, pp 5–16
13. Dinelli C, Racette J, Escarcega M, Lotero S, Gordon J, Montoya J, Hassanalian M (2023) Configurations and applications of multi-agent hybrid drone/unmanned ground vehicle for underground environments: a review. Drones 7(2):136. https://doi.org/10.3390/drones7020136
14. Edelman H, Stenroos J, Peña Queralta J, Hästbacka D, Oksanen J, Westerlund T, Röning J (2023) Analysis of airport design for introducing infrastructure for autonomous drones. Facilities 41(15/16):85–100. https://doi.org/10.1108/F-11-2022-0146
15. Eun Y, Park J, Jeong Y, Kim D, Park KJ (2022) A resiliency coordinator against malicious attacks for cyber-physical systems. In: 2022 22nd international conference on control, automation and systems (ICCAS). IEEE, pp 1698–1703. https://doi.org/10.23919/ICCAS55662.2022.10003922
16. Gill SS (2024) Quantum and blockchain based serverless edge computing: a vision, model, new trends and future directions. Internet Technol Lett 7(1):e275. https://doi.org/10.1016/j.teler.2024.100116
17. Gugan G, Haque A (2023) Path planning for autonomous drones: challenges and future directions. Drones 7(3):169. https://doi.org/10.3390/drones7030169
18. Gugueoth V, Safavat S, Shetty S, Rawat D (2023) A review of IoT security and privacy using decentralized blockchain techniques. Comput Sci Rev 50:100585. https://doi.org/10.1016/j.cosrev.2023.100585

19. Hadi HJ, Cao Y, Nisa KU, Jamil AM, Ni Q (2023) A comprehensive survey on security, privacy issues and emerging defence technologies for UAVs. J Netw Comput Appl 213:103607. https://doi.org/10.1016/j.jnca.2023.103607
20. Karbasishargh K, Moghimi Esfandabadi MH, Esmaeili A (2024) Innovative approaches to UAV performance: enhancing safety, reliability, and flexibility. Int J Reliab Risk Saf Theory Appl 7(2):28–39. https://www.ijrrs.com/article_206237_cff5a38dbb2ab850a2d2764bc975d210.pdf
21. Lian B, Kartal Y, Lewis FL, Mikulski DG, Hudas GR, Wan Y, Davoudi A (2022) Anomaly detection and correction of optimizing autonomous systems with inverse reinforcement learning. IEEE Trans Cybern 53(7):4555–4566. https://doi.org/10.1109/TCYB.2022.3213526
22. Limnaios G (2022) Cybersecurity considerations for aerial networks. https://repository.ihu.edu.gr/xmlui/handle/11544/29950
23. Liu W, Zhang T, Huang S, Li K (2022) A hybrid optimization framework for UAV reconnaissance mission planning. Comput Ind Eng 173:108653. https://doi.org/10.1016/j.cie.2022.108653
24. Mohsan SAH, Othman NQH, Li Y, Alsharif MH, Khan MA (2023) Unmanned aerial vehicles (UAVs): practical aspects, applications, open challenges, security issues, and future trends. Intel Serv Robot 16(1):109–137. https://doi.org/10.1007/s11370-022-00452-4
25. Nagel L (2023) Types of drones and UAVs (2024). https://www.tytorobotics.com/blogs/articles/types-of-drones
26. Niyonsaba S, Konate K, Soidridine MM (2023) A survey on cybersecurity in unmanned aerial vehicles: cyberattacks, defense techniques and future research directions. Int J Comput Netw Appl 10(5):688–707
27. Omolara AE, Alawida M, Abiodun OI (2023) Drone cybersecurity issues, solutions, trend insights and future perspectives: a survey. Neural Comput Appl 35(31):23063–23101. https://doi.org/10.1007/s00521-023-08857-7
28. Parrot (2024) Parrot ANAFI USA—professional drone made in USA. https://www.parrot.com/en/drones/anafi-usa
29. Parrot (2024) The largest software ecosystem for drones. https://www.parrot.com/us/partner-ecosystem
30. Pasdar A, Koroniotis N, Keshk M, Moustafa N, Tari Z (2024) Cybersecurity solutions and techniques for internet of things integration in combat systems. IEEE Trans Sustain Comput. https://doi.org/10.1109/TSUSC.2024.3443256
31. Quy NM, Ngoc LA, Ban NT, Hau NV, Quy VK (2023) Edge computing for real-time internet of things applications: future internet revolution. Wirel Pers Commun 132(2):1423–1452. https://doi.org/10.1007/s11277-023-10669-w
32. Radoš K, Brkić M, Begušić D (2024) Recent advances on jamming and spoofing detection in GNSS. Sensors 24(13):4210. https://doi.org/10.3390/s24134210
33. Rugo A, Ardagna CA, Ioini NE (2022) A security review in the UAVNet era: threats, countermeasures, and gap analysis. ACM Comput Surv (CSUR) 55(1):1–35. https://doi.org/10.1145/3485272
34. Rupanetti D, Kaabouch N (2024) Combining edge computing-assisted internet of things security with artificial intelligence: applications, challenges, and opportunities. Appl Sci 14(16):7104. https://doi.org/10.3390/app14167104
35. Sabour MH, Jafary P, Nematiyan S (2023) Applications and classifications of unmanned aerial vehicles: a literature review with focus on multi-rotors. Aeronaut J 127(1309):466–490. https://doi.org/10.1017/aer.2022.75
36. Sentil (2024) VOXL® 2 AI & 5G development drone—Sentinel. https://www.modalai.com/products/sentinel-development-drone?variant=47564491227440
37. Sindiramutty SR, Jhanjhi NZ, Tan CE, Yun KJ, Manchuri AR, Ashraf H, Hussain M et al (2024) Data security and privacy concerns in drone operations. In: Cybersecurity issues and challenges in the drone industry. IGI Global, pp 236–290. https://www.igi-global.com/chapter/data-security-and-privacy-concerns-in-drone-operations/340079
38. Smith K (2024) Australia selects quantum-systems, SYPAQ to provide new drones to ADF. https://govconexec.com/contract-awards/australia-selects-quantum-systems-sypaq-to-provide-new-drones-to-adf/

39. Teixeira K, Miguel G, Silva HS, Madeiro F (2023) A survey on applications of unmanned aerial vehicles using machine learning. IEEE Access. https://doi.org/10.1109/ACCESS.2023.3326101
40. Thomas D, Jain D (2024) Precision mapping and obstacle detection with LIDAR-equipped nano drone. In: AIP conference proceedings, vol 3044, no 1. AIP Publishing
41. Vanitha N, Rakshun Selvaa SK (2023) Cyber security investigation of GPS-spoofing attack in military UAV networks. In: International conference on internet of everything and quantum information processing. Springer Nature Switzerland, Cham, pp 116–124. https://doi.org/10.1007/978-3-031-61929-8_17
42. Wang C (2020) Investigating the threats of unmanned aircraft systems (UAS) at airports. Doctoral dissertation, Purdue University
43. Wei X, Ma J, Sun C (2024) A survey on security of unmanned aerial vehicle systems: attacks and countermeasures. IEEE Internet Things J. https://doi.org/10.1109/JIOT.2024.3429111
44. Yang W, Wang S, Yin X, Wang X, Hu J (2022) A review on security issues and solutions of the internet of drones. IEEE Open J Comput Soc 3:96–110. https://doi.org/10.1109/OJCS.2022.3183003
45. Yang Z, Ying J, Shen J, Feng Y, Chen QA, Mao ZM, Liu HX (2023) Anomaly detection against GPS spoofing attacks on connected and autonomous vehicles using learning from demonstration. IEEE Trans Intell Transp Syst 24(9):9462–9475. https://doi.org/10.1109/TITS.2023.3269029
46. Yaseen A (2023) The role of machine learning in network anomaly detection for cybersecurity. Sage Sci Rev Appl Mach Learn 6(8):16–34. https://orcid.org/0009-0002-8950-0767
47. Yu Z, Gao H, Cong X, Wu N, Song HH (2023) A survey on cyber-physical systems security. IEEE Internet Things J 10(24):21670–21686. https://doi.org/10.1109/JIOT.2023.3289625
48. Yu Z, Wang Z, Yu J, Liu D, Song HH, Li Z (2023) Cybersecurity of unmanned aerial vehicles: a survey. IEEE Aerosp Electron Syst Mag 39(9):182–215. https://doi.org/10.1109/MAES.2023.3318226
49. Zaitseva E, Levashenko V, Mukhamediev R, Brinzei N, Kovalenko A, Symagulov A (2023) Review of reliability assessment methods of drone swarm (fleet) and a new importance evaluation based method of drone swarm structure analysis. Mathematics 11(11):2551. https://doi.org/10.3390/math11112551
50. Zeng Y, Zhang R, Lim TJ (2016) Wireless communications with unmanned aerial vehicles: opportunities and challenges. IEEE Commun Mag 54(5):36–42. https://doi.org/10.1109/MCOM.2016.7470933

MIX
Papier aus verantwortungsvollen Quellen
Paper from responsible sources
FSC® C105338

If you have any concerns about our products,
you can contact us on
ProductSafety@springernature.com

In case Publisher is established outside the EU,
the EU authorized representative is:
Springer Nature Customer Service Center GmbH
Europaplatz 3, 69115 Heidelberg, Germany

Printed by Libri Plureos GmbH
in Hamburg, Germany